£65 · 00

POLYMERS AS MATERIALS FOR PACKAGING

ELLIS HORWOOD SERIES IN POLYMER SCIENCE AND TECHNOLOGY

Series Editors:
T. J. Kemp, Professor of Chemistry, University of Warwick
J. F. Kennedy, Professor of Chemistry, University of Birmingham

This series, which covers both natural and synthetic macromolecules reflects knowledge and experience from research, development and manufacture within both industry and academia. It deals with the general characterisation and properties of materials from chemical and engineering viewpoints and will include monographs highlighting polymers of wide economic and industrial significance as well as particular fields of application.

POLYMERS AS MATERIALS FOR PACKAGING

J. ŠTĚPEK, M.Sc., Ph.D., D.Sc.
Associate Professor
Prague Institute of Chemical Technology
V. DUCHÁČEK, M.Sc., Ph.D.
Associate Professor
Prague Institute of Chemical Technology
D. ČURDA, M.Sc., Ph.D.
Associate Professor
Prague Institute of Chemical Technology
J. HORÁČEK, RNDr, Research Fellow
Institute of Hygiene and Epidemiology
M. ŠÍPEK, M.Sc., Ph.D.
Associate Professor
Prague Institute of Chemical Technology

Translator:
J. PANCHARTEK, M.Sc., Ph.D.
Associate Professor
Institute of Chemical Technology, Pardubice

Translation Editor:
Dr. G. E. J. REYNOLDS
Polypack Consultancy, Haslemere, Surrey

ELLIS HORWOOD LIMITED
Publishers • Chichester
Halsted Press: a division of
JOHN WILEY & SONS
Chichester • New York • Brisbane • Toronto

668.42
STE

This English-language edition first published in 1987 by

ELLIS HORWOOD LIMITED

Market Cross House, Cooper Street, Chichester, West Sussex, PO 19 1 EB, England

in co-edition with

SNTL – Publishers of Technical Literature, Prague

The publisher's colophon is reproduced from James Gillison's drawing of the ancient Market Cross, Chichester.

Distributors:

Australia, New Zealand:
JACARANDA WILEY LTD.,
G.P.O. Box 859, Brisbane, Queensland 40001, Australia

Canada:
JOHN WILEY & SONS CANADA LIMITED
22 Worcester Road, Rexdale, Ontario, Canada

Europe, Africa.
JOHN WILEY & SONS LIMITED
Baffins Lane, Chichester, West Sussex, England

East European countries, China, Northern Korea, Cuba, Vietnam and Mongolia:
SNTL – Publishers of Technical Literature
Spálená 51
120 00 Praha 1, ČSSR

North and South America and the rest of the world:
HALSTED PRESS, a division of
JOHN WILEY & SONS
605 Third Avenue, New York, N.Y. 10158, U.S.A.

© 1987 SNTL—Publishers of Technical Literature/Ellis Horwood Limited
Translation © J. Panchartek, 1987
Translated from Czech: Polymery v obalové technice

BL **British Library Cataloguing in Publication Data**
Štěpek, J.
Polymers as materials for packaging. –
(Ellis Horwood series in Polymer Science and Technology)
1. Polymers in packaging
I. Title
668.4'2   TS198.3.P6
ISBN 0–85312–677–1 (Ellis Horwood Limited)
ISBN 0–470–20720–5 (Halsted Press)
LIBRARY OF CONGRESS Card No. 86–27424

6-6-88

Table of Contents

6   Table of Contents

Foreword

The development of science and technology adds to traditional disciplines new knowledge and its application for practical purposes of increasing importance. Packaging technology is such a discipline. Its modernization, initiated by increasing demands on package quality and by economic aspects, was undoubtedly made possible by the development of macromolecular chemistry and the technology of the production and processing of plastics.

Although plastics will not be the only packaging materials in future, in many cases they are replacing the less convenient or more expensive classical materials such as metal, glass, and wood. In other cases plastics enable reduction of consumption of such classical materials where their use is indispensable, and plastics confer refinement of these packaging materials. This applies, e.g., to laminated aluminium foils or laminated paper which can be welded to produce large, easily shaped, and airtight packages suitable, e.g., for sterilization of some foodstuffs in a more economical way than in the traditional cannery tins or bottles. The large variety of physical and chemical properties of plastics extends their applicability as

far as the protective function of packages is concerned.

Of course, the given advantages of plastics in
packaging technology are far from being devoid of
problems. The physical and chemical properties of
the materials must be understood before use for
packing foodstuffs to avoid doing more harm than
good. In particular it is necessary to consider in
advance what will be the effect of the packaging
material with respect to its transparency for light
or radiation, permeability to gases, or extractability of some of its components, any of which could
affect the food quality.

This comprehensive basic handbook, written by a
team of specialists in the relevant fields, is
therefore welcome; it deals with the above-mentioned
and some other problems of the applications of plastics in packaging in a series of well defined chapters,
so that all those involved with the production or use
of this significant group of packaging materials may
readily obtain reliable information.

May the book be a success in this technologically
and economically important mission.

Prof.Dr.Ing. Vladimír Kyzlink, DrSc.

1

General Problems
of Packaging Technology

D. ČURDA

In packaging technology there are many materials available for the production of packages of different types. Besides plastics the main materials are wood, paper and cardboard, textiles, metals, glass, and, for
packing food, some edible substances, too.

An important point is the fraction of the individual
materials in total packaging consumption. The greatest
part (usually 40-50%) is paper and cardboard. Thus a
developed packaging industry is characterized by relatively large consumption of paper and cardboard, metals (especially in the form of thin metal sheet), and
plastics, and by a distinct decrease in the consumption
of wood, partial retreat from glass and, to some extent,
from textiles. However, the importance of the individual materials is not always determined by the mere proportions in total consumption.

1.1 TYPES OF PACKAGING IN RELATION TO PLASTICS

Packaging is a summary technical term denoting the packaging materials and packages made of them. Thus it includes
auxiliary components which complete the function of the
packages themselves (e.g. sealing, fixing, binding).

Packages can be classified not only according to material composition, but also according to other criteria. First of all there exists a basic functional classification:transport packages, multi-unit packages, and consumers'packages.

Another classification takes into account the mechanical properties of the packaging which determine the method of application and, to a considerable extent, the machinery suitable for the packing operations.

Soft packaging includes paper, plastic films and metal foils, and their various combinations. The soft materials serve either for unit packing or for bulk packages usually of the sack type.

Semi-rigid packaging includes cardboard and paperboard for folding boxes, also plastic sheets and metal foils usually in the form of dishes and cups.

Rigid packages are predominantly made of glass, metals, paperboards and plastics, the typical representatives being bottles, cans, tins and boxes.

Consumers'packages are bags, boxes, bottles, jars, tins, tubes, cups, dishes, etc. Transport packages are cribs, barrels, sacks, transport cans, carboys, etc.

For a summary of packaging it is advantageous to adopt the criterion of material composition which provides links to package production and the relation between package and the goods packed.

As far as the relation between plastics and alternatives is concerned, the classic packaging materials may be improved or complemented by plastics, or they may be replaced by plastics if functionally more suitable.

1.1.1 PACKAGING MADE OF WOOD

Thanks to its availability and good workability, wood belongs to the oldest packaging materials. Even now, though with a decreasing trend, wood is used for transport packages, such as cribs, barrels, and large containers. However, small packages such as chip baskets, small boxes, and basketry products still find applications.

The chemical resistance of wood is good: wood resists acetic, and other organic acids.

Sulphuric acid of concentrations above 10% damages wood markedly, and concentrated sulphuric acid causes carbonization. Wood exhibits no resistance to nitric acid. Dilute (about 10%) solutions of hydrochloric acid affect wood only slightly at room temperature.

Most mineral salts do not damage wood, on the contrary, they conserve it against microbial attack.

Soap solutions containing alkali hydroxides have little effect and wood is fully resistant to ammonia up to medium concentrations.

Methanol, ethanol, and other alcohols do not act on wood, but high-percentage ethanol causes a volume reduction. Vegetable oils and mineral oils have no effect while drying oils which penetrate the wood substance have a conserving effect /1,2/.

The physical properties of wood which are appreciated in packaging are: relatively good mechanical strength combined with low average density, elasticity and vibration damping ability, good heat insulation properties and small thermal expansion.

The drawbacks of wood include water absorption, volume changes depending on the moisture content, susceptibility to microbial attack, anisotropic nature with consequently mechanical properties depending on the direction of fibres. The properties of wood also vary according to its origin.

Wooden cases are either nailed crates or collapsible crates. They differ in size, purpose, as well as in construction (e.g. full-wall cases, cleated boxes, all-wooden boxes, and cases of wood combined with other materials).

Boxes and crates are used for packing various goods ranging from articles of 20,000 kg weight (usually machinery products) to fruit, vegetables, and beverage containers.

To reduce the consumption of wood in packages, various types of light-weight constructions are used which only have a wooden frame of substantial timber while the sides are formed of thinner materials. Two groups of these modern packaging materials are made by upgrading some lower-quality wood substance:

a) Sheets formed by combination of several layers of thin wooden veneers or layers of other materials. This group comprises the usual, normal or speciality plywoods, veneers coated on both sides with paper or cardboard, or a layer of regenerated rubber between two lower-quality veneers (this material offers good protection against weather effects, but it is not suitable for packing foodstuffs due to its rubber odour).

b) Sheets produced by pressing torn or crushed wood substance, even that of annual plants. This group comprises the hard type of pressed fibreboard which exhibits lower water absorption than chipboard, and sheets made of tyre cord waste.

Up to 80% of wood substance is saved in the production of these light-weight cases as compared with the all-wooden boxes.

It is the sheet materials which exemplify the use of plastics in combinations with wood substance. Various plastics are employed in the production of sheet materials incorporating wood. Urea-formaldehyde adhesives are favoured for their light colour, rapid hardening at low temperatures, hygienic acceptability, and low prices.

The boxes and crates are typical transport packages whose main function is to protect the content against any mechanical damage.

Steel strips or wires are used as closures and for security. Binding strips from plastics (especially oriented polypropylene) have proved to be very popular, with strength as much as 60% of that of steel and thus sufficient for binding the majority of packages. Also they are light, weldable, resistant to corrosion, safe during handling, and attractive. The polypropylene binding strips manufactured are 12-16 mm wide and 0.3-0.8 mm thick, and their colour can meet the customers' wishes /3/.

When the requirements for mechanical protection of the goods are greater, especially in the packing of sensitive apparatus, glass articles, etc., protective packing is often used in combination with wooden packages. Here a number of plastics can be employed advantageously (e.g. foam polystyrene, foam polyurethane, sponge rubber or expanded rubber).

For better protection against moisture, wood is combined with polyethylene film or plasticized poly(vinyl chloride) film which can be sealed by glueing or welding.

In many instances, wooden packages have been completely replaced by packages from plastics. A typical example is the plastic transport crates used in the foodstuff industry (e.g. for fruit and vegetables, bakery products, beverage bottles). Until recently, wooden, metal-sheet, and plastic packages competed in this field, and the transport crates from high density polyethylene proved most useful, as can be seen in Table 1.

However, plastic packages not only successfully replace   wooden boxes and crates and especially

Table 1. Comparison of production costs of 1000 beer
bottle crates made of three different materials /4/

Material	Energy (fuel) consumption	Labour consumption	Weight of one crate	Service life
	t	h	kg	years
HDPE	5.36	66.0	2.00	8
metal	3.38	253.5	4.12	8
wood	2.48	728.1	5.83	5

cardboard boxes, but they also meet the requirements
of multi-unit packages and solve a lot of problems
in this field. Application of the so-called shrink
films (especially those made of polyethylene) and
stretch films to multi-unit packing and in connection
with the palletization meant a revolutionary step in
the direction of rationalizing of multi-unit and transport packing of some kinds of piece goods.

The second large group of wooden packages used in
the foodstuff industry includes barrels, casks, vats,
and tubs. The production of barrels used to be one of
the most exacting crafts, and even after the introduction of mechanization into coopery, it required considerable special skill particularly in construction
of large storage containers. The labour-intensive
manufacture of coopery products also has the consequence
that other materials are increasingly preferred in the
construction of transport containers. The drawbacks
of wood, such as the danger of microbial infection
and the associated difficulties of cleaning, and the
greater weight contribute to this trend. On the other
hand, the advantages of wood, especially in cases
where moderate permeability for gases (during the
fermentation process), and where good thermal insulation during transport is required, are still appreciated.

Plastics are used in this group of packages, first
of all, for a complete substitution of traditional
coopery products, i.e. casks, barrels, vats, and tubs,
mostly made of high density polyethylene. It must be
stressed here that besides low weight, hygiene, and
low noise, the substantial advantage of these plastics
lies in their versatility in offering much more varied
shapes and closures, so that further requirements can
be met such as e.g. better stacking and palletization or
better storability of the empty containers. Larger
coopery products, mainly freestanding containers, are
replaced by containers made from glass-reinforced plastics.

1.1.2 PAPER, CARDBOARD, AND PAPERBOARD FOR PACKAGING

Paper products are widely used for consumers' and transport packages, due to relatively good availability of
the raw material, i.e. mainly wood, the wide assortment
of the paper products, the possibility of refinements
by impregnation and combinations with plastics, the
recycling of paper wastes, and finally, also to the
relatively low price of this kind of packaging material,
determined, inter alia, by considerable maturity of the
paper industry.

The properties of paper, inclusive of those which
are significant in packaging, such as tensile strength,
overpressure resistance, impermeability to fats and
oils, resistance to water, are determined by the starting raw material and, to a considerable extent, by the
manufacturing process.

The principal raw material for paper production is
coniferous soft-wood (spruce and fir) and, to a lesser
extent, hard-wood. A significant process in paper production is mechanical or chemical pulping. Mechanical
pulping is accomplished by grinding on rapidly rota-

ting discs in the presence of water to give an aqueous
suspension of wood fibres called groundwood pulp. This
pulp is used directly for the manufacture of paperboards and cheaper sorts of paper. Groundwood pulp
may be added to other sorts of pulp, e.g. cellulose,
waste paper, and, to a lesser extent, waste textiles.
Accordingly, wood-containing paper or wood-free paper
are produced. If scalded billets are submitted to the
grinding, the resulting pulp is brown in colour and
contains relatively long fibres. It is also used in
the production of paperboard.

Chemical treatment of wood can be accomplished by
the action of acids or alkalis. This process serves
to degrade the lignin and hemicelluloses.

Roughly ground wood pulps must be refined before
converting into paper. This stage also affects substantially the final structure of paper, and the
attainment of the required properties, i.e., an
adequate quality, stiffness, brittleness, porosity,
etc.

The paper pulp is mixed with some additives, e.g.
sizing agents (which reduce water absorption and
ink-feathering, fillers (which serve mainly, for
attaining paper opacity), and sometimes dyestuffs.

Polymer additives such as melamine formaldehyde
resins or urea formaldehyde resins improve the wet
strength of the final paper. The wet tensile strength
reaches 25-50% of the dry tensile strength, whereas
the non-modified paper exhibits 5-15%.

The paper pulp is felted on wires usually arranged
as a continuous belt. When the pulp is poured on the
wire, the fibers are oriented in the direction of
its movement, which, causes a higher strength or paper in this direction. This fact must be taken into
account in production of paper packages, and the
paper used must be oriented in accordance with the
direction of the maximum tensile stress (e.g. in the

case of paper bags).

The last operation consists in the passage of paper
between cylinders to obtain machine glazed (M.G.)
paper, paper glazed on both sides, or unglazed
paper.

The production of cardboard and paperboard goes, in
principle, through all the operations outlined above with
the only difference that several layers of the felted
paper pulp are combined (usually by wet pressing) to
obtain the required thickness or required number of
layers.

The paperboards and cardboards can also be produced
by pasting together dry paper sheets. This method is
used for the production of corrugated board. The wavy paper (of the surface density about 120-220 g cm^{-2})
is combined with cardboard usually by means of water
glass (sodium silicate) adhesive to produce light but
relatively strong material especially if several such
layers are combined. The so-called single-faced corrugated board is obtained by combining only one wavy
layer with one flat layer. If the wavy layer is combined
with two flat layers - one on each side - the product
is called the double-backed corrugated board, whereas
the double-wall corrugated board is composed of two
wavy layers and three flat layers.

A great boon to packaging technology are paper products modified by impregnation and/or by coating. Paper,
paperboard, or cardboard are - in these cases - responsible for mechanical strength and insulating properties, whereas the impregnation or coating usually imparts
water resistance, impermeability to water vapour or
other gases, increased chemical resistance, heat-seal-
ability and sometimes an improved appearance. The substances used for paper impregnation or coating are
usually classified according to the state in which
they are applied, viz. hot-melts, aqueous dispersions,

and organic solutions.

The impregnation or coating of paper with hot-melts
is advantageous in that no dilutent needs to be evaporated. In this category are the substances of hydrocarbon type, paraffin wax and microcrystalline waxes,
i.e. solid products obtained from crude oil. Common
paraffin wax contains a large proportion of aliphatic
hydrocarbons, it has relatively low melting point
(about $50-60^{\circ}C$), is rather brittle, and is characterized by relatively large flaky crystals. The microcrystalline waxes contain a large amount of isoalka-
nes, isoalkenes, and naphthenes, have higher melting
points ($63-90^{\circ}C$), are more plastic and sometimes even
sticky, and they crystallize in fine needles. They are
useful where higher flexibility at lower temperatures
is required, e.g. for collapsible boxes used for packing frozen food products. The considerable adhesive
power in the melted state is also useful e.g. for making combined packaging materials (two paper layers or
paper with metal foil).

The improvement of wax compositions is directed to
greater flexibility, greater adhesive power or weldability. A better appearance can be achieved with the
modification, especially of the microcrystalline waxes,
by the addition of macromolecular compounds, viz. polyethylene, polyisobutylene, and mainly ethylenevinyl
acetate copolymers (cf. Sec. 4.2.1.1.1) which impart
a great deal of the character of plastics to these
materials after solidification, the good workability
due to low viscosity of the melt being maintained.
The term "hot-melt" is used for this type of coating
materials (melting adhesives). They have found broad
applications as binding agents or coatings for varied
paper products and proved extremely useful in boxes
for frozen foodstuffs, where the packages are highly
impermeable to water vapour and possess an excellent
appearance (lustre) with the guarantee of uninterrupted
sealing.

The hot-melt may penetrate through the whole structure of the paper or it may provide a coating at the
surface (on one side or on both sides), or it may adhere
two layers of packaging material (laminating, backing).
The first way, penetrating impregnation, is realized
by passing the pre-heated porous paper through the
melt at enhanced temperature. At lower temperatures,
the melt remains more at the surface. The production of
a continuous surface layer is especially desirable in
cases where the aim is a slight permeability to gases
and/or vapours. If contact between the package content
and the hot-melt is undesirable (e.g. if the package
contains some biologically active compounds such as
insecticides, rodenticides, etc.), application as an
interlayer is preferred.

The applications of melts to paper also include
extrusion coating. This method serves for the application of some plastics to paper (mainly polyethylene,
less frequently polypropylene, poly(vinylidene chloride),
and polyamides). The extruded polyethylene layers vary
from 17 to 30 g m^{-2}. The polypropylene coatings improve
both some barrier properties (e.g. resistance to oils)
and thermal resistance of the packages (110°C or even
150°C for a short time). A still higher thermal resistance is reached with paperboard and cardboard coated
with polymethylpentene (up to about 200°C) facilitating
use in food roasting and baking where aluminium previously predominated.

If the impregnation agents are applied as aqueous
dispersions or organic solvent solutions, it is necessary to evaporate the dispersion medium (solvent) during
the process. Aqueous dispersions (e.g. poly(vinyl acetate) or poly(vinylidene chloride)) offer the advantage of safer handling, whereas organic solvents necessitate an efficient exhaust of vapours with regard to
their toxicity, and fire hazards, and solvent recycling
dictated by economic reasons.

Besides the impregnation modifications already mentioned, some specific chemical modifications are used.
They include the hydrophobic modifications of paper and
paperboard with complex salts of higher fatty acids and
the production of non-adhesive (release) papers through
modification with silicones.

In the former case the most frequent method is
application of chromium(III) complexes of stearic acid.
The hydrophobic properties are provided by the alkyl
group, whereas the chromium(III) has an affinity for
the surface of the fibers, through the functional
groups possessing excess negative charge and is able to
occupy the free coordination sites of chromium /4/.

The impregnation of paper with silicones at $1-3$ g m^{-2}
imparts not only the hydrophobic but also lyophobic
properties and general non-adhesivity to a wide variety
of materials. This fact allows paper impregnated with
silicones to be used with a wide assortment of products.
Such products as sweets, dates, figs, frozen foods, meat,
and also asphalt, waxes, adhesives, whose adherence to
paper earlier presented considerable difficulties, can
now be packed with inserted layers of the impregnated
paper (e.g. between cake portions and other bakery
products, cut cheese, etc.). In the field of self-ad-
hesive materials (labels, tapes) the release paper is
indispensable as a support and interlayer. Besides
plastics being used in combination with paper, cardboard
and paperboard, aluminium foils are also laminated with
hot-melts as the adhesive interlayer. Such packaging
materials afford the best protection against penetration
of moisture.

Packages from paper products use the sheet material
for wrapping or comprise complete containers which are
flexible (sacks, bags, etc.) or rigid (collapsible boxes, cases, etc.).

Especially suitable for packing food is greaseproof
paper used as bags or sheets and produced by extensive
refining or beating.

Of the greaseproof papers, parchment paper possesses
outstanding properties: it is impermeable to fats and
resistant to water (even boiling water). It is a chemically modified paper made of high-quality cellulose.
The treatment consists in passing through 50% sulphuric
acid, which causes swelling of the cellulose fibres and
transformation into a membrane of amyloid character,
the original fibrous structure of paper being lost.
The product serves for packing of fatty and wet food
(meat, cheese, fats) either directly or as a lining
or covering membrane.

Parchment substitute and glassine have some greaseproof character, though much less than that of parchment,
and are redispersed in water. The characteristics are
obtained by the slow beating of paper pulp. Glassine
differs in being glazed on both sides. Both are used
for packing goods with higher fat content but not too
wet; glassine, having relatively low permeability to
organic vapours, serves also for packing seasoning,
coffee, and spices. The Havana paper produced by
half-slow beating resembles these greaseproof papers;
its wood-free modification is called beya and is used
in the same way as the foregoing types and in combination with aluminium foils for packing butter and
chocolate. It is also impregnated with wax. Greaseproof paper is being replaced by biaxially oriented
high density polyethylene film /5/.

Wrappers and bags for packing foodstuffs which do
not require special protection against moisture or
fat (e.g. pulse and cereals, etc.) are made of sulphite papers with a range of wood content and surface
density.

Bags, either flat or gusseted, may be manufactured
directly in the paper mills, but modern automatic

packaging machinery produces bags directly from a roll.

Paper shipping sacks are not substitutes for jute sacks, but are advantageous for loose materials, their weight being negligible as compared with the weight of the content, and volume of the empty sack being small. The large paper sacks are made mainly of high-quality strong sulphate paper of about 70 g m^{-2}, the sulphite paper (which has lower strength, particularly if it is too dry) being only used as a substitute. To increase their strength, the paper sacks are stitched or stuck together from several plies of paper: usually three plies for inland transport and five plies for export purposes and for more exacting goods. To increase the resistance of these packages to moisture (from outside or from inside - if the content is wet) or to more aggresive chemicals, one of the plies can be impregnated with bitumen or paraffin. One type of sack is composed of two plies of sulphate paper both 50 g m^{-2}, laminated with asphalt of 30 g m^{-2}; the maximum filling temperature of the sacks corresponds to the softening point of asphalt, i.e. 45°C.

Another material used for paper sacks which is outstanding with regard to low permeability to water vapour is composed of two layers of sulphate paper, each 70 or 50 g m^{-2}), laminated with 60 g m^{-2} of microcrystalline wax; sacks made of this material withstand a filling temperature of 60°C and they are also suitable for packing foodstuffs.

Paper sacks can also be made of oiled paper, e.g. sulphate or sulphate-sulphite 75 g m^{-2} paper impregnated with slack wax to 82 g m^{-2} giving a maximum filling temperature of 45°C.

The paper sacks with the impregnated interlayers must be marked (usually with a strip), so that the resulting waste paper may be treated separately. Besides the impregnated papers, plastic interlayers (polyethylene) are used to improve the performance of the sacks.

Plastics are used not only as materials for laminated paper sacks, but also in their own right for
self-supporting plastic sacks which become, both in
quantity and performance, a significant alternative
to paper sacks. The three main thermoplastics used
are low density polyethylene, high density polyethylene, and plasticized poly(vinyl chloride). The properties of sacks made of low density polyethylene and of
plasticized poly(vinyl chloride), do not differ greatly,
but heat and frost resistance, method of manufacture,
and material prices favour sacks of extrusion blown
polyethylene tube /6/. High density polyethylene has
much better impact resistance, especially at low temperatures. Strong cross-laminated films of oriented
high density polyethylene are used successfully for
heavy duty sacks /7/.

Rigid paper packages include folded or collapsible
boxes of wide-spread use for consumer products. They
are usually made of paperboard or cardboard of
230-400 g m^{-2} with the outer layer of better quality
for printing. If used only to protect contents mechanically, no special requirements are imposed, but if
designed for pasty, liquid, or frozen foodstuffs, they
must be waterproof and/or vapour-proof. The cardboard
is then modified with paraffin wax or a hot-melt coating or laminated with plastics, often polyethylene or
polypropylene /8/). These materials not only ensure
water impermeability but provide for machine assemb-
ling of the individual flat blanks or edges by heat-sealing.
Shapes and methods of folding and closing can be very di-
verse. The boxes are either of one part or of two parts
or hermetically closed. For machine assembly, the closures
are made mainly by folding and sealing flaps.

Loose materials may be packed in bags before being
placed in the boxes. Other types of closures such as
tuck end, ear, and lock end, can be reclosed after

opening and, therefore, are used mainly for goods
which are shown to consumers in shops or which are
taken from the boxes by parts (e.g. the lock-end closure for cigarette boxes).

Two-part boxes have either tuck-end or flap closures.

Air-tight collapsible boxes are used for granular,
liquid, or pasty contents, especially in the foodstuff
industry. This group adopts shapes which enable ration-
al mechanization of both the packing process and storage (prismatic, tetrahedral, cylindrical, conical
shapes and flat welded closures, Fig. 1).

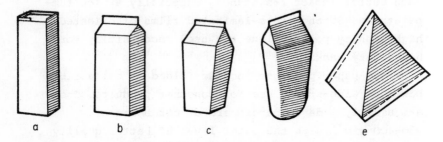

Fig. 1. Various types of water-tight sealable packages made of cardboard backed with plastic
a - prismatic shape, b, c, d - various shapes with roof
closures, e - tetrahedral shape

These packages offer good possibilities for the application of impregnated papers or papers laminated
with plastics.

The prismatic shape (Fig. 1a) is used for one of the
oldest packages of this sort, the *Eko box* used in
freezing plants. A newer version is the *Hermeted box*
with an inner bag which is heat-sealed simultaneously
with the flaps.

Other packages of prismatic shape are known under
the names *Zupack, Expresso*. The Expresso package has
an inner surface covered with a layer of a thermo-

plastic material, which after filling is sealed to the
paper exterior when the flaps are folded.

Other packages are characterized by the "roof"
closure (Fig. 1b-d), are used predominantly for liquid
or pasty products and provide a simple way of closing.

Prismatic packages with roof closures are used in
the *Pure-Pack* system, conical shapes are known under
the names *Perga* and *Satona* (Fig. 1d). An innovative
four-walled package (Fig. 1e) is the *Tetra-Pak* which
is manufactured from a web of modified cardboard which
is heat-sealed both longitudinally and transversly
(Fig. 2).

This type of package, especially when used for
aseptic filling and for longer storage life necessitates excellent barrier properties. Thus the *Tetrabrik
Aseptic* package (used for long-life milk) is composed
of the following layers (from outside to inside):
1) polyethylene, 2) printing, 3) duplex board, 4) polyethylene, 5) aluminium foil, and 6) polyethylene.
Figure 3 represents schematically the production of
this package.

An important group of consumers paper packages are
the wound cartons of cylindrical or conical shape
produced by winding and gluing one or several layers
of cardboard or paper. Such cups used for spreads,
mustard, honey, etc. must be waterproof, and are
therefore internally coated with paraffin wax or
latex. They are being replaced by plastic tubs and
cups made by injection moulding or by thermoforming
which are superior in all respects. These can be
produced in an unlimited range of shapes and sizes,
are resistant to moisture and meet other barrier requirements, so that they are suitable for pasty,
semi-liquid, and liquid food products.

Compared with the simple cardboard cups which
usually have simple plug closures (unless they only

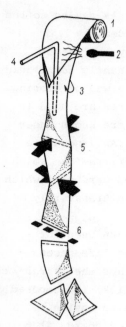

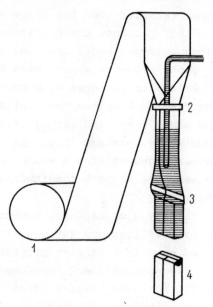

Fig. 2. Schematic
representation of
production and filling
of Tetra-Pak packages
1 - a roll of packaging
material, 2 - steriliza-
tion of the polyethylene
layer, 3 - winding and
longitudinal sealing of
the tube by means of
rotating rollers,
4 - the filling nozzle,
5 - formation of a tetra-
hedron by welding the tube
with parallel jaws, 6 - cutting off of the final product

Fig. 3. Schematic re-
presentation of pro-
duction and filling
of Tetra-Brik type
packages from one roll
of laminated card-
board
1 - a roll of the lami-
nated cardboard,
2 - production of a
tube by longitudinal
sealing and its filling,
3 - transversal sealing
and completing operation,
4 - the final package

serve as drinking cups), the plastics tubs can be provided with various closures according to the requirements for tightness, reclosing, etc. Figures 4 and 5
give examples of the press-on lids with the corresponding edges of the tubs made by injection moulding and
thermoforming, respectively.

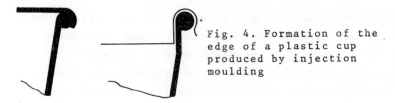

Fig. 4. Formation of the
edge of a plastic cup
produced by injection
moulding

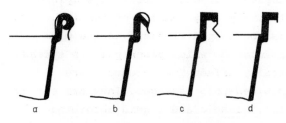

Fig. 5. Shapes of edges of plastic cups made by drawing
a - rolled edge with a press-on lid, b - sharp edge with
a press-on lid, c, d - the edges suitable also for
welding-on lids

The two pictures also show the characteristic difference between the tubs made by injection moulding
and those made by thermoforming in the uniformity of
wall thickness. Control of wall thickness can be
attained selectively by injection moulding. For firm
fixing of the lid, the decisive parameter is the space
between the outer edge of the tub and the lid edge.
Such strenghtening is achieved much less easily by
thermoforming. The problem is usually solved by shaping
the tub edge as in Fig. 5. The rolled edge (Fig. 5a)
is more exacting and, hence, less common than the
other variants /9/. A fundamental advantage of plastic tubs over cardboard ones lies in the possibility
of tight sealing by welding.

Suitable lidding materials for poly(vinyl chloride)
tubs are: paper or aluminium foil both with a thermoplastic coating, or cellophane with a thermoplastic
coating and with a layer of vinyl chloride - vinylidene

chloride copolymer (*Saran*) as water-vapour barrier. The
lids for polystyrene tubs are usually made of aluminium foil coated with butadiene-styrene latex or
polystyrene film without coating. Polypropylene tubs
are usually covered with lids made of a poly(ethylene
terephthalate)-polyethylene laminate.

Aluminium foils or poly(vinyl chloride) films with
a lining of poly(ethylene terephthalate)-polyethylene
laminate enable relatively easy removal of the lid.

Paper products can be shaped directly on the wire
mesh of suitable profiles on which paper pulp is spread
by suction and partially dried. Such products are
frequently called moulded-pulp packages. They are strong
and, at the same time, elastic and liquid-absorbing
and are suitable mainly as trays for packing eggs,
meat, instruments, etc. However, also in this field,
as far as mechanical protection of goods is concerned, the
moulded packages are being gradually replaced by plastics,
most often by expanded polystyrene. Moulded-pulp packages retain their value in special cases which require
water absorption as in the packing of fresh meat,
where it is desirable to remove the liberated juice.

Cardboard cases for transport packages are increasingly important and predominate over the wooden
transport packages because of low weight (about 25% of
the wooden cases) and the connected material economy,
the small volume of the folded empty package (about
5-10% as compared with the filled package), lower production costs as compared with wooden cases, easy
recycling, and other advantages such as printability,
easy closing, etc.

Joints of cardboard cases are made with wire staples,
and more recently also by hot-melt adhesives.

The cardboard cases are easily closed with adhesive
tapes, textile tapes, and also plastic and steel tapes.

Cylindrical cardboard barrels with suitable plastic inserts (e.g. of polyethylene) can serve for the transport of pasty and more liquid materials.

All-cardboard drums are produced within the volume range from 10 to 90 litres; when combined with top and bottom of plywood or metal sheet 200 l volume is practicable. The all-plastic drums (especially those of high density polyethylene) represent a functionally very attractive alternative to cardboard barrels.

1.1.3 TEXTILE PACKAGES

Fabric sacks and bags are well-tried, still used as transport packages, and valued for their strength, flexibility, and low weight. Also consumers´ packages of net construction become increasingly important.

The raw materials for sacks and bags are mainly jute, the Chinese grass meshta, tow (especially flax tow), and for packing nets, also cotton. More recently, sacks from woven tapes formed by longitudinal splitting of unidirectionally oriented films of high density polyethylene or polypropylene /10/ have been employed. They are much lighter than jute fabrics, have higher strength, and are resistant to microbial attack.

Woven sacks are often returnable and are especially useful where good wet strength is required (e.g. for potatoes). Powders can penetrate the fabric, or loose fibers from the fabric can contaminate the contents. These drawbacks can be overcome by combining the woven sacks with paper sacks (crepe inserts) or with polyamide film. The combinations with plastics can protect the contents from moisture.

Bales represent transport packages for use with pressed materials (cotton, hay, straw, feathers, hops, tobacco, etc.). They are made of jute fabric and, sometimes of paper. The bales are tightened with steel bands. They are heavier than sacks, and require mechanical handling.

Nets for consumers packages are suitable for fruit
and vegetables where ventilation is required. They are
produced either in the form of bags of appropriate
size, or, for modern packaging machinery, as sleeves
which are filled with goods, closed at intervals and
separated by cutting.

Plastic net is produced without weaving from polyethylene by special extrusion technology.

1.1.4   METAL PACKAGES

Metal is an important material for production of numerous
kinds of consumers´ and transport containers including
metal foils and tubes, cans of thin metal sheet, canisters and drums of various sizes up to several cubic
metres volume. The most widely used metals are steel
(i.e. technical iron with carbon content up to 1.7 %)
and aluminium with various surface treatments in which
plastics play a significant role.

Tin, being subject to world shortage, is no longer
used as an independent packaging material (true tin
foil), but it is still indispensable for surface protection. Some transport packages made of steel are
protected by zinc. Lead is used in the manufacture of
tubes for some technical products.

The properties of metals which are especially
appreciated from the point of view of packaging include
their considerable strength, impermeability, and good
thermal conductivity. The main drawback of metal packages lies in their susceptibility to corrosion by some
products or by the atmosphere.

Cannery tins are the most important packages made
of thin steel sheet of 0.20-0.25 mm thickness usually
tin plated or fire tinned.

For some food products, the tinned steel sheets
without further surface treatment are satisfactory,
but if the contents is corrosive (acidic and contain-

ing oxygen or other depolarizers) or if it causes blackening due to the formation of sulphides (see below), a
varnish layer is needed. Cannery tins are varnished
internally, but sometimes external varnishing is
required also for protective and decorative purposes.
The tinned steel is varnished on one or both sides,
either in the rolling mill or in the cannery by the
process of roller coating. The process is carried out
in a continuous line with a capacity of 5000-7000
sheets per hour and controlled by one operator. Each
automatic line incorporates a tunnel for drying and
curing the varnish.

The inner surface is coated with non-toxic, non-tain-
ting, sterilizable varnish. These varnishes are based
on natural resins and drying oils or synthetic polymers such as phenol-formaldehyde resins, epoxides,
vinyl compounds, and polybutadiene, and often mixtures
of natural and synthetic resins. For lower sterilization temperatures (which are usual for fruit and
some vegetables) oleo-resinous varnishes are sufficient,
whereas for more exacting sterilization conditions (which
are common for meat products) varnishes based on phenolic
resins are often used. Special varnishes with slip
additives, based usually on higher hydrocarbons, form
a thin layer at the surface of the baked varnish film
which reduces friction and prevents damage of the varnish
during can fabrication.

The varnishes are usually applied to give 4-8 g m^{-2}
when dry. Varnish performance, mainly good adhesion to
the substrate and protection of the tin, depends on
complete curing by baking.

The outer surfaces of cans may be provided with
varnishes applied to the sheets and chosen according to
sterilization temperatures. For temperatures below 100°C
the varnishes based on alkyds and drying oils are sufficient, whereas higher temperatures (up to 120°C)
require boil-resistant compositions based on epoxy-esters.

Table 2. Important properties of the main types of lacquers and varnishes used for cannery tins /11/

Lacquer or varnish type	Taste	Flexibility	Adhesion	Colour	Price	Resistance against			
						Soldering	Sterili-zation	Acids	Alkalies
acrylic	3	3	1	1	3	1	1	1	1
alkyd	3	2	1	2	1	2	2	2	3
epoxy-amine	1	1	1	1	3	1	1	1	1
epoxy-ester	2	1	1	2	2	2	1	1	2
epoxy-phenolic	1	1	1	3	3	1	1	1	2
phenolic	2	3	1	3	2	1	1	1	3
oil-resin	2	2	1	2	1	2	1	1	3
polybutadiene	2	2	1	2	1	1	1	1	1
vinyl	1	1	2	1	3	3	2	1	1
polyester	2	2	1	1	2	2	2	2	3

The evaluation scale: 1 - good, 2 - moderate, 3 - worst

Table 2 presents important properties of the main types of varnishes used for inner and outer surfaces /11/.

If the can is to be decorated by printing, then, as a rule, the operation is carried in line with varnishing using a special cylinder and flat bed machine or offset web-fed rotary machine. Clearly the priming coat, the printing ink, and the top coat must all fulfil the same requirement with respect to heat-resistance, as for varnishes.

Tinning and varnishing belong to classical ways of surface protection of cannery tins. The trends in tin economy led to electrolytic tin plating and, more recently, to the introduction of other surface treatments

of steel sheet which use chromium or aluminium in place
of tin.

The production of cans uses blanks which are formed
into cylindrical shape and soldered in a bodymaker.
At both ends of the cylindrical tube flanges are formed
for application of the lids by rolling-on.

The lids are cut from metal sheets and profiled ready
for the closing machine.

On the periphery of the lids thus prepared, rubber
sealing must be applied. For *Bax* cans the flanges of the
cylinder and lid are butt soldered using rings of rubber
compound which are vulcanized into a saddle shape. *Bliss*
cans are seamed and soldered and rubber latex is used
as the seal which is finally dried.

The alternatives to tinned steel, especially chro-
mium-plated steel cannot be soldered and require a new
technology. Thus the *Mira Seam System* /12/ employs an
adhesive based on polyamide; the adhesive-coated components are heated, pressed together and rapidly cooled.
The strength of such a bond is said to be higher than
that of the soldered joint.

In use, it is necessary to know what corrosion may
take place on contact of acid foodstuffs with tin-plated
or varnished steel. Steel is dissolved by acid foodstuffs with evolution of hydrogen so it must be protected with tin or tin plus varnish.

Ideally the varnish should separate the contents
from the corrosible metal, but such coating has some
porosity. According to Nehring /13/ the relationship
between the number of pores and the thickness of the
varnish is exponential: if the thickness of the varnish
layer is reduced, the number of pores may be doubled.

A solution of the problem of corrosion could be laminates of metal sheets and plastic film. The literature
refers to laminates of steel foil rather than metal
sheets. Such foils, produced by rolling, have a thickness
as small as 0.05 mm and can be laminated with paper,

cardboard, and plastic /14, 15/ for the production of
e.g. collapsible boxes.

Among large metal packages there are steel barrels
which are returnable containers designed for transport
and storage of industrial liquids, especially oil and
petrol, and steel drums, light transport packages with
corrugated walls designed for powder and granular or
semi-solid materials. Further types of metal containers
are steel tight-head or open-head drums and pails,
transport cans and canisters.

In all these cases there are corresponding plastic
packages, especially those made of high density polyethylene.

The large-volume packages also include vessels for
liquids of all sorts. Fruit juices and still beverages
are stored and transported also in vessels made of
glass-reinforced plastics.

Aluminium is the second important metal in packaging
technology. Its importance increased considerably when
it was found to be a good substitute for scarce tin in
the production of foils and tubes.

Aluminium is manufactured by electrolysis of aluminium
oxide in the presence of fluxing agents which requires
much electrical energy. For packaging its advantages are
low density (2.70 g cm^{-3}) and relative softness which
allows cans and tubes to be made by cold drawing or
pressing, or foils as thin as 0.005 mm by cold rolling.
On the other hand, aluminium packages exhibit lower mechanical strength than steel packages, and aluminium is
less chemically resistant than tinned steel to acid media.

Surface protection of aluminium against corrosion
consists in electrochemical surface oxidation (eloxal
coated or anodized aluminium) which, for more severe
conditions, is followed by varnishing.

Aluminium is used for the production of consumers´
packages such as cans, tubes, aerosol containers, foils,

and various types of semi-stiff items, mainly dishes,
and functional components such as lids. Aluminium also
forms a significant component of various laminated
packaging materials.

Tins are made of aluminium of 99.5% purity or aluminium with low content of manganese, from sheets by
cold drawing (mainly low shapes as boxes for sardines)
or by pressing. The closures of aluminium tins are
made in the same way as those of tin-plated steel.

Interesting stiff or semi-rigid packages, mostly
of dish shape, are produced by deep drawing from thin
aluminium sheets ($>$ 0.2 mm) or aluminium foil ($<$0.1 mm)
backed with a plastic film ($\sim$0.05 mm) of high density
polyethylene, polypropylene, or polyamide. The lids
are made of a similar laminate and attached to the
dish (which has a suitable edge made by pressing) by
heat sealing, so that the final packing is air-tight
and can be sterilized in a counter-pressure autoclave
temperatures up to 120°C. These "semi-rigid tins" can
become significant among sterilizable packages. Their
first advantage lies in the possibility of production
directly from a web of the aluminium laminate in a
deep-drawing press which can form part of the packaging
line. Secondly, ready-made packages take little space
in transport, because they can be shaped to provide
a nesting feature. Other advantages lie in the production
facility for various volumes (30 ml to 400 ml, the optimum
being 250 ml) and shapes, the relatively thick layer of
polymeric film, as compared with the usual varnish
coating, affording better corrosion protection, and
easy opening - the lid may be cut off or peeled off
manually using a projecting tag.

Tubes are made of high purity aluminium from annealed
pre-forms by deep-drawing.

Foils are made of un-refined aluminium containing
silicon, iron, traces of copper, etc. by rolling to

less than 0.1 mm. The aluminium strip is annealed to
obtain the necessary softness before final rolling.
The rolling process can produce holes (pores) at some
places (especially with thin foils) due to imperfect
polishing of the rollers or to the presence of in-
clusions and crystals in the aluminium. With foils of
0.008 mm thickness the perforations are visible when
viewed against light; at 0.012 mm the permeability is
reduced to about one half, and at 0.030 mm the foil
can be considered practically impermeable. Hence,
thin foils are not air-tight, and for exacting uses
should be provided with a varnish layer, or combined
with other materials to close the pores; also the foil
becomes more resistant to corrosion and stronger.

Rolled foil can be modified by stamping (design
forming) using profiled rollers to provide the foil
with various patterns, relief inscriptions, etc. Uncoated aluminium foils are used for non-aggressive
foodstuffs such as chocolate and tobacco products,
and for the cups for make-up creams, etc. (thickness
0.008 to 0.012 mm), for bottle closures (0.04 to
0.08 mm), and dishes for pre-cooked foods in freezing
plants (up to 0.1 mm).

Aluminium foils can be varnished by roller-coating
and printed; cellulose nitrate lacquer and baking
varnishes based on epoxy resins are used for foils
for packing cheese. Thermoplastic varnishes can provide
an air-tight closing of aluminium foil packages by
heat-sealing.

The printing of foils can be easily carried out
but is not practical in cases where the aluminium is
expected to provide protection against thermal radiation.

Aluminium foils on their own are usually too soft
and easily creased even after varnishing, and their
strength is insufficient for packaging purposes unless
they are combined with other packaging materials,

particularly paper, plastic films, or fabrics. Such
laminates provide not only flexibility and strength,
but also a barrier against gases, light and to some
extent against thermal radiation. If aluminium foil
is combined with paper using microcrystalline wax or
some water-proof adhesive, then the intermediate layer
contributes to protection.

There are numerous laminates of aluminium foils and
plastic films, with thermoplastics alone or with
thermoplastics and cellophane, and also with thermoplastics plus paper plus cellophane /16-22/.

This wide assortment of laminates can be classified
according to various criteria: sterilizability, printability, and mechanical performance. The vacuum depo-
sition of aluminium onto various materials including
plastic films has become significant. The process is
carried out in an evacuated horizontal vessel in which
a roll of the web material is unwound over a rotating
cooled drum and rewound on another roll at a speed of
30-40 m min^{-1}. Below the cooled drum there are small
crucibles with aluminium heated at 1700°C by resistance
heating. The evaporated aluminium is condensed at the
cooled surface of the web. This method can produce material of 1600 mm width and 12-60 μm thickness with
the metal deposited on one or both sides at a thickness
of 0.01 to 0.05 μm. Most often, this method of metallization is applied to films of biaxially oriented
polypropylene, polyamide, polyester, and polyethylene,
also laminated films are made with the aluminium layer
sandwiches. Metallized plastic films have an attractive
appearance and excellent barrier properties, and they
are economically interesting packaging materials.

Aluminium is also suitable for the production of
transport containers such as cans and barrels, but
plastics are advantageous in some cases.

1.1.5 GLASS CONTAINERS

Glass is a well-established packaging material which
gets along with newer materials very well. Its advantages include high chemical resistance, good washability,
easy sterilization, transparency, returnability and
availability of raw materials for production. The
main drawbacks of glass containers are their brittleness and considerable weight. Sometimes the low heat
conductivity of glass and its lower thermal shock
resistance are drawbacks. The high energy consumption
in glass production is a serious disadvantage.

Physical and chemical properties are determined by
the composition of glass. Glass is an amorphous substance formed by annealing the melt of oxides of silicon,
sodium, calcium, and metal oxides. High silicon dioxide
content lowers the thermal expansion of glass and thus
improves its resistance to temperature changes, but
such glass is less easily melted and processed. Calcium
contributes to high chemical resistance whereas sodium
lowers the melting temperature but increases thermal
expansion. The composition is also dictated by the
production method.

The required shape of the package is achieved by
the direct moulding of molten glass in the case of low
flat shapes; the press-and-blow process, comprising
pre-moulding of the vessel and blowing into final
shape in a mould, is used in the case of wide-necked
jars; a third method consists in blowing a portion
of molten glass into a mould by means of compressed
air for narrow-necked bottles.

All these methods can be fully automated. However,
glass containers larger than 15 l and short runs of
bottles of complex shapes are still made by the old
manual method; in this the glass-blower blows, by
means of a blowing iron, the parison is then

enclosed by a cast-iron mould where the blowing is
completed.   •

   The quality of glass containers can be impaired by
defects in the process of glass melting or hot forming.
The main defects are bubbles, blisters, internal stress,
stones, or fissures which can be especially troublesome
in the stressed regions of containers such as the necks
of bottles. Such defects are subject to quality standards.

   The heat resistance of glass containers is characterized by the temperature difference which must not cause
any damage when transferred from a bath of higher
temperature to another bath of lower temperature.
Bottles for sterilization must withstand a thermal
shock of $40^{o}C$, and other beverage bottles of $35^{o}C$
(from $55^{o}C$ to $20^{o}C$). The direction of temperature change
is important because glass is much more sensitive to
cooling than to heating. As far as resistance to internal
pressure is concerned, bottles for pasteurization must
withstand at least 1.2 MPa for one minute, other bottles
must withstand 0.8 MPa, unless higher requirements are
dictated by practical considerations (e.g. bottles for
Champagne require 2.0 MPa). A linear dependence between
resistance to internal pressure and glass weight in
grams per ml of the bottle volume was found for bottles
of circular cross-section /12/. The high compressive
strength of glass enables containers to withstand a
large perpendicular load (about 10 kN), thus facilitating stacking. On the other hand the brittleness of
glass causes sensitivity to impact. Protection against
scratching of the glass surface prevents loss of its
mechanical strength.

   The current trends in the production of glass containers are weight reduction and improved scratch
resistance. At first, surface coatings based on polyoxyethylene stearate and silicones were applied on
annealed bottles. A more efficient method appears to
be a very thin coating of metal oxides, especially of

titanium and tin but also vanadium, zirconium, or
aluminium. These "inorganic" surface treatments are
applied at the "hot" end of the annealing lehr where
the annealing phase of hot formed glass containers
(600-700°C) starts. The methods or surface treatments
by organic compounds applied at the "cold" end at the
annealing lehr still continue to be developed; the
majority are polyethylene and poly(vinyl chloride)
coatings. The polyethylene layer is usually of
0.15-0.20 mm thickness, but that of poly(vinyl chloride)
is thicker. These plastic layers increase the impact
resistance by about 40% (PVC) and 70% (PE) as compared
with the surface treatments with metal oxides. The
resistance to internal pressure is increased also by
35%(PVC) and 50% (PE) /23/. Coatings of epoxide resins
are said to be more efficient even than those of
polyethylene /24/. It is obvious that glass containers
with surface coatings must meet the same requirements
as uncoated glass in resistance to heat and to alkaline
washing baths.

A further development of glass containers utilizes
plastics protection in a different way: a thin-walled
pear-shaped glass bottle with a wide opening is supported by a covering of expanded polystyrene /23/.
This combination represents the further advance in
weight reduction of glass containers (Fig. 6).

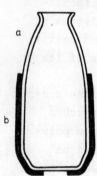

Fig. 6 A bottle of very thin glass (a)
in a covering made of expanded polystyrene (b)

The chemical durability of glass /25/ is excellent, especially in comparison with other packaging materials; nevertheless, corrosion may also occur. Acids do not damage glass by attacking the silicon-oxygen bonds (which form the main molecular structure of glass) but only replace Na^+ and Ca^{2+} ions by protons. Thus the alkali metals are extracted without any substantial damage to the glass. In contrast, alkalies cleave the silicon-oxygen bonds, so that the glass surface is dissolved in the alkaline medium. A similar phenomenon can be observed if water vapours condense at the glass surface and the condensate is not removed; at first, water acts as an acid and extracts alkali metals from the glass surface, and when the content of alkali metals in the water layer increases, alkali corrosion becomes significant.

The choice of colour of glass containers is influenced by any requirement for filtering radiation and by the need to enhance appearance.

Glass containers are mainly intended for liquids, but also for pastes, powders or granules, and pickled and dry products. These are mainly consumers' packages, but larger glass containers, such as carboys, may be used for fruit juices, concentrates, chemicals, etc. For foodstuffs, the glass containers are usually divided into two main classes, for beverages and for preserves. The former includes milk, beer, wine, fruit juices, soda waters, squashes, syrups, spirits, mineral waters, vegetable oil, soup-stock, etc. Narrow neck bottles are used for medicines, and a special type of medicine containers is the ampoule.

Bottles are usually cylindrical ending with a neck of diameter below 30 mm in most cases. For minimum material consumption and minimum weight it is desirable for bottles to have the minimum surface area for a given volume, which is best achieved by low and wide shapes. This also meets the requirements of mechanized

filling lines where beverage bottles must have good
stability on quickly moving conveyers. A long and
narrow neck increases the weight of bottles and, therefore, the shortest possible necks are made, especially
with one-way bottles. The economical shape must, of
course, meet aesthetic and performance requirements.
For example, the shape of the neck must ensure easy
emptying.

These criteria largely apply also to the design
of plastic bottles and the main principles governing
the choice of closures apply equally to both glass and
plastic packages.

Bottle closures belong to the most important functional
parts of glass containers. They must be tight and hygienic. Perfect tightness is essential for sterilized
products and carbonated beverage. For pasteurization
of the latter, a crown cap must withstand a pressure of
1.2 MPa. The hygienic requirements are that the closure
should cover, and protect against contamination, the
whole bottle mouth as is the case with the crown cap,
aluminium cap, tear-off seal, or screw-on cap. In
contrast, and unsatisfactory in this respect, were the
press-on cardboard lids previously used for milk bottles
and now replaced by aluminium tear-off caps. A bottle
closure must be pilfer-proof. This requirement is met
by the crown cap, tear-off seal, and sealed cork.
Sometimes bottles must be reclosable, which need not
conflict with the previous requirement, and may be
realised with cork stoppers (wine, spirits), especially
screw plugs, screw-on caps, or quick-turn closures
(lug type) secured only by a part of the thread (bottles
for ketchup, syrups, etc.). The lock-lever closures
formed by a porcelain plug and rubber gasket and used
earlier for soda-water bottles also enable easy reclosing
but did not meet hygienic requirements nor were they
pilfer-proof. The latter drawback can be overcome by

applying a printed adhesive paper strip over the closure;
screw-on caps can be secured by turning their lower
edge over a lug on the bottle neck, which necessitates
removing the lower part before unscrewing the closure
of thin aluminium sheet.

The requirement for quick mechanical closing is
met with crown caps, but mechanization for other
closures is well developed also, but with lever-lock
types is complicated.

The crown cap is forced on to the bottle neck and
is held there by the elasticity of its edge, the aluminium cap is usually press-formed on the bottle neck.
Plastic heat shrinkable caps made mostly of poly(vinyl
chloride) are an alternative. Screw caps are closed
and opened by one or more turns or, sometimes, a fraction of a turn for quick-turn closures, e.g. one third
(the "*Flavor Lok*" closure) or a quarter (the "*Twist-Off
Cap*"). The twist-off cap is used for wide-necked jars
and is secured by the vacuum produced during processing.
Each of the closures mentioned needs a specific design
of bottle mouth.

Plastics are used in conjuction with glass for
packing and for closures. Their importance in packaging
can be exemplified by crown caps, where the original
inserts of solid cork or of chippings moulded with a
binder are increasingly replaced by plastic inserts
based on PVC pastes or PE or its copolymers. Plastics
also find use in protecting glass containers from
mechanical damage, as mentioned at the beginning of
this chapter. Plastics, more significantly, compete
with glass for packaging, so that nowadays there is
a broad assortment of plastic bottles. Table 3 gives
a survey of their properties for comparison with those
of glass containers /11/ described at the beginning
of this chapter.

Table 3. Properties of bottles made of various plastics /12/

Properties	Acrylic mul-tipolymers	Nitrile polymers	Low-density polyethylene	High-density polyethylene
density (g cm^{-3})	1.09-1.14	1.10-1.17	0.91-0.925	0.95-0.96
transparency	transparent	transparent	cloudy	cloudy
water absorption	medium	medium to low	low	low
permeability for				
water vapour	high	medium	low	very low
oxygen	low	very low	high	medium to low
carbon dioxide	medium	very low	high	medium to high
resistance to				
acids	low to good	low to good	good to very good	good to very good
ethanol	sufficient	sufficient	good	good
bases	low to good	good	good	good
mineral oils	good	very good	low	sufficient
solvents	low	sufficient	good	good
heat	good	sufficient	low	sufficient
cold	low	low	excellent	excellent
sunlight	good	good	sufficient	sufficient
high humidity	sufficient	excellent	excellent	excellent
strength	medium to high	medium to high	low	medium
impact strength	low to good	good	excellent	good
world market prices	medium	high	low	low
applications	foodstuffs, medicines, cosmetics	foodstuffs, medicines, carbonated beverages, cosmetics, household chemicals	cosmetics, foodstuffs	detergents, milk and other food-stuffs, me-dicines, cosmetics

Table 3. Properties of bottles made of various plastics /12/

Properties	Poly-propylene	Poly-styrene	Styrene-acrylo-nitrile	Poly(vinyl chloride)
density (g cm^{-3})	0.89–0.91	1.0–1.1	1.07–1.08	1.2–1.4
transparency	transparent	transparent	transparent	transparent
water absorption	low	medium to high	high	low
permeability for				
water vapour	very low	high	high	medium to low
oxygen	medium to high	high	high	low
carbon dioxide	medium to high	high	high	low
resistance to				
acids	good to very good	good	good	very good
ethanol	good	low	low	very good
bases	very good	good	good	good
mineral oils	sufficient	sufficient	sufficient	good
solvents	good	low	low	good
heat	good	sufficient	sufficient	sufficient to low
cold	low to good	low	low	very low
sunlight	sufficient	sufficient to low	sufficient to low	low
high humidity	excellent	excellent	excellent	excellent
strength	medium to high	medium to high	medium to high	medium to high
impact strength	sufficient to good	low to good	low to good	sufficient to excellent
world market prices	low	low	medium	medium
applications	medicines, cosmetics, syrups, juices	dry goods, medicines, vaselines	dry goods, medicines	shampoos, oils, detergents, whisky, wine, vinegar, edible oils, floor waxes

An indisputable advantage of plastic bottles is their
low weight, and, as a rule, they have better impact
resistance. In addition there is the facility to produce
much more varied shapes than are possible with glass
technology. Sometimes, however, their lower chemical
resistance and permeability to gases must be taken into
account. The main applications are included in Table 3.

Since ecological problems arise with bottles (particularly those for beverages), attemps have been made
to develop packages which could be dissolved in water
after use. In case of glass, soluble silicates (water
glass) could be suitable if coated with a polymer on
both sides. A perfect bond of the water glass to the
polymer would be achieved by first spraying the bottle
surface with poly(vinyl hydrogen phtalate) as an anchor
coat /26/.

The wide neck glass containers are primarily represented by preserve jars produced predominantly in
cylindrical shapes of 150 ml to 5 l capacity. They
present problems with respect to closures of required
air-tightness (which is less easily achieved than with
narrow neck bottles), combined with easy closing and
opening.

The main criteria applied to the evaluation of
these closures are the method of attaching the closure
to the neck, the method by which air tightness is ensured, and the method of opening. Other criteria include
the position of the seal in the cap and the material
of which the cap is made.

As far as the material is concerned, the majority
of caps are produced from tinned steel sheets, only
the *Omnia* and *Pano* closures being made of aluminium.
Also patented /27/ and successfully tested are sterilizable closures of plastics (mainly high density
polyethylene, polypropylene, or polyamide). Such a
cap can be applied to a bottle with the Omnia type of
neck by pressing, and is able to vent.

The sealing compound is usually placed on the cap
in such way that it seals the glass at the edge side
of the neck (e.g. the cap types *Omnia, Pano, Eurocap,
Twist-Off*) or at the side of the neck (*Libby, White-
-Cap-Pry-Off, Silavac, SKN, SKO*). The former seal can
more easily come into contact with the contents,
which is usually undesirable, particularly if some
components (mainly fats) attack the seal or can
themselves be contaminated, for example by an odour.
Such problems are especially connected with the use
of natural rubber, which therefore, has been largely
replaced by synthetic rubber (especially butyl rubber
or butadiene-acrylonitrile rubber). Plastics are also
used particularly for compositions based on poly(vinyl
chloride). For seals the main functional requirements
are hygienic acceptability and consistent performance
during sterilization and closing.

Another type of consumers´ packages used for spreads,
mustard, and other nonsterilized products are slightly
conical "tumblers" which are closed with snap-on caps
made of un-plasticized poly(vinyl chloride).

Cosmetics products and medicines are often packed
in wide neck glass jars which can easily be replaced
by plastics.

An innovation for labelling glass containers are
labels made of water-soluble polymers /26/, which
solves the problems of removal of the residues of paper
labels from washing machines for returnable bottles.

Although the glass containers are used primarily
in the field of consumers´ packages, they have also
been used for storage and transport of liquid chemicals,
aromatic substances, concentrates, fruit juices, and
other substances, in larger volume bottles, demijohns,
carboys, and duplex carboys.

1.1.6 PACKAGING AND COATING MATERIALS MADE OF EDIBLE SUBSTANCES

Edible packaging materials are used in the food industry. They are mainly derived from saccharides,
proteins, and lipoids. In addition, some synthetic
compounds can also be involved. All are produced in
the form of films or coatings. Coatings bring the
material into intimate contact with the foodstuff,
and as complete removal prior to use cannot be ensured,
edibility of the coating is essential. The edible
packaging materials can be applied as a protection
against contamination. In this case they should be
removed from the foodstuffs prior to use, and their
edibility only guarantees hygienic acceptability.
When the package is eaten with the foodstuff either in
the original state or after suitable modification, e.g.
dissolution, boiling, or application of enzymes, protection against contamination must be used, e.g., for
portion packing of beverage concentrates, soup cubes,
or other foodstuffs, the external packaging then comprises a conventional material. Where the edible packages can be used they improve the goods-to-package
weight ratio and may afford better protection than
earlier simple packages by sealing the surface of the
foodstuff.

The ice layer used as a coating for some frozen
foodstuffs to minimize dehydration or oxidation represents a simple edible package. This ice layer is
produced by brief immersion of the frozen foodstuff
in water prior to packing e.g. in cardboard boxes.

Coatings of simple saccharides or disaccharides
have a limited protective effect, because they are
hygroscopic, their use being also limited by their
sweet taste (they are specially suited for "glazing"
candied fruit).

Amylose films and coatings are often recommended
as edible packaging material. Amylose, which is characterized by linear chains of glucose units, is present
in common starch to about 20%, whereas 80% is amylo-
pectin with branched chains. However, special maize
varieties have been developed for industrial purposes
in which the starch contains 60% or more amylose. Amylose is isolated by fractionation based on the formation
of complexes of starch with various alcohols (mainly
butanol). Amylose films show excellent resistance to
organic solvents and fats, being also resistant to
weaker acids and bases. Their permeability to water
vapour and other gases is comparable with that of
cellophane. The presence of amylopectine detracts from
the mechanical properties of amylose film.

Amylose can be plasticized with aqueous glycerol
for casting or extrusion of films. Such films can be
thermally welded (at about 105-120°C), although the
material is not a true thermoplastic /28/. Solubility
and mechanical properties can be modified by esterification (especially acetylation) /29/ or by etherification,
as in the case of hydroxypropyl derivatives. Amylose is
recommended for packing frozen meat, poultry and fish,
and for special cases of sauces, soup preparations,
or even sausage casing. Amylose is also recommended
for surface modification of paper to which it imparts
impermeability to fats.

Cellulose films are indigestible in contrast to
fibrous cellulose in plant tissues which are a desirable
component of diet. The skins of some fruits present a
good example of natural packaging with cellulosic,
pectinic, and lipoid components of plant tissues.

Water-soluble derivatives of cellulose, such as
methylcellulose and carboxymethylcellulose, are defined
as edible and can be used for transparent films and
coatings /30/. Ethylcellulose (mainly in combination

with oils), cellulose acetobutyrate (mainly in combination with acetylated fats /31/), and hydroxypropylcellulose are edible constituents of a number of coatings
for foodstuffs.

Pectin is structurally close to polysaccharides
and is used in the form of calcium(II) salts - mostly
as coatings. It consists of linear macromolecules
composed of galacturonic acid units whose carboxylic
groups are partially esterified with methanol. Calcium
yields a three-dimensional structure with the divalent
cations cross-linking adjacent chains.

Alginates, which are also used as edible packaging
materials, are obtained commercially from brown algae
(sea-weeds). They are similar structurally and in
colloid properties to pectins including their ability
to form three-dimensional structures with calcium(II)
salts. Alginate coatings applied to processed poultry
ensure retention of quality of the surface layers by
preventing diffusion of salts into the meat during
its immersion in sub-zero brine as a freezing medium.

The calcium alginate can also be used for artificial
skins, but if it comes into contact with salty contents,
an undesirable exchange of sodium and calcium ions
occurs. This exchange is prevented by using macromolecular bases including synthetic ones such as
melamine-formaldehyde condensates. Alginate coatings
under gelatin skins promote removability of the skin
from meat products.

The second large group of edible packages is
derived from proteins, of which gelatine is the most
important. The advantages of gelatine, mainly as capsules, were appreciated in pharmacy; however, this
material is used in numerous branches of the food
industry. Gelatine capsules are formed from strips
whose composition most often is 40% gelatine, 30%
glycerol, 30% water. Filling, heat-sealing, and drying
are performed in line /32/. For coating, gelatine is

combined with pectin, alginates, sugar, casein, starch,
and (for coating of sausage) with addition of meta-
phosphoric acid.

Artificial skin from glue stock is an established
edible package. It replaces in a more hygienic way
sausage skin made from sheep, goat, or pig intestines.
Its production starts from the flesh split of hide
(the best is cow hide) from which collagen fibers are
obtained and pasted together with the collagen
hydrolysate - glutin.

An advance in the technology consists in dissolving
collagen and recrystallizing into fibrillar form.
The American firm Ethicon Inc. Dewro introduced a
procedure starting with a mixture of fibrous collagen
and collagen solution, which is shaped, coagulated
with salt, and dehydrated in acetone. This gives
products of high quality.

The third group of edible packaging materials is
derived from lipids and related substances. These are
mostly hydrophobic and thus resistant to water and
show relatively low permeability for water vapour,
properties relevant to their natural function, for
example the waxes on the surfaces of fruits and leaves.

Waxes can be used as edible coatings, because they
are chemically stable and physiologically acceptable.
They are esters of higher fatty acids (e.g. hexacosanoic
and montanic acids) with higher primary alcohols (e.g.
cetyl, stearyl, and myristyl alcohols. The natural
waxes also contain alkanes (e.g. nonacosane from cabbage
leaves and heptacosane from tobacco leaves).

Additional emulsion or solution coatings (e.g. from
carnauba wax or beeswax) are applied to fruit and
vegetables to prevent dehydration, evaporation of aromatic
substances, and microbial infection, while permitting
sufficient transpiration.

Mixed esters of lower fatty acids with glycerol,
especially saturated acetoglycerides /33/, for the
packaging have melting points from 27°C to 47°C.
Acetoglycerides resemble fats in a number of physical
and chemical aspects; permeability to water vapour is
comparable with that of polyamides, whereas permeability
for permanent gases (particularly oxygen and nitrogen)
is about the same as that of polyethylene. They are
hygienically acceptable and digestible, and are applied
as coatings on cheese (to prevent surface mildew),
meat, fish, poultry, and fruit, especially to prevent
dehydration and undesirable changes during freezing
and cold storage.

Commercial coatings based on acetoglycerides are
sometimes modified with other substances, e.g. cellulose
acetobutyrate /34/.

A fully synthetic material which is suitable for
edible packages is poly(vinyl alcohol) giving water-soluble films or coatings.

It is also possible to combine edible materials in
the form of two-layer coatings to attain optimum protection.

1.2   PROTECTION OF GOODS BY THE PACKAGE

The main function of a package is to protect the article
packed between production and consumption or use. The
vast variety of the articles packed leads to differing
requirements for protection. Comparing the two main
groups of goods packed, foodstuffs and industrial products,
the requirements of foodstuffs are usually much greater.
Industrial products need protection against mechanical
damage and/or chemical and physical changes (e.g.,
corrosion of metals, caking of powdery chemicals). The
biochemical complexity of foodstuffs makes possible

partial or complete deterioration even within a relatively short period.

There are two modes of protection, the package
acting as a barrier to external influences or as a
barrier to prevent the goods contaminating the surroundings. The former is exemplified by barriers to
moisture, oxygen, or other gases, to light or other
radiation, to heat, or to pests. With regard to
protection against mechanical damage, the package
can be considered as a barrier to pressure, or,
possibly with protective inserts, against impact.

The protective effects of packages may be complemented
by some chemical means, e.g. drying agents used to
absorb moisture, or impregnants with antioxidant,
microbiocidal, or anti-corrosion effects.

In many cases, the protective function of a package can be seen in that it provides a microclimate
which otherwise would have to be created in a con-
trolled-atmosphere store, e.g. correct humidity, oxygen
content, light intensity, or sterility. Many of these
requirements can be met within the package better
than in a room.

There are three main influences which should be
controlled by the package, viz. mechanical, climatic,
and biological effects.

1.2.1 PROTECTION OF GOODS AGAINST MECHANICAL EFFECTS

Mechanical damage can be encountered with many industrial products as well as with foodstuffs.

Mechanical stress arises mainly during transport
and the associated handling. A specific type of mechanical stress consists in pressure changes during
pasteurization and sterilization of foodstuffs in
packages.

During storage the most important mechanical factor is the pressure of upper layers on lower ones.

The characteristics of deformation during mechanical stress are particularly important in the case of plastic packages, and must be taken into account when these packages are exposed to a static load for a long time. Bottle crates are exposed to high stress, because 35 such crates may be piled on pallets, which corresponds to a load of about 7 kN for the lowest crate, and the stress can operate for a period of several weeks or even months. Such packages require careful attention both to the choice of material, and also to design (stiffening ribs, reinforcement of corners, etc.). Figure 7 gives an example of the time dependence of deformations of milk bottle crates made of various polyolefins and exposed to a stress of 2.5 kN /35/.

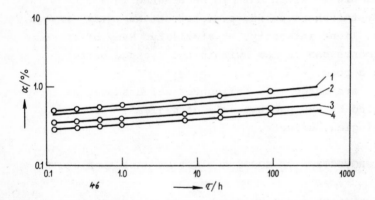

Fig. 7. Time dependences of the deformation α of milk bottle crates made of various polyolefins and exposed to a stress of 2.5 kN at the temperature of 23°C /35/

1 - polyethylene (density 0.958 g cm^{-3}), 2 - polyethylene (density 0.965 g cm^{-3}), 3 - copolymer of propylene with high content of ethylene, 4 - copolymer of propylene with low content of ethylene

Prediction on the behaviour of plastic packages
during mechanical stress is also obtained from the
curves derived from short-term testing by compression
at a constant rate and plotting resistance (force)
against deformation. Examples of such curves are given
in Fig. 8 /36/. Figure 8 also shows that the contents
of a package act as a reinforcement to a certain extent.

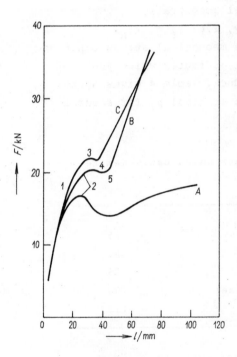

Fig. 8. The force-path
diagram of compression
of a 120 l wide-necked
barrel made of linear
polyethylene (direction
of the compression = the
vertical axis, rate 10 mm
min^{-1}, temperature 23ºC)
/36/
A - the empty barrel,
B - the same barrel filled
with a granulated plastic
and C - filled with water;
1 - the beginning of de-
formation of the barrel in
its middle section,
2 - the beginning of the
deformation at the lower
section of the barrel
(at the boundary of the
edge and the bottom),
3 - formation of a fold
in the same section of the
barrel, 4 - the beginning
of the increase of the
internal pressure and
blow-out of the barrel
wall, 5 - the lid bears
against the solid filling

With transport crates it is advantageous if they
can be stocked cross-ways (particularly for high
stocks) which improves both stability and distribution
of forces.

With round and spherical objects piled up on one
another, the pressure stress is restricted to a small
surface area of mutual contact, which can easily

result in mechanical damage. The problem can be solved
by packaging the product in single layers each in an
adequately strong container or by using shaped inserts
which distribute the stress over the largest possible
surface area of the product.

During handling the goods are often exposed to
impacts and vibrations. In this case the intensity
of mechanical stress is expressed by the g_f factor
expressing the ratio of acceleration a (m s^{-2}) of the
respective mass to the normal acceleration
(g_n = 9.81 m s^{-2}) of gravity, i.e. $g_f = a/g_n$.
The ability to withstand the mechanical stress can then
be expressed by the critical g_f factor which just
causes no damage to the product. Table 4 gives approximate numerical values of the critical g_f for several
groups of articles.

Table 4. Approximate g_f values which cause damage
to some groups of goods /37/

precision instruments	15
electronic instruments	25
typewriters, cash registers	50
television sets	75
refrigerators, washing machines	100
less sensitive instruments	125

If a package has to provide protection against
impacts and vibration, firstly the package must resist
the corresponding stress and, secondly it must absorb
sufficient of the kinetic energy of the impact. Fixing
is the method of placing and securing goods inside
the package to prevent external impacts damaging the
goods by attaining the maximum possible absorption
of kinetic energy. Two types of fixing are usually

differentiated, viz. rigid and flexible.

Rigid fixing means that the package and the article form one solid unit: in this case any shifting of the article and impacts within the package are prevented. This type of fixing is suitable for solid articles.

Flexible fixing means that the package and the goods are connected, but do not form one solid unit. The fixing materials after an impact allow the article packed to continue in its inertial movement for a short time, until it is brought to rest by the fixing material (which is either elastic or non-elastic) absorbing the impact kinetic energy.

In general, plastic packages withstand both impacts and vibrations, and also plastics are useful as fixing materials.

The efficiency of various materials may be characterized by the dependence between the critical g_f factor and the static load for a given thickness of the fixing material, and drop height.

The curves exhibit characteristic minima, as can be seen in Fig. 9 /37, 38/.

Figure 10 gives these curves for polyurethane foam, and shows the calculation of the optimum (economical) thickness of foam. If we know the weight of the article packed (e.g. 10 kg) and the surface area which should be protected (e.g. 25 cm x 25 cm), we deduce a static load of about 1.5 kPa. If we know the critical g_f factor of the article (e.g. 50) and the potential fall height is 75 cm, then we can find the point of intersection of the g_f factor and the static load, and all the thickness values found below that point will satisfy the expected conditions of mechanical stress. The most economical curve, of course, is the nearest one, i.e. thickness of 5 cm in this case.

Plastics offer a number of possibilities for effective protection by packing. Examples are the in-situ

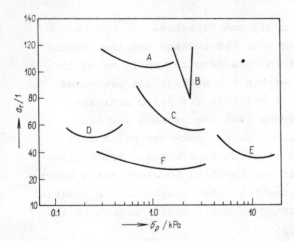

Fig. 9. Rough data on fixation efficiency of some
materials against dynamic stress (dependence of the
relative limit deceleration a_r of the product packed
on the compressive strength σ_p^r of the fixation material)
for a fall from 750 mm height at the temperature of
21°C /37/

A - cellulose stuffing of 75 mm thickness,
B - double-wall corrugated board, C - expanded
polyethylene (mean density of 19 kg m^{-3}) in a
layer of 50 mm thickness, D - the so-called rubber
horsehair in a layer of 50 mm thickness, E - expanded
polyethylene (mean density of 16 kg m^{-3}) in a layer
of 50 mm thickness, F - expanded polyethylene (mean
density of 23 kg m^{-3}) in a 100 mm layer

formation of polyurethane foam around articles /39/,
and "pneumatic packing" which makes use of plastic
cushions filled with air (dunnage bags) /40/.

A special case of mechanical stress is encountered
with the goods packed in shrink-films or stretch-films,
on pallets or in smaller units. The tension in the
film acts on the goods packed and can cause deformation
of non-rigid articles: therefore, the shrinking tension
should not exceed about 3 MPa, otherwise it is necessary
to use inserts, protective frames, etc.

Whereas the mechanical stress due to impact or
vibration arises usually during distribution, that
due to pressure changes within the package is normal

during production processes involving heat treatments of foodstuffs. In these cases the stress affects the packages because of thermal expansion of the content.

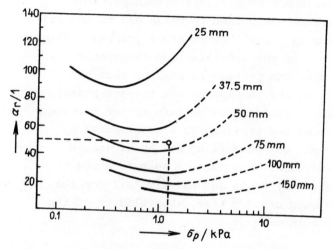

Fig. 10. Fixation efficiency of expanded polyurethane layers (mean density of 19 kg m^{-3}) of various thickness for a fall of a fixed object from 750 mm height /37/. The graph also shows the fixation thickness for an object with critical factor g_r = 50 and compressive strength σ_p = 1.5 kPa, presuming a possibility of fall from 750 mm height. a_r – relative deceleration

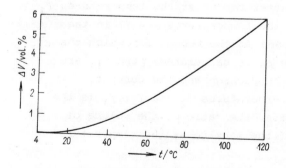

Fig. 11. Volume changes ΔV in dependence on temperature t

Figure 11 shows the thermal expansion of water (the
volume at $4^{\circ}C$ is taken as 100%) and as water is a substantial component of foods which require sterilization
or pasteurization, the curve in Fig. 11 can serve as
a basis for assessment of pressure within packages.
The curve indicates the volume increase due to heating
the content from an initial temperature (during filling
or after closing) to the sterilization temperature. If
the free volume in the package is smaller than the
volume increase of the incompressible aqueous phase,
the inner pressure would cause rupture of the package
or of the closure. The pressure is also affected by
gaseous components of the filling and by expansion
of the package itself. This fact is seen from the
equation suggested by Dikis and Malskiy /41/ for the
pressures in closed cannery tins, the validity of the
relation being more general.

$$p_2 = p''_p + (p_1 - p'_p)\left(\frac{1 - f_1}{\gamma - \gamma_1 f_1} \cdot \frac{T_2}{T_1}\right)$$

In this equation p_1 and p_2 stand for the pressures
(in Pa) in the package after its closing (at temperature
T_1) and at T_2, respectively; $p^{\wedge}_p$ and p''_p are the partial
pressures (in Pa) of water vapour at the temperatures T_1
and T_2; T_1 and T_2 are the temperatures (in K) in the
package after closing and at the moment for which the
corresponding pressure p_2 is calculated; γ and γ_1 are
the volume ratios of the package and the content,
respectively, at the temperatures T_2 and T_1; f_1 is the
filling coefficient (i.e. the ratio of the volume of
the content to that of the package at the temperature
of filling).

Although the expansion of packages decreases the
risk of extremely high pressures, it must be noted

that the different types of package vary in their
resistance to rupture. Glass packages designated for
pasteurization should withstand internal overpressure
of at least 1.2 MPa (for one minute), whereas cannery
tins are irreversibly deformed and, sometimes, perforated at about 0.28 MPa, flexible plastics packages are
damaged at overpressures as low as 0.11 MPa (depending
on the plastics used and the quality of the weld).

From the given formula it is clear that higher
filling temperatures will lead to smaller expansion of
the contents and lower maximum pressure. From this point
of view it would be ideal to fill the package at the
sterilization temperature. Filling at higher temperatures also has the advantage of preliminary removal
of gaseous components contained in the foodstuffs
either in dissolution or as occluded bubbles which
would otherwise increase the maximum internal pressure.
However, filling at higher temperatures is not possible
with carbonated beverages, beer, mineral waters, etc.;
on the contrary, the beverage must be filled at the
lowest possible temperature to ensure the highest
solubility of carbon dioxide. Consequently, the pasteurization of carbonated beverages results in the highest
pressures.

Carbonated beverages also exemplify the case where
permanent inner pressure stress must be expected to
affect the package during storage. Whereas no problems
are connected with the traditional packages, glass
bottles or cannery tins, the long-term pressure stress
can cause trouble in the case of plastic packages.

The plastic package introduced under the commercial
name *Rigelo Pak* by the Swedish firm of the same name
is one of the first for carbonated beverages. From
Fig. 12 it can be seen that this package has a spherical convex bottom, which best withstands the pressure;
the middle part is cylindrical and (being most stressed)

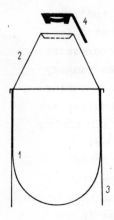

Fig. 12. The Rigelo Pak package for beer and beverages saturated with carbon dioxide

1 - the bottle body with spherical bottom (poly(vinyl chloride) lined with poly(vinylidene chloride), the wall thickness 0.1 mm), 2 - the upper conical part of the vessel made of poly(vinyl chloride), 3 - the re-inforcing cardboard collar (usually backed with printed aluminium foil), 4 - the closure made of branched polyethylene

is reinforced with a cardboard collar (which also ensures stability of the package); the upper part of the package (which is welded to the middle part) has a conical shape which best withstands down-ward pressure during filling. The closure is made of polyethylene as one unit by injection moulding, it is of the plug type and has a projection to facilitate removal by hand. The volume of the package is 280 ml, and its weight is 20 g. As the package itself is made of poly(vinyl chloride), it does not withstand sterilization, but, according to the producer, beer pasteurized before filling can be kept in it for six weeks.

The problems of internal pressure arise also with aerosol packages. The light-weight packages made of metal sheets, which are most usual for these products, are considered safe up to 0.28-0.30 MPa. Plastic pack-ages have not found any significant applications in this field so far. There exist, however, production units for smaller aerosol vessels made of high-density polyethylene, polypropylene, polycarbonates, or poly-amides /42/.

1.2.2 PROTECTION OF GOODS AGAINST CLIMATIC EFFECTS

1.2.2.1 Protection of Goods against Moisture

Humidity is one of the most important climatic factors, and relevant to a broad range of products /43/. In principle the following groups of products can be differentiated:

1. Products whose necessary water content is continuously influenced by humidity and temperature changes. This group includes especially foodstuffs, pharmaceuticals, textiles, paper products, wood, etc. With these products either water absorption or dehydration may be undesirable depending on the type of product. The corresponding sorption and desorption isotherms show smooth curves without abrupt changes.

2. Products which may contain chemically bound water. The reaction of such products to humidity variation is not continuous, but abrupt changes take place beyond certain specific limits. This group includes especially crystalline chemicals. The corresponding sorption and desorption isotherms are characterized by sudden jumps corresponding to the various hydrates.

3. Products which are anhydrous and are degraded or corroded by water. This group includes many metals as well as some other materials with sensitive surface (optical instruments). The corrosion of metals begins after humidity reaches a specific limit.

4. Products containing little or no water and showing no substantial changes due to humidity. This group includes glass products, some metals and metal alloys, some plastics, ceramics, etc.

5. Products consisting of various components given in the previous groups.

In the first group the products are characterized
by the equilibrium relative humidity which means the
humidity reached above the product in a small enclosed
space. The quality of this type of product is optimum
at a certain water content. If the goods are kept at an
air humidity lower than the optimum equilibrium humidity, they lose water by evaporation; on the other
hand, the foods kept in more humid conditions absorb
water. Hence, the sorption isotherm represents graphically the relation between the water content of goods
and the relative humidity, and it forms a basis for
considerations about packaging goods for protection
against humidity changes. In Fig. 13 there are typical
sorption isotherms of some bakery products /44/.

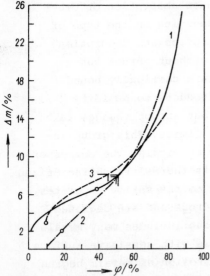

Fig. 13. The sorption isotherms for some sorts of
bakery products (durable pastry) /44/.

Δm - the water sorption expressed as the weight
change, φ - relative humidity. 1 - sponge-biscuits,
2 - wafers, 3 - crackers. The empty circles denote
the initial humidity, and the arrows point at the
maximum admitted humidity of the products sold

Similar considerations apply to the second group of products, the only difference being that the water content is changed suddenly after a certain relative air humidity is reached: as a rule, such abrupt changes in water content cause sticking, caking, or even dissolving. A sorption isotherm characterizing this group is given in Fig. 14.

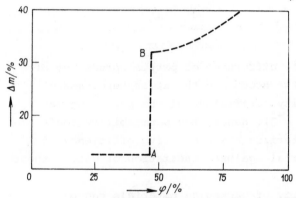

Fig. 14. The sorption isotherm of disodium dichromate /43/ Δm - the water sorption expressed as the weight change, φ - relative humidity, AB - the jump change of humidity

The main problems in the third group are those of metal corrosion following from condensation of water at the surface products for the engineering and electrical industries.

The products of the fourth group usually need no special protection against climatic factors.

Considering the fifth group, it is obvious that the requirements for protection must be judged depending on which component of the product is most endangered, and whether it can be separated and packed alone.

As the greatest requirements with regard to protection against climatic effects are those of the products of the first two groups, in the following text they are given special attention.

If a product which is characterized by the partial
pressure p_1 (in Pa) of water vapour is separated from
the surrounding atmosphere (with partial pressure p_2
(in Pa) of water vapour) by a package material of the
surface area A (m^2) and thickness t (mm), then the
volume Q (m^3) of the water vapour which diffuses through
this package during the time τ (d) is given by the
relation

$$\dot{Q} = PA \, \Delta p \, \tau t^{-1}$$

where Δp means the difference of partial pressures of
water vapour of the goods and the atmosphere, and p
is the permeability coefficient of the packaging material (g cm d^{-1} m^{-2}). Hence, the permeability coefficient P characterizes the protective efficiency of
the package material against penetration of water vapour
(Table 5).

The permeability of packaging materials can be
affected to various degrees by mechanical effects. If
accidental damage of the package is excluded, these
mechanical effects include, first of all, the systematic weakening of packaging materials by shaping and
folding in the production of the packages. Various
materials exhibit differing degrees of resistance to
folding, as it can be seen in Table 6 by Heiss /44/,
where the permeability coefficients for water vapour
are compared for the flat unfolded materials and those
folded twice along perpendicular axes; in the latter
case the permeability of the material was tested after
flattening, when the folds appeared as a cross on the
sample area of 42 cm^2.

Closures present similar problems with regard to
permeability to water vapour. The permeability of
glued closures is calculated on the basis of surface
area and thickness of the glued region and permeability

Table 5. Coefficients of permeability P (g m^{-2} d^{-1}) of some
packaging materials of the given thickness t (mm) or surface
density ϱ (g m^{-2}) (surface densities of the coating are given
in brackets) for water vapour at 25 ^{o}C at the water vapour
pressure difference 231 Pa (corresponding to the difference
between relative humidities of 0% and 75%). $P_{0.1}$ means the
corresponding coefficient for 0.1 mm thickness

Packaging material	t	ϱ	P	$P_{0.1}$
aluminium foil with polyethylene		110 (50)	0	
aluminium foil backed with cellophane bonded with microcrystalline wax	0.01		0.14	
aluminium foil laminated with sulphate (kraft) paper	0.01		0.4	
aluminium foil with vinyl chloride	0.009		1.0	0.1
vinylidene chloride (Saran) copolymer	0.035		0.3	0.1
low density polyethylene	0.03		1.1	0.3
polypropylene	0.04		1.3	
paper with vinyl chloride-vinylidene chloride copolymer	0.03		2.0	
paper with polyisobutylene	0.03		3.0	
cellophane with low density polyethylene		60 (24)	3.5	
rubber-coated fabrics	0.8		3.8	
rubber hydrochloride (Pliofilm)	0.04		4.5	1.8
raw candy wrap with polyethylene		70 (20)	5.6	
unplasticized poly(vinyl chloride)	0.04		9.0	3.6

Table 5 - Continued

Packaging material	t	ϱ	P	$P_{0.1}$
slack wax impregnated paper			11.4	
vapour-tight varnished cellophane			17.0	
plasticized poly(vinyl chloride)	0.13		17.4	22.6
polyamide (Silon)	0.1		17.5	17.5
tar paper			31.4	
paraffin-wax-coated paper		60 (20)	39.0	
container board		1250	44.0	
polystyrene	0.04		60.0	23.0
sulphite paper with lacquer			70.0	
spruce wood	4		84.0	33.6
paper with poly(vinyl acetate)	0.03		100	
cellulose acetate	0.04		100	40
parchment paper		60	251	
cellophane (common sort)	0.03		252	76
artificial skin Cutisin (75 gauge)	0.068		330	224
parchment substitute			332	
pergamyn	0.03		446	133
sulphite wrap	0.07		649	454
air	0.1			5.1×10^5

of the glue used. If the closure is only folded or
rolled in, the resulting resistance to penetration
of water vapour is usually lower.

Moisture transfer between the goods packed and
atmosphere does not take place at a constant difference
of partial pressures of water vapour outside and inside
the package (or at constant difference of relative

Table 6. Permeability coefficients (P) of some folded and unfolded packaging materials for water vapour

Packaging material	P ($g\ m^{-2}d^{-1}$)		
	unfolded	folded	difference
paper with paraffin wax coating one side	7.5	35.2	27.7
bag paper coated with bitumen (150 $g\ m^{-2}$)	7.3	11.8	4.5
paper with paraffin wax coatings at both sides	1.8	6	4.2
unplasticized PVC film (0.04 mm)	8.2	8.5	0,3
pergamyn paper with LDPE coating	5.1	5.4	0.3
LDPE film (0.07 mm)	1.4	1.4	0.0
paper with Diofan coating one side	1.3	0.9	-

PVC - poly(vinyl chloride), LPDE - low density polyethylene

humidities at the same temperatures), but represents
a non-steady process in which the difference between
the partial pressures tends to zero during storage.

This applies especially to the calculation of the
admissible storage period of dried packed products
or to the calculation of the required permeability
of the packaging material to protect the product
against water uptake for a specified time.

For practical purposes it is usually sufficient
to make approximate calculations, because of the
variability of some factors (e.g. atmospheric relative
humidity): instead of the gradually diminishing
difference between the partial pressure of water vapour
in the atmosphere and that over the product it is
possible to use the difference between the partial

pressure in the atmosphere and the average value of
the partial pressure over the product protected.

1.2.2.2 Protection of Goods against Oxidation

Oxidation processes can be considered the most important chemical changes taking place during storage of a
number of products, particularly foodstuffs; the
requirements of the latter for protection by packaging have been investigated most.

Atmospheric oxygen takes part in most oxidation
processes in foodstuffs, hence protection can be
secured with a package which reduces the access of
oxygen. This protection is often enhanced by removing
oxygen from the package either by evacuation or by
displacement with an inert gas. In this connection,
it must be mentioned that packaging materials may be
permeable to gases other than oxygen, e.g. nitrogen,
carbon dioxide; the latter and water are the main
gaseous products of some oxidation-reduction processes.
Antioxidants are sometimes incorporated in the packaging materials to complement the protective action.

In some kinds of foodstuffs the restriction of
oxygen can disturb the normal oxidation-reduction
processes and cause undesirable by-products or even
complete spoilage. This applies especially to fresh
fruit and vegetables.

Some packaging materials can react with the contents
and produce a reducing atmosphere which can be favourable for some components but undesirable for others
in foodstuffs. Such reduction can result from partial
dissolution of metal in the content of the package.

As for water vapour, there are materials of various
permeabilities for oxygen and other gases, as e.g.
completely air-tight packages made of glass or metals,

plastic packages exhibiting selective permeability for
gases, and permeable porous packages made of paper.

With regard to exchange of oxygen and gaseous products
of oxidation-reduction processes between the outer
atmosphere and foodstuffs, these can be classified in
two groups. The first includes products which require
the exchange in order to maintain optimum quality; the
main representatives are fresh fruit and vegetables
and some products in which fermentation processes take
place. The second group includes products in which
chemical changes resulting from exchange of gases
(especially with absorption of oxygen) are undesirable;
the main representatives are sterilized, frozen, or
dried products, especially those containing an oxidable
component, as e.g. vitamin C, some dyestuffs and flavouring additives. Some foodstuffs form a transition between
the two groups, i.e. those which in one phase need oxygen or removal of some gaseous products of oxidation-
-reduction processes, but in subsequent phases it is
advantageous to minimise such processes by separation
of the foodstuff from the atmosphere; this is the case
with meat.

A package must meet the requirements for exchange
of gases, exclusion of oxygen, etc. according to the
characteristics of individual food products.

The permeability of packaging materials to oxygen
and other gases is usually expressed by the gas volume
which passes through a sample of defined surface area
and thickness under specified conditions of time,
temperature and partial gas pressure differential
(cf. Section 3.3.1).

Water vapour and/or water may affect the permeability
to gases in two ways. If water causes swelling of
packaging media, like plasticizers, it increases the
permeability. If the medium does not absorb water,
water vapour can be absorbed at the package surface,
whereby the permeability is somewhat decreased.

Therefore, the more precise data on permeability also specify the relative humidity during testing.

Table 7 summarizes the permeability coefficients of several packaging materials for some gases.

The permeability to oxygen is a criterion for comparing individual packaging materials, but may be misleading in the case of a specific oxygen-sensitive product.

If the package contains a solution of some oxygen-labile substance (e.g. ascorbic acid) which reacts so rapidly that the oxygen passing through the wall of the package is immediately consumed in the chemical reaction, then the barrier function is of prime importance.

Obviously, the effective protection with oxygen-sensitive foodstuffs is determined by the relation between package permeability and the rate of consumption of oxygen in the foodstuff.

For a liquid the oxygen transfer is convection-dependent, and the ratio between the rates of oxygen supply and consumption can be characterized by a dimensionless factor Sl suggested by Slavíček /45/:

$$Sl = DAt^{-1}k^{-1}c^{-1}V^{-1}$$

where D is the diffusion coefficient of oxygen for the packaging material $(m^2 s^{-1})$, A is the package surface area (m^2), t means thickness of the film (m), k is the second order rate constant of the reaction of oxygen with the reactive component $(1 \, mol^{-1} s^{-1})$, c is the concentration of oxygen in the foodstuff $(mol \, 1^{-1})$, and V is the package volume (m^3).

This factor denotes the actual protective efficiency of the package, provided that the oxidation process taking place is a bimolecular reaction of the second order as for the oxidation of ascorbic acid /46/.

Table 7. Permeability coefficients (P) of several packaging materials for some permanent and technically important gases (measured at 133.3 Pa partial pressure difference, area 1 m², and thickness 1 mm)

Packaging materials	Test conditions T, °C	r.h. %	Permeability coefficient P (ml m^{-2} d^{-1}) O_2	CO_2	N_2	H_2	SO_2	NH_3
low density polyethylene	25	0	11–26	47–120	4–9	43–69	258	241
PVC (unplasticized)	25	0	1.0	80		112		
poly(vinylidene chloride)	25	0	0.88	0.017–0.17	0.013	1		
poly(vinyl alcohol)	21	50	0.8	1				
polyamide (Nylon)	25	50	0.17	1.5				
polystyrene	20		17–206	86–285	67	774		
rubber hydrochloride	25	0	0.4–103	2.6–602	4–86	17		
poly(ethylene terephthalate)	25		0.17			5		
cellophane (unimpregnated)	25	0	0.018	0.037	0.03	0.054	0.015	0.14
	25	43	0.06	1.2	0.06	0.14		
	25	100	0.09	2.2	0.15	0.65	366	1522
cellulose acetate	25	25	3.6	48–104				
pergamyn paper (t = 0.03)	21	75	1.2–62	2.5–52				
" " (t = 0.06)	21	0	254	172				
sulphite paper (t = 0.045)	22	0–75		700–1400				

t – thickness (mm), T – temperature, r.h. – relative humidity

If $Sl = 1$, then the rate of introduction of oxygen
into the material is the same as the rate of its
consumption in the chemical reaction. If $Sl \ll 1$, then
the introduction of oxygen is slow compared with its
consumption, hence the rate of the oxidation process
is limited by the low permeability of the packaging
material. If $Sl \gg 1$, the introduction of oxygen is
rapid compared with the reaction rate, and the package does not function as an effective barrier.

If the initial concentration of the oxidized
substance and the solubility of oxygen in the foodstuff
are known, it is possible to calculate the kinetics
of the oxidation, i.e. the time dependence of the
concentrations of both the oxidized substance and
the dissolved oxygen. It is thus possible to determine
the storage life of the product, provided it is known
at what oxidation level the product becomes unaccept-
able. Such calculations are time consuming, therefore
a computer program was set up for this purpose /46/.

For solid foodstuffs and technical products, the
effective protection of the package also depends on
the ratio between the permeability of the package for
oxygen and the rate of consumption of oxygen in the
product; in addition account must be taken of the
diffusion coefficient (for oxygen) of the product.
It is useful to introduce the concept of "equivalent
thickness," t_e, i.e., such a layer of the product as
would have the same diffusion resistance to oxygen as
the packaging medium. The value is easily obtained
from the relation

$$t_e/t_f = D_p/D_f$$

where t_f is the thickness of the packaging medium, D_p
and D_f are the diffusion coefficients of oxygen in
the foodstuff and in the packaging medium, respectively.

It is calculated that jam covered with a polyethylene

film of 0.05 mm thickness is protected against penetration of oxygen as little as by a 0.9 mm layer of
the jam itself, which is negligible protection. In
contrast, if a poly(ethylene terephthalate) film of
0.03 mm thickness is used instead, the protective
effect is equivalent to a 4.8 cm layer of the jam.

When calculating the oxidation kinetic for solid
foodstuffs, it is possible to start from the mass
transfer equation suggested by Slavíček, using the
conduction conditions of oxygen transfer /47/. It
is necessary to know the values of the relevant parameters; for the packaging material its thickness
and its diffusion coefficient for oxygen; for the
product its diffusion coefficient for oxygen, the
initial concentration of the substance oxidized, the
solubility of oxygen, and the rate constant of the
oxidation reaction. The treatment necessitates that
the oxidation proceeds a second order bimolecular
reaction. This condition is fulfilled quite well in
the case of ascorbic acid, but does not apply fully
to the more complex oxidation of fats.

1.2.2.3 T h e P a c k a g e a s B a r r i e r
        t o O r g a n i c V a p o u r s

The barrier function of a package to penetration of
vapours of various organic compounds is important both
where changes of taste or smell of foodstuffs must
be avoided and when packing volatile organic compounds.

The risk of escape of fragrant substances is especially great with foodstuffs containing very small
amounts of readily volatile components as in some fruit
juices. Apple juice contains about 0.15% volatile components which escape during evaporation together with
about the first 10% water. Pear juice is similar in
this respect, whereas morello juice and raspberry juice

are deprived of their typical aroma by evaporating 15
and 30% water, respectively. Therefore packages for
fruit juices (especially those designated for long-term
storage) must possess low permeability to esters of
lower fatty acids, carbonyl compounds, alcohols, and
volatile acids. In citrus fruits the characteristic
taste and smell are due to less volatile components
like terpenes, sesquiterpenes, unsaturated aliphatic
alcohols of terpenic type (e.g. linalool, geraniol,
citronellol) and aldehydes (citral); as these components are sensitive to oxidation changes, the package must also possess low permeability to oxygen.

With some vegetables (e.g. garlic, onion,
horse-radish) their distinct aroma (due especially to
organic sulphur compounds, like allyl sulphide, allyl
isothiocyanate) can penetrate into other products,
which must be prevented.

Good impermeability to aromatics is required in
the case of various spices with the aim of preventing
both the escape of volatile components from the spice
and the absorption of foreign smell.

The risk of absorption of alien odours (either
from foodstuffs kept side by side or from various
cleaning agents, disinfectants, etc.) is especially
marked with foods of higher fat content (e.g., butter
or chocolate), because most odours substances are
soluble in fats. Fats are affected most readily
by spices, flowers, perfumes, coffee, fish, kerosene,
soap, paints and varnishes, etc. The taste of foods
rich in fats is also considerably affected by oxidation of the fat, so that impermeability to aromatics
must also be accompanied by impermeability to oxygen.

Changes of taste and smell are frequently encountered
with products having large surface areas, and are due
to easy oxidation of the organoleptic components or
to adsorption of alien odours at the large surface or

to both. Roasted coffee is an example of such products
with high demands on packaging materials.

Meat is also vulnerable to odour contamination but
some kinds of fish must be sealed in a suitable package in order not to contain the typical fish smell
(usually ascribed to trimethylamine).

The main problem in packing aromatic foodstuffs
(and other products) is to attain the necessary impermeability of the packaging material to the volatile
organic components. In contrast to impermeability to
water vapour and gases, this problem is complicated
by the large variety of organic vapours.

Beyond the airtight packages made of metals or
glass, the permeability of other materials (i.e. either
non-porous materials like plastic films or porous ones
exemplified by paper) to vapours of organic compounds
is controlled by the same laws as the permeability
to water vapour and gases.

The experimental determination of permeability to
vapours of organic compounds is based on quantitative
measurement of the compound passing from one side of
the packing medium to the other using chemical methods
(a selective reaction or a group reaction of the organic
vapour), or physical methods (e.g. interferometry), or
physico-chemical methods (gas chromatography); organoleptic methods (olfactometry) may be employed also.

Table 8 lists some more important packaging materials
arranged according to increasing permeability to ethanol
and propyl acetate vapours /47/. From the table it can
be seen that the order of single materials (polyethylene
cellophane, glassine, polyamide) is practically reversed
when going from ethanol to propyl acetate, whereas
certain combined materials (Celothen, i.e. a combination
of cellophane and polyethylene, and lacquered cellophane)
do not show such deviations.

Table 9 gives the permeability coefficients of low

Table 8. Permeability coefficients (P) of some packaging materials for organic vapours (measured at
133.3 Pa partial pressure difference, temperature
$25^{\circ}C$, surface area 1 cm^2)

Material	Thickness (mm)	$Px10^9$ (ml s^{-1})
For ethyl alcohol vapour:		
Low density polyethylene	0.09	2.54
pergamyn paper	0.025	6.33
Celothen (cellophane + polyethylene)	0.09	6.63
lacquered cellophane	0.025	12.08
cellophane	0.03	20.0
polyamide	0.06	38.5
plasticized poly(vinyl chloride)	0.135	81.6
For propyl acetate vapour:		
cellophane	0.03	0.023
polyamide	0.06	0.84
Celothen	0.09	0.91
lacquered cellophane	0.025	1.36
glassine	0.025	6.46
low density polyethylene	0.09	31.3
plasticized poly(vinyl chloride)	0.135	144.0

density polyethylene for ten different organic vapours
while Table 10 provides a comparison of barrier efficiency for five packaging films against various
organic vapours /48/.

Comparison of the suitability of packaging material can
also be based on specific organoleptic properties; the times
needed for the concentration of some volatile compounds
(which diffuse from the inside of the package to the
outside) to reach the level which can be detected

Table 9. Permeability coefficients (P) of low density polyethylene for organic vapours (measured at 1 cm thickness, partial pressure difference 133.3 Pa, temperature $35^{\circ}C$, surface area 1 cm^2)

Penetrating vapours	$Px10^{13}$ (ml s^{-1})
ethanol	2.49
methanol	3.3
propanol	4.1
acetone	4.4
water vapour	4.5
butyric acid	14.2
acetic acid	22.1
ethyl acetate	22.9
butylamine	75
tetrachloromethane	170

olfactometrically are given in Table 11 /49/.

From the examples in Table 10 and 11 it is seen that packaging films based on poly(ethylene terepthalate) possess generally a low permeability to organic vapours. Poly(vinylidene chloride) and polyamide films are usually a little more permeable, which is also true of un-plasticized poly(vinyl chloride). Cellophane and some varieties of greaseproof paper (parchment and pergamyn papers) can often be advantageously used. On the contrary, polyethylene finds rather limited application due to its considerable permeability to most organic vapours.

Table 10. Permeability coefficients (P) of five types
of packaging films for some organic vapours (thickness
1 mm, temperature $20^{\circ}C$, partial pressure difference
133.3 Pa)

Organic vapour	P $(g$ $m^{-2}d^{-1})$				
	1	2	3	4	5
petrol	0.016	0.062	0.14	66.8	0.052
carbon disulphide	0.41	1.44	150	1148	22
chloroform	2.8	8.7	29	306	1536
methanol	45	0.026	0.005	0.46	218
ethanol	5.1	0.19	0.007	0.46	107
diethyl ether	0.84	1.6	0.8	130	40
acetone	0.23	0.34	26	2.6	x
ethyl acetate	0.19	0.28	10.9	6.6	x
ammonia	1.4	0.29	0.009	0.58	128
diethylamine	0.0012	3.15	xx	80	200
pyridine	0.005	0.006	x	6.8	x

1 - polyamide, 2 - poly(ethylene terephthalate),
3 - unplasticised poly(vinyl chloride), 4 - low density
polyethylene, 5 - cellophane, x - the film is dissolved, xx - the film is attacked

Table 11. Olfactory determination of various organic
vapours penetrating packaging films (time in hours
to odour detection)

Volatile substance	Poly(ethylene terephthalate) (Mylar)	Cellophane	Vinyl chloride - vinylidene chloride copolymer Saran
	t = 0.025 mm		t = 0.06 mm
vanilin	141	20	24
extract from spruce needles	140	20	1
methylsalicylic acid	116	24	24
ethyl butyrate	20	0.5	0.016
propylenediamine	92	1	4

t - thickness of the film

1.2.2.4 P r o t e c t i o n   a g a i n s t
R a d i a t i o n

Electromagnetic radiation of very broad wavelength
range (i.e., gamma radiation, X-rays, ultraviolet
radiation, visible light, infrared radiation) and
also corpuscular radiation (beta radiation and
alpha radiation) can have immediate effects, both
favourable and unfavourable, in packaging numerous
products, especially foodstuffs, medicines, some
chemicals, sanitary materials, etc. The favourable
and technologically useful effects include especially
sterilization by short-wave radiation (particularly
gamma radiation, but also X-rays and ultraviolet
radiation) and by corpuscular radiation (mainly beta
radiation); the insecticidal effects of radiation,
and other specific effects, e.g. reduction of
germination of potatoes by gamma radiation, or
formation of vitamin D from its provitamins by UV
light. A special use is the X-ray determination of
amount and/or condition of products in opaque metal
packages. Some unfavourable effects are known for
almost any type of radiation. Initiation of secondary
radioactivity by high doses of radioactive radiation
can be neglected as a rare case, but oxidation-reduction changes in foodstuffs initiated by radioactive,
ultraviolet, and visible radiation of shorter wavelengths
and undesirable changes in some kinds of packaging
materials by radiation are quite important.

Of the corpuscular radiation, in packaging, alpha
radiation, i.e. positively charged helium ions, is
unimportant. It is practically completely absorbed
by a sheet of paper or aluminium foil even if energy
of the particles is relatively high preventing
technological use.

In contrast, beta radiation, or cathode glow, having

a higher penetrating power, could be applied to "cold"
sterilization of foodstuffs and packages. Usually, the
maximum range of beta particles is expressed by the
thickness (surface density) of the aluminium layer
causing complete absorption. The penetrating power
of this radiation is determined by the energy of the
particles.

Absorption of beta particles obeys quite well the
exponential relation

$$I = I_0 e^{-\mu t}$$

where I is the activity of the radiation after passage
through the absorbing layer of the thickness t (cm),
μ is the overall absorption coefficient (cm^{-1}), and
I_0 is the activity measured in the absence of the
absorbing material.

To a first approximation, the absorption coefficient
μ depends linearly on the density of the absorbing
material. Obviously, the requirement of maximum penetration in the "cold sterilization" of foodstuffs is met
best by packaging materials of low surface density and
minimum possible thickness.

The same principles apply to radioactive radiation,
gamma rays, of very short wavelengths (0.5 to 40 μm).
Being the most penetrating of the radiation, gamma
radiation is most suitable for radio-sterilization
of foodstuffs.

In the radio-sterilization of foodstuffs, the sealed
package not only prevents subsequent contamination of
the products packed, but it also restricts the entry
of oxygen thus reducing undesirable oxidation processes
which proceed very rapidly under ionizing radiation.

For this purpose aluminium, because of its low
density (2.70 g cm^{-3}), is more suitable than steel
(7.90 g cm^{-3}). If the foodstuffs had to be protected

against radioactive radiation, the order would, of
course, be reversed. In such case lead would also
be suitable for outer protection (11.30 g cm^{-3}).

Glass is usually characterized by a high chemical
resistance, but it is not very suitable for foodstuffs
which are to be sterilized by radiation, because the
glass thickness usually absorbs a large part of the
radiation; in addition, glass turns dark and opaque
under radiation.

Plastics, having very low densities (e.g. polystyrene
1.05 g cm^{-3}, cellophane 1.50 g cm^{-3}), would be suitable
as packaging materials for products which are to be
sterilized by irradiation, but under high doses of
radiation, they undergo changes consisting either
in cross-linking or in degradation of the macromolecules
(cf. Section 3.3.5). These chemical changes induce
physical changes, especially in colour, mechanical strength,
and also permeability to water vapour and gases.

A number of packaging materials have been studied
from the point of view of the changes caused by the
sterilization doses of radiation. Proctor and Karel
/50/ studied the changes in polyethylene, rubber
hydrochloride, poly(vinylidene chloride), coated
cellophane and other combined types of packaging materials,
caused by relatively high doses (30-40 kJ kg^{-1}), of
cathode rays. The characteristic "packaging" properties,
such as permeability to oxygen and fats, remained
practically unchanged. Small changes were observed in
the weldability of some films. Darkening occurred with
poly(vinylidene chloride).

Holečková, Čekalová, Lanzer /51/ and Holečková,
Kyselová and Lanzer /52/ sterilized a wide range of
packaging materials by irradiation with doses of 5 to
50 kJ kg^{-1}. They arrived at the conclusion that the
most suitable packaging material, with regard to
radiation resistance, is poly(ethylene terephthalate)

(*Melinex*), followed by polyethylene. The odour produced
by irradiation could be prevented by a suitable stabilizer.
The radiation doses used for treatment of foodstuffs
vary from about 0.15 kJ kg^{-1} (e.g. to prevent germination of potatoes), through 0.1-0.5 kJ kg^{-1} (as applied
against insects and parasites in corn), 1-5 kJ kg^{-1}
(pasteurization), up to 30-50 kJ kg^{-1} (sterilization).
Most materials withstand doses up to 10 kJ kg^{-1}, suitable for radio-pasteurization. For higher doses the
selection of plastics is restricted because of various
adverse effects.

Radiosterilization can, in principle, be achieved
by application of another type of short-wave radiation,
X-rays of wavelengths 0.01 to 15 nm. The demands
which must be met by packaging materials are the same
as in the previous case. X-rays are also used in some
cases for the detection of defects in opaque (especially
metal) packages. The penetrating power of X-rays is
sufficient for radiography of cannery tins and their
content. Resolution of individual components of the
contents and/or contaminants necessitates different
absorption of these rays in the components (which is
determined primarily by density differences).
Therefore, it is possible to detect free volume (air)
in the tin, air bubbles in the content, bones and
stones or metal impurities, etc. X-ray radiography
made it possible to solve the problem of colour changes
of sausages not completely immersed in juice (in tins)
and the problem of deterioration of sausages accompanied
by the formation of gas bubbles /53/. X-ray radiography
was also used to study the internal corrosion of
cannery tins.

Sun light reaching the earth contains ultraviolet
radiation of wavelength 286-400 nm, whereas shorter
wave-lengths are absorbed by the atmosphere. The shorter
wavelength ultraviolet radiation from artificial sources

is used for the surface sterilization of foodstuffs
and/or packages but its penetrating power is very
little. Both ultraviolet radiation and the short-wave
part of visible light have serious undesirable
oxidation-inducing effects. They stimulate oxidative
decomposition of fats, because the energy required
for formation of the fatty acid residues is of the
same order of magnitude as the energy of the radiation.
Therefore, products sensitive to light require packages
especially opaque to the short-wave region of the
spectrum. Metals are completely opaque. Colourless
glass is transparent to wavelengths of about 380 nm
but the transparency decreases with increasing iron
content. Most plastics which are transparent and
colourless also transmit ultraviolet radiation to a
considerable extent. This can be seen from Fig. 15
which shows that polyethylene transmits short-wave
radiation, whereas poly(vinylidene chloride) exhibits
the lowest transparency of the materials considered.

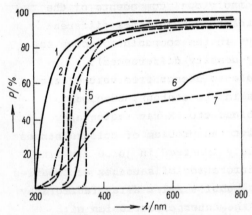

Fig. 15. Dependence of optical transmittance P of some
packaging materials on the wavelength λ of the radiation used
1 - polyethylene, 2 - cellophane, 3 - cellulose acetate,
4 - rubber hydrochloride, 5 - poly(vinylidene chloride),
6 - poly(vinyl chloride), 7 - transparent paper made of
bleached cellulose

The incorporation of suitable absorbers of ultraviolet radiation can impart considerable or complete UV
opacity to quite transparent packaging materials. A
number of compounds exhibiting this effect have been
patented. Gluck and Barany /54/ have reported effective
substances having the general formula RCH=NH=CHR´ in
which R and R´ are substituted or unsubstituted hetero-
cyclic or aliphatic groups. Ellickson and Hasenzahl /55/
used substituted benzophenones as filter agents and
applied them in wax as a coating on cellophane; 0.25%
of the agent in a wax coating showed very good protection
of cheese against rancidity. The authors quote further
compounds useful as the UV filter agent, e.g. 3-methoxy-4-
-hydroxybenzoic acid for hydrophilic films. Other very
efficient substances are diphenylglyoxal, diphenyl-
thioglyoxal, and butyl p-dimethylaminobenzoate. Various
compounds with absorption maxima in the short-wave
region can serve as UV filters, and it is possible to
use substances which have additional properties, viz.
antioxidant, microbiocidal, etc. (e.g. sorbic acid).

Sometimes it is desirable to remove also the
short-wave or other part of visible radiation from the
sunlight impinging on foodstuffs. As the colour of
the transmitted light is complementary to that of the
absorbed light, it is useful to choose the colour of
the transparent or translucent packaging material so
that it is complementary to the colour of the undesirable
portion of the spectrum.

Absorption of radiation, especially of longer
wavelengths, results in transformation of the energy
into heat. Intense heating occurs particularly with
infrared radiation of 760 to 50 000 nm, which is
technologically useful in drying foodstuffs. In contrast,
the thermal effect of solar radiation on packed products
is usually undesirable.

The amount of heat from solar radiation impinging

upon a horizontal surface area of 1 m^2 reaches
3.65o kJ h^{-1}, if the sun is 90° above the horizon,
however, mean values of the radiant heat passing through
windows or shopwindows vary about the value of O.830
J $m^{-2}h^{-1}$ and if the windows are shaded by trees, the
value decreases to about O.315 J m^{-2} h^{-1} /56/. Glossy
aluminium foils provide the most efficient protection
by reflecting a substantial part of the radiant heat.
The problem of absorption of radiant heat is very
important with both consumers´and transport packages
and, in particular, with transport containers for cooled
and frozen foodstuffs. Therefore, materials exhibiting
the lowest possible absorptivity of radiant heat are
preferred in these cases.

1.2.2.5 Role of the Package in Temperature Changes

Heat transfer from outside through the package wall
into the product packed, or in the opposite direction,
is a process which is frequently either intentionally
initiated (as in thermal sterilization or freezing of
foodstuffs) or spontaneous (i.e. temperature equalization
between the goods and their surroundings in accordance
with climatic conditions).

The spontaneous temperature equalization sometimes
has unfavourable effects on the product: not only in the
case of frozen and cooled foodstuffs (whose temperature
should be maintained) but also in the case of some more
sensitive goods which must be protected against short-term
changes of temperature during transport (against freezing
or undesirable heating).

In all these cases, both the consumers´ and transport
packages can be considered to be acting as thermal
resistances, of magnitude determined by thermal conductivity
and thickness of the material used. The thermal conductivity

values of packaging materials and, for comparison, some
other important materials are given in Table 12. The
Table also gives numerical values of specific heat
capacity which can play an important part in some
technological operations in which the heat capacity
of packages has a significant effect (e.g. the energy
expended in heating and cooling packages during ther-
mosterilization).

The density and thickness of packaging materials
vary within broad limits; the thermal insulation of
packages is therefore expressed more reliably by their
thermal resistance, i.e. ratio of thickness to thermal
conductivity. Table 13 gives numerical values of these
quantities for some packaging and/or insulating materials.

Estimating the magnitude of heat transfer from
outside through the package wall into the product packed
or in the opposite direction, we must take into account
not only the conductive resistance of the package, but
also the convective resistance at the inner and the
outer surfaces of the package wall which is given by
the reciprocal value of the heat-transfer coefficient
from gas (or liquid) into the wall or vice versa.
The value of this heat-transfer coefficient does not
express a material property of the packaging material,
but is connected, in a complex way, with convection of
the gaseous or liquid medium.

Insulating properties are required for some transport
packages, especially for frost-sensitive goods, as
fruit, vegetables, eggs. The protection of potatoes
in jute bags and in paper bags has been investigated
/57/. The potatoes were cooled from the initial temperature of 6.7°C to 0°C in the two types of bags, and
the time of cooling was longer by 4-7 h in the paper
bags than in the jute bags in stationary air, the
difference being even greater in an air flow. The temperature of unpacked eggs kept at -17°C decreased from

Table 12. Specific heat conductivity (λ), specific heat capacity (c_p), and density (ϱ) of some packaging and other important materials

Materials	λ ($W\ m^{-1}\ K^{-1}$)	c_p ($kJ\ kg^{-1}\ K^{-1}$)	ϱ ($g\ cm^{-3}$)
copper	5112.8	0.38	8.9
aluminium	2653.2	0.89	2.7
steel	601.2	0.45	7.7
stainless steel (18% Cr, 8% Ni)	194.4	0.46	
glass	9-15.12	0.79	2.5
porcelain	10.08	0.79	2.3
bricks	6.012-10.44	0.83	1.6
concrete	10.44-16.56	0.88	2.2
bitumen, asphalt	9	0.92	2.1
hardenable phenolic materials	6.012	1.09-1.25	1.7
polyamides	2.988-4.5	1.67-2.08	1.13
low density polyethylene	4.5	2.30	0.92
poly(methyl methacrylate)	2.376	1.46-1.67	1.18
poly(vinyl chloride)	1.944-2.232	0.96	1.3-1.4
polystyrene	1.8	1.33	1.05
expanded polystyrene	0.504		0.02-0.2
wood	1.8	1.46-2.72	0.8
paper	1.512	1.34	0.8-1.2
cork	0.432-0.504	1.67	0.016-0.05
water	7.056	4.18	1
ice	28.62	2.05	0.9
vegetable oils	1.8-2.232	1.96	0.93
air (dry)	0.288		0.0012

Table 13. Thermal resistivity $(t\lambda^{-1})$ of some packaging materials with specific heat conductivity λ for the layer thickness t (mm) or the layer of surface density $\rho(g\ m^{-2})$

Material	t (mm)	ρ (g m^{-2})	$(t\lambda^{-1})$ x10^4 (m^2 K W^{-1})
steel sheet for cannery tins	0.25		0.0504
parchment paper		64	13.71
paraffin-coated half-parchment paper		60	36
wall of a bottle	3		36.83
"Eco"-type packages for freezing plants	(two cardboard layers with an aluminium foil between them)		68.54
paraffin-coated cardboard	0.625		82.08
cork insulating material	150		42.84

24°C to -3.5°C within two hours, whereas that of eggs packed in boxes with inserts only decreased to 9°C /58/. The second result is attributable to the insulation of the package itself and to that of the air gaps involved.

Very interesting possibilities are offered by smaller, thermally insulated containers in cases where it is necessary to transport and maintain low or high temperatures of foodstuffs for several hours. Commercially available transport packages for frozen foodstuffs have a volume of about 1 m^3 and can hold up to about 680 kg, the weight of the package itself being 150 kg. With insulation of 65 mm thickness the temperature of the foods packed should not increase more than by 3°C within 15 hours /59/. On the other hand,

also commercially available are packages made of expanded polystyrene which are said to maintain the temperature of hot beverages at 70-85°C for 5 hours /60/.

In Czechoslovakia, a thermally insulating (so called
isothermal) container was developed for a content of
130 l with a polystyrene or polyurethane insulating
layer /61/, the latter being more effective. The container is suitable for transport and short-term storage
of frozen foodstuffs and for storage of solid carbon
dioxide.

Kraft et al. /62/ described a number of applications
of expanded polystyrene in the form of thermally insulating packages and transport containers.

Also interesting is the possibility of utilizing
the thermal insulating properties of transport crates
made of expanded polystyrene; for the transport of
pre-cooled fruit and vegetables it is possible to make
use of the morning coldness when gathering the product.

The storage time in thermally insulating packages
depends on the initial temperature and also on the
amount and temperature of any cooling agents added.
If a low temperature is to be maintained for several
days, solid carbon dioxide (dry ice) is often employed.

A prerequisite for such exposure of the package to
heat or cold is resistance of the packaging material
to that temperature. The temperature range is from
-40°C (cooling media in freezing plants) to about
120°C (sterilization in autoclaves).

1.2.3 PROTECTION AGAINST BIOLOGICAL SPOILAGE

The biological factors which can cause damage to the
packed products include microorganisms, insects, and
rodents, which are hazards especially for foodstuffs.

1.2.3.1 P r o t e c t i o n a g a i n s t M i -
      c r o o r g a n i s m s

Contamination with microorganisms is a major cause of
damage to various products,especially foodstuffs, and
is manifested by mould, fermentation, and rotting to
various extents.

Packages play an important role in the protection
of foodstuffs against microbial infection. This protection can function by the package forming a barrier
separating the product from the atmosphere (which is
usually rich in microbial species), or the package
can contain substances with microbiostatic or microbiocidal effects and thereby contribute to the elimination of superficial microflora.

Besides the character of the foodstuff packed, i.e.
its natural stability or method of conservation
(preservation), the microbial resistance of the packaging materials is also important. Non-hygienically
produced, improperly treated and unsuitably applied
packages can themselves be a source of infection.

Metallic and glass materials cause no substantial
troubles in this respect. Although they can be infection
carriers, they are easily sterilized, so that they
meet the highest requirements. Metal packages, especially
returnable containers (as e.g. milk canisters, beer
barrels made of aluminium), sometimes present corrosion
problems, if unsuitable chemical means are used for
cleaning and sterilization.

Wood, paper, and textile packages present a much
more serious problem from the microbiological point of
view, as they can themselves serve as substrates for
a number of microbial strains.

Significant growth of moulds and some bacteria
strains on these materials usually needs a relative
humidity above 65% and, hence, proportionally increased

moisture content in the material (the critical value
for paper is about 8%), and, of course, adequate temperature. Wooden packages are often attacked by ligni-
perdous fungi e.g. wood decay fungus (*Merulis lacrimans*),
various mildews which can even grow through the package wall into the products packed, and besides that,
improperly stored and badly treated packages can become
carriers of pathogenic microbes which are transferred
by e.g. rodents (*Salmonella typhi murium, Escherichia
coli*, etc.). Returnable wooden packages, especially
those which come into close contact with the foodstuffs
packed (e.g. in packing fats or vegetables) must meet
increased hygiene requirements in this respect. Besides
possible impregnation of wooden packages with bactericidal
substances (see below) it is sometimes desirable to wash
the wooden packages. This washing as well as application
of some cleaning agents or conservation agents (e.g.
sulphur dioxide) present no substantial problems in the
case of hard wood with a smooth surface (wooden casks
and vats). More roughly processed packages made of soft
wood are not so easily washed. To increase the washability
of crates, it is recommended to cover the surface with
a plastic layer (e.g. poly(methyl methacrylate)) by coating
or spraying.

   A similar problem is encountered with paper and
cardboard packages, where the risk of infection is
greater, if the material is made of waste paper, because
the production process does not always ensure inactivation
of the microflora thus introduced. Such materials are
excluded from use in packages for direct contact with
foodstuffs. Parchment paper, containing lower saccharides due to hydrolysis of cellulose by the action of
sulphuric acid during production, is also a substrate
on which most microbes and especially mildews can grow.
As parchment paper comes into close contact with a number
of moist foodstuffs, its rigorous microbiological control

is recommended. It is also not always possible to avoid
microbial growth on paraffin-coated paper and
cardboard, which can cause trouble in packing sweets.
Therefore, it is necessary to ensure hygienic conditions
and to exclude any possibility of reinfection during
both production and storage of paper packaging materials.

Textile packages (made of cotton, hemp, and jute)
are often exposed to moisture and (during potato
harvest) contamination with soil, which produces
favourable conditions for growth of microbial strains.
Of the textile packaging materials, jute has the
highest natural microbial resistance.

Cork stoppers and cork inserts in crown caps are
frequent sources of infection especially with sterilized
beverages. As both natural cork and the pressed cork
have good heat insulation properties, it is difficult
to destroy various mildews (especially the thermally
resistant ones) by heating. The sterilization procedure
is much more difficult in the case of strongly infected
materials. Besides the preventive microbiological
control it is recommended to soak the cork material in
sulphur dioxide or ethanol solutions and to apply
intermittent thermal sterilization. For disinfection
of crown caps it is recommended to soak the closures
in 0.25% solution of dimethyllaurylbenzylammonium
bromide.

With plastics the danger of microbial infection
depends primarily on the type and composition of the
packaging material (cf. Section 3.3.7).

The polymers based on cellulose can be denoted as
non-resistant: in this respect they belong to the
previous group of cellulosic packaging materials. The
main representatives are cellophane and some cellulose
esters, particularly the nitrate and acetate.

There are different opinions about the microbiological resistance of plastics such as poly(vinyl

chloride), poly(vinylidene chloride), polystyrene,
polyamide, poly(methyl methacrylate), and polyesters
and also synthetic rubbers and phenol-formaldehyde
resins. The reason is that the resistance of these
products also depends on the content of additives
(cf. Section 3.3.7). Under normal conditions, however,
these materials can be considered sufficiently resistant.

Of the thermoplastic and thermosetting resins,
polyethylene and urea-formaldehyde resins, respectively,
are considered to be unequivocally resistant to microbial
attack.

The sterilization of plastics is limited by their
thermal and chemical resistance. Practically only
the materials resisting temperatures above 100°C can
be sterilized by heat (e.g. high density polyethylene,
polypropylene, polyamides, poly(ethylene terephthalate),
polycarbonates, and fluorinated polyolefins). Surface
treatment of plastics with aqueous solutions of various
sterilizing chemicals is limited by the hydrophobicity
and, hence, non-wettability of plastics. The application of cationactive surfactants of 0.1 to 0.4%
concentration in 65% aqueous ethanol was fairly
successful /63/. Sterilization with ultraviolet radiation
is especially advantageous for films, and in some cases
sterilization with gaseous ethylene oxide is also useful.

The polyethylene-coated cardboard for use in packing
long-life milk by the *Tetrapak Aseptic* method is
successfully sterilized with hydrogen peroxide solution.

It should be noted that in most cases plastic films
produced and supplied in rolls exhibit high microbial
cleanliness or even sterility (especially the inner
surface of extrusion blown tube).

The protection of goods against microbial deterioration by packages is thus limited in the case of a
number of materials by specific conditions, e.g. for
cellulosic materials by humidity and temperature limits.

Under more severe conditions, e.g. for transport to
highly humid tropical zones, these materials must be
impregnated by microbiocidal or microbiostatic substances.
Such modification should not only protect the packaging
material itself but in case of consumers' packages
should also be effective on the surface microflora
of the foodstuffs packed.

The selection of preserving agents for impregnating
of packages not used for foodstuffs is much easier than
that of harmless, hygienically unobjectionable and,
at the same time, efficient compounds for impregnating
of foodstuff packages.

Several groups of efficient compounds can be identified
as technically useful; first of all those based on
phenol and its derivatives - p-nitrophenol, biphenyl,
cresol, thymol, xylenol, resorcinol and a number of
efficient chlorinated phenols such as pentachlorophenol
and its salts with sodium, zinc, cadmium, and copper,
p-chloro-m-cresol, and p-chloro-m-xylenol. Mercury
compounds are very efficient, especially arylmercury(II)
derivatives such as phenylmercury(II) acetate, naphthenate, oleate, stearate, and salicylate. Also highly
efficient are organotin compounds possessing very low
toxicity for man, so that some applications in the
foodstuff industry are conceivable; possible representatives are tributyltin oxide or benzoate.

For transport packages exposed to rain it is also
important whether or not the impregnation agent is
extracted with water. Such extraction is significant
e.g. with sodium pentachlorophenoxide which is very
soluble in water. In contrast, organomercury(II)
fungicides are very soluble in hydrocarbons but sparingly soluble in water. Also 8-hydroxyquinoline and
its copper(II) salt are almost insoluble in water,
and the organotin compounds are resistant to extraction
as well. Better permanence of the conservation agent
is achieved by mixing it with a composition which is

coated on the packages.

New and little developed fungicidal coating compositions involve sporicidal and self-sterilizing compounds;
formaldehyde polymers or some halogenated compounds
(chlorosuccinimide, bromosuccinimide, bromoacetamide)
would suit this purpose.

Interesting results were obtained from the applications
of some textile fibres possessing antimicrobial effects.
Palishey et al. /64/ describe antimicrobially active
acetate fibres prepared by acetalization of partially
saponified triacetate fiber with aldehydes of 5-nitro-
furane. The modified fibers inhibit the growth of
Staphylococcus aureus and some other representative
pathogenic microflora.

The impregnation of packages which come into contact
with foodstuffs can be carried out firstly with the
preservatives currently used in the foodstuff industry,
e.g. benzoic acid, sodium benzoate, *p*-hydroxybenzoic
esters, sorbic acid, but also salicylic acid and its
sodium or calcium salts, dehydroxyacetic acid, dimethyl
dichlorosuccinate, diethyl or dipropyl succinate, and,
to a lesser extent, *p*-nitrobenzoic acid.

1.2.3.1 P r o t e c t i o n   a g a i n s t
        I n s e c t s

Macroscopic biological pests affecting packed goods,
particularly foodstuffs (pulses, cereals, dried fruits
and vegetables etc.) include insects of many kinds
selective in their attack on materials and under
specific conditions. Besides the insects attacking
stored goods in temperate climates there are more
dangerous kinds which must be taken into account if
the goods are exported. Termites (white ants) are
dreaded because almost no organic material (even
impregnated) can resist them. Insect varieties known

to attack both packaging materials and foodstuffs
include: cadelle (*Tenebriodes mauritanius*), drugstore
beetle (*Stegobium paniceum*), larder beetle (*Dermetes
lardarius*), granary weavil (*Calandra granaria*), confused
flour beetle (*Tribolium confuseum*), sawtoothed grain
beetle (*Oryzaephilus surinamensis*), yellow mealworm
(*Tenebrio molitor*), caterpillars of the Indian meal
moth (*Plodia interpunctella*), the mediterranean flour
moth (*Ephestia Kühniella*), and the tobacco moth
(*Ephestia elutella*).

Insects prefer loose and unprotected foodstuffs,
but packed goods are not protected either. Their attack
is either in search for food or to become pupae or to
lay eggs (some *Tenebriodes* kinds and e.g. *Calandra*,
respectively). Some insects attack packaging materials
directly, as do termites and, in moist locations
silverfish (*Lepisma saccharina*) /65/.

As any attack on a package by insects depends on
physiological and morphological factors at the time,
the degree of resistance by a packaging material
depends largely on the conditions. Apart from glass
and metals which are perfect insect barriers, all
other materials are liable to attack. Penetration of
the package depends on the quality, thickness, and
surface properties of the material. It is obvious that
the package should have no holes through which the
insect could gain direct access.

If the requirements are not too exacting, then
sufficiently thick wooden or cardboard packages,
multilayer paper bags, and thicker plastic films are
quite satisfactory. Cellophane was found relatively
good above 0.04 mm thickness, but polyethylene of
thickness above 0.1 mm was attacked by more active
species. Considerable resistance against insects was
found with combined materials and with vinyl chloride-
-vinylidene chloride copolymer film or polycarbonate

film alone. Also satisfactory are cardboard boxes with
thermoplastic coatings, but paraffin wax coating alone
gives no substantial protection /66/.

In more exacting situations it is necessary to
impregnate the less resistant packaging materials (paper
and cardboard) with insecticides.

1.2.3.3. Protection against Rodents

Another group of pests are rodents. Although there are
not many kinds, they are numerous and cause trouble
by eating foodstuffs, damaging packages, and sometimes
acting as carriers of infections, which constitutes
a serious health problem.

The most widespread rodent in Europe is the house
mouse (*Mus musculus*), making it widely necessary to
protect stored foodstuffs against this pest. The brown
rat (*Rattus norvegicus*) prefers damp conditions and
a meat diet. Black rats (*Rattus rattus*) cause damage
especially in sea and river ports.

As with insects, animals prefer foods loose rather
than in a package. Physiological factors and the
construction of the package determine whether the
animal will attack the packed food. Important factors
are the mechanical strength of the package and those
which make the package and its content attractive,
especially holes in the package, and its permeability
to odours. Therefore, cellulose-based packages and
those made of some synthetic materials provide no
reliable protection against rodents.

The protection afforded by these materials can be
improved by impregnation with chemical agents with
toxic or repellent effects for rodents.

It must be stressed that the prevention of access
of rodents to foodstuffs, which is provided by packages, must form part of more extensive measures,
including proper waste disposal, cleanliness, and

prevention of access by rodents into store-rooms and vehicles.

REFERENCES

/1/ Nikitin V.M., Chémia dreva a celulózy, Slovenské vydavatelstvo technickej literatúry, Bratislava, 1956.

/2/ Polyakov K.A., Nekovové chemické materiály, SNTL, Prague, 1956

/3/ Majerik J., Chem. revue, 1975, (4), 19.

/4/ Belšán L., Obaly, 1966, 12, 16.

/5/ Weinert M., Köhler F.J., Verpack. Rdsch., 1971, 22, 261.

/6/ Planeta A., Obaly, 1971, 17, 188.

/7/ Kaltenbacher E.J., Mod. Packag., 1971, 44 (10), 54.

/8/ Gustavssen C.E., Verpack. Mag., 1973, 17, 14.

/9/ Kühne G., Neue Verpack., 1969, 22, 1413.

/10/ Stoeckert K., Kunststoffe, 1973, 63, 700.

/11/ Modern Packaging Encyclopedia and Planning Guide 1972-1973, McGraw-Hill, New York, 1972.

/12/ Anonym, Packaging, 1976 (6), 15.

/13/ Nehring P., A lecture at the firm Obal n.p., Prague, 1967.

/14/ Anonym, Food Process., 1964, (6), 202.

/15/ Barbieri J., Malanese G., Rosso S., Imballagio, 1970, 21, (175), 22.

/16/ Anonym, Rev. Cons., 1969, (11), 126.

/17/ Anonym, Verpack. Rdsch., 1965, 16, 16.

/18/ Anonym, Verpack. Ber., 1969, 13 (1), 22.

/19/ Anonym, Emballages, 1969, 39 (261), 118.

/20/ Anonym, Food Eng., 1973, 45 (3), 85.

/21/ Anonym, PIRA Packag. Abstr., 1971, 28 (2), 63.

/22/ Anonym, Emballages, 1965, 35 (231), 198.

/23/ Anonym, Emballages, 1970, 40 (275), 84.

/24/ Sampsel D.F. Mod. Packag., 1971, 44 (2), 78.

/25/ Paine F.Q., Paine H.Y., A Handbook of Food Packaging, Leonard Hill, Glasgow, 1983.

/26/ Löfflerová J., Šiman J., Manipul. sklad. bal., 1974, 3 (4), 16.

/27/ Austrian pat. 247 126, 1966.

/28/ Mamma C.E., Cereal Sci. Today, 1967, (1), 4.

/29/ Mark A.M., Mehltretter C.L., Stärke, 1969, 21 (4), 92.

/30/ Anonym, Verpack. Rdsch., 1962, 13, 676.

/31/ Clark W., Shirk R., Food Technol., 1965, 19 (10), 105.

/32/ Widmann A., Fette, Seifen, Anstrichm., 1965, 58, 553.

/33/ Alfa-Slater R.B., J. Amer. Oil Chem. Soc., 1958, 35, 122.

/34/ Anonym, Food Eng., 1966, 38 (2), 136.

/35/ Morris A.C., Whittal A.G., Plastics, 1966, 31 (349), 1421.

/36/ Frank H., Stephen H.E., Verpack. Rdsch., 1975, 26, 169.

/37/ Hanlon J.F., Handbook of Package Engineering, McGraw-Hill, New York, 1971.

/38/ Reif G., Penzkofer J., Heiss R., Verpack. Rdsch., 1975, 26 (2), Techn. Wiss. Beil. 9.

/39/ Anonym, Bessere Verpack., 1973, (11-12), 55.

/40/ Schweizer Verpackungs- und Transportkatalog 1958-59, 91.

/41/ Dikis M.J., Malskiy M.A., Oborudovaniye konzervnych zavodov, Pishtchepromizdat. Moscow, 1953.

/42/ Anonym, Verpack. Rdsch., 1973, 24, 1416

/43/ Granzer R., Proceedings of Euro Food Pack, Vienna, 1984, 16.

/44/ Heiss R., Verpackung feuchtigkeitsempfindlicher Güter, Springer, Berlin, 1956.

/45/ Slavíček E., Sborník Vysoké školy chemickotechnologické v Praze, E 17, 1967, 27.

/46/ Čurda D., Slavíček E., Kyzlink V., Die Nahrung, 1967, 11, 70.

/47/ Slavíček E., Sborník Vysoké školy chemickotechnologické v Praze, E 17, 1967, 63.

/48/ Hanousek J., Žák J., Obaly, 1960, 6, 183.

/49/ Koch W.T., Massengale J.T., Food Eng., 1962, 34 (2), 60.

/50/ Proctor B.E., Karel M., Mod. Packag., 1955, 28, 137.

/51/ Holečková L., Čeláková B., Lanzer O., Prům. potrav., 1970, 21, 43.

/52/ Holečková L., Kyselová B., Lanzer O., Prům. potrav., 1971, 22, 73.

/53/ Malčič B., Rapič S., Fleischwirtschaft, 1960, 12, 23.

/54/ Gluck B., Barany H.C., Brit. pat. 591 275, 1965.

/55/ Elickson B.E., Hasenzahl V., Food Technol., 1958, 12, 577.

/56/ Bäckström M., Technika chlazení, SNTL, Prague, 1959.

/57/ Anonym, Eur. Potat. J., 1958, 1 (3), 66.

/58/ Orel V., Vejce, jejich zpracování a ošetření, SNTL, Prague, 1959.

/59/ Plank R., Handbuch der Kältetechnik, Springer, Berlin, 1960.

/60/ Anonym, Packaging, 1969, 40 (475), 94.

/61/ Horáček V., Prům. potrav., 1970, 21 (9), 133.

/62/ Kraft W., Müller G., Weiss G., Verpack. Rdsch., 1970, 21, 103.

/63/ Havlíček J., Šilhánková L., Prům. potrav., 1962, 13, 70.

/64/ Palishey B.V., Khim. Volok. 1970, (4), 247.

/65/ Schelhorn M., Packstoffe und Verpackungen, Keppler, Frankfurt am Main, 1959.

/66/ Highland H.A., Jay E.G., Mod. Packag., 1965, 38 (7), 205.

2

Basic Principles of Chemistry and Technology of Production of Plastics

J. ŠTĚPEK, V. DUCHÁČEK

2.1 DEVELOPMENT OF STRUCTURAL THEORY OF MACROMOLECULAR COMPOUNDS AND THEIR CLASSIFICATION

Concepts of the structure of macromolecules developed through several stages /1-6/. Natural macromolecular substances, such as natural rubber, cellulose, proteins, were investigated by a number of scientists during the 1920s. Polanyi /7/ proposed a theory of small parallel pipeds - cells, according to which a macromolecular substance represents a complex of low-molecular compounds linked by special associative forces. Meyer and Mark /8-10/ supported the theory which considers macromolecular substances to be composed of micelles (crystallites), i.e. sets of relatively small molecules containing 50-100 basic structural units linked mutually by associative forces which are so strong that each micelle behaves as an independent unit.

Baekeland /11/ suceeded in synthetizing a macromolecular phenol-formaldehyde resin which became (under the name *Bakelite* in 1920) the first polymer produced industrially.

Staudinger (who was awarded the Nobel Prize in 1953)

began in 1920 to investigate the structure of paraformaldehyde and polystyrene /12/, proved the existence of
covalent bonds in these compounds, denoted them as
high-molecular compounds, and coined the term macromolecule /13/. He supported the idea that macromolecules
are thread-like chains composed of small low-molecular
units connected through the main valencies. According
to his theory, the physical properties depend on the
chain length, and the chemical properties on the reactive
groups present in the chain /14-18/. His results did
not agree with the micellar theory; investigation of
the relationship between the viscosity of solutions
and the relative molecular weight of the polymers
allowed him to determine the degree of polymerization.

Brilliant papers by Carothers /19, 20/ appeared
in 1930s followed by those of Flory /3, 21/ (the latter
was awarded the Nobel Prize in 1974) and completed
the foundations for further development of modern
macromolecular theory. In the field of polymerization
processes, a substantial change was induced by the
works of Ziegler /22/ and Natta /23/ (they were
awarded the Nobel Prize in 1963) dealing with coordination catalysts as polymerization initiators. These
allowed the development of polymers with regular
steric arrangement of structure - stereo-regular
polymers (cf. Section 2.2).

Plastics are fully or partially synthetic organic
materials. Their main components are high-molecular
compounds - polymers whose molecules are composed of
a large number of basic building units - mers - linked
by chemical bonds. Structurally they comprise chains or
a three-dimensional network composed predominantly of
carbon atoms and, to a lesser extent, of oxygen, silicon,
nitrogen, or sulphur atoms. The length and size of
these macromolecules can be different in a single
sample, ranging from several atoms to 10^4 atoms.

The exceptional property of plastics - thermoplasticity
at technically easily attainable conditions - allows
their shaping at some stage of their manufacture,
facilitating mass production (often continuous) of
consumer's articles or intermediate components.

Plastics are often not worked in the pure state,
but with various additives, such as fillers, lubricants, plasticizers, light and heat stabilizers,
antioxidants, and pigments, which either improve their
workability or modify their properties in various
ways.

There are several schematic classifications of
plastics. According to the starting materials used
for their production /24/ plastics are divided into:

- semisynthetic materials formed by chemical
or physical transformations of natural polymers, such
as the materials obtained by transformation of natural
rubber, cellulose, and proteins.

- fully synthetic materials produced by synthesis
from low-molecular organic compounds.

Some authors classify /25/ plastics as

- condensation products like phenolformaldehyde
or aminoaldehyde resins, and polyesters.

- polymerization products such as poly(vinyl
chloride), polyacrylates, polyolefins, and polystyrene.

- natural products, viz. compounds of cellulose,
proteins, bitumens, and plant oils.

Plastics can also be clasified according to their
behaviour during heating:

- materials fusible on heating - thermoplastics,
i.e. those which soften and finally melt on heating
and solidify again on cooling, the cycle being
repeatable several times without changing substantially
their properties

- materials hardened by heating - thermosetting
resins, i.e. those which melt initially but, on further

heating, are transformed by chemical reaction into an
infusible state with properties substantially changed.
Many transition types exist between these two extremes
with various degree of branching or cross-linking.

Besides large-scale basic materials, the modified
types of plastics /26/ become increasingly important.
They are obtained by chemical or physical modifications
of the basic polymers and have substantially different
properties.

2.2   THE BASIC PROCESSES OF FORMATION OF SYNTHETIC
       POLYMERS

The synthetic polymers are formed predominantly by
three basic reactions: polymerization, polyaddition,
and polycondensation.

Polymerization is a chain reaction in which a large
number of simple organic unsaturated compounds (monomers) add on to each other to yield long macromolecules.
This reaction does not give side products, but the
macromolecules incorporate virtually all the molecules
of the monomer. Hence, polymerization does not change
the chemical composition. The macromolecular chain
grows to its final length in a very short time which
depends on the reaction conditions, so that at any stage
of the reaction the mixture contains the unreacted
monomer and completed macromolecules of the final
polymer, a consequence of the free-radical character
of the reaction.

The overall polymerization rate and the length of
the macromolecules formed are determined by the rates
of the individual kinetic processes which constitute
the whole polymerization mechanism, i.e. initiation,
propagation reaction, and termination.

For polymerization to take place it is necessary
to activate the monomer molecules. This can be ac-

complished by heat, light, radiation, or by a substance
which is easily activated, the initiator, sometimes
called the catalyst. If the activated molecule is
denoted as a radical R$^\bullet$ and the monomer by the symbol
M, then the individual kinetic processes can be expressed
as follows /27/:

initiation $R^\bullet + M \longrightarrow RM^\bullet$

propagation $\begin{cases} RM^\bullet + M \longrightarrow RM_2^\bullet \\ RM_x^\bullet + M \longrightarrow RM_{x+1}^\bullet \end{cases}$

termination $RM_x^\bullet + RM_y^\bullet \longrightarrow RM_{x+y}R$ by recombination

termination $RM_x^\bullet + RM_y^\bullet \longrightarrow RM_x + RM_y$ by disproportionation

During the growth reaction a side reaction may occur,
called the transfer, in which the activity of an activated molecule is transferred to a molecule of the
monomer or final polymer or even another compound AB
(e.g. a solvent molecule):

transfer $RM_x^\bullet + M \longrightarrow RM_x + M^\bullet$
$ RM_x^\bullet + AB \longrightarrow RM_x A + B^\bullet$

These processes represent a termination of growth
of one macromolecule concurrent with the formation of
another activated particle. Hence, this reaction does
not change the number of activated particles in the
polymerizing system, but it reduces the average length
of the macromolecules formed.

The polymerizability of monomers is governed by
their chemical structure, i.e. their molecular skeleton
(whether it contains one or several double bonds), the
character of any substituent groups present, and their
position with respect to the double bond. If the
unsaturated compound contains several substituents,
their relative positions affect the polymerization:
the symmetrically substituted derivatives polymerize
with difficulty or not at all, whereas unsymmetrical
substitution increases the polymerizability of monomers,

such substituted derivatives being even more reactive
than the parent olefinic hydrocarbons which polymerize
with difficulty. So e.g. /24/:

ethylene	$CH_2=CH_2$	is not easily polymerized
unsymmetrical dichloro-ethylene (vinylidene chloride)	$CH_2=CCl_2$	are polymerized much
vinyl chloride	$CH_2=CHCl$	more easily
symmetrical dichloro-ethylene	$ClCH=CHCl$	can hardly be polymerized (only a low-molecular polymer is formed)

The size of the substituent must also be taken into
account. Chlorine atoms are bulky and in a symmetrical
arrangement, sterically hinder the mutual approach of
the molecules at the distances necessary for the reaction to take place. If the volume of the halogen is
very small, as e.g. in the case of fluorine, then even
the substitution of all four hydrogen atoms of ethylene
by fluorine will not reduce the polymerizability of
the resulting tetrafluoroethylene ($CF_2=CF_2$). Steric
effects are very important aspect of polymerization.

Electronegative (unsymmetrical) substituents, as
e.g. $-F$, $-Cl$, $-CN$, $-COOR$, $-C_6H_5$, in the ethylene molecule increase polymerizability, but alkyl substituents
have little effect.

The polymerizability of diolefine also depends on
the position of substituents. The most important
dienes are the conjugated ones which can polymerize
entirely by 1,4-addition or, to a lesser extent, by
1,2-addition (or even by a mixture of the two).

According to the initiation and the course of the
reaction, polymerizations are classified as radical or
ionic.

2.2.1   RADICAL POLYMERIZATION

Radical polymerization can be induced by heat,
initiators, or ultraviolet or ionizing radiation.

The polymerization may be influenced by certain
compounds such as these containing conjugated double
bonds in the form of substituted aromatic systems. As
a rule, such compounds react readily with a reactive
radical to give one of lower reactivity, so that they
slow down or completely stop the polymerization (polymerization retarders and inhibitors, respectively),
however, a compound acting as the inhibitor for one
monomer can be a retarder for another.

Simultaneous polymerization of two or several
monomers is called copolymerization. Copolymerization
takes place with such monomers whose homopolymerizations
are not very different energetically.

Copolymerization has great technical importance,
because it allows modification of the properties of
homopolymers or the generation of quite new properties.

2.2.2   IONIC POLYMERIZATION

Ionic polymerization is a reaction in which the growing
ends or the active centres of the macromolecule formed
represent a more or less strongly ionized or polarized
grouping X and Y able to attract and incorporate
monomer molecules.

If the carbon atom connected to the active bond
carries a positive charge, the reaction is called
cationic polymerization /28/:

$$YCH_2-\overset{(+)}{\underset{R}{CH}} X^{(-)}$$

The growth reaction is realized by carbocations.

If, on the contrary, the carbon atom connected to
the active centres carries a negative charge, the
reaction is called anionic polymerization /2, 29, 30/:

$$XCH_2{-}\overset{(-)}{CH}\ Y^{(+)}$$
$$\underset{R}{|}$$

The growth reaction is realized by carbanions.

In contrast to radical reactions, ionic reactions
usually require lower activation energies. As a rule
they need low temperatures (-80 to -130°C) and proceed
much faster.

The catalysts used for the cationic polymerization
are strongly acidic compounds, such as BF_3 or its
hydrates or etherates, $TiCl_4$, $SnCl_4$, $AlCl_3$ (the
Friedel-Crafts catalysts), H_2SO_4, etc. Cationic polymerizations are characteristically affected by the
presence of co-catalysts (e.g. traces of water, hydrogen
chloride, alcohols, ethers, etc.) which accelerate the
polymerization process in different ways depending on
the monomer and temperature. Inhibitors of cationic
polymerization are generally compounds of basic
(nucleophilic) character, i.e. those able to attach
a cation.

The catalysts used in anionic polymerizations are
strong bases and compounds of basic reaction, such as
e.g. alkali metal amides, alcoholates, alkyl-metallic
compounds, and alkali metals.

2.2.3 POLYCONDENSATION

Polycondensation represents a repetition of the reaction
of functional groups of the starting compounds. Formation
of a macromolecular product requires the proper number
of functional groups in the starting components, i.e.
at least two in each of the reacting partners: in

such a case the macromolecular product has a linear
structure. If some of the reaction components contains
more that two functional groups, then the products are
cross-linked and have threedimensional structure
/24, 2/.

Schematicaly, polycondensation can be expressed as
follows:

$$aAa + bBb \longrightarrow aABb + ab$$
$$aABb + aAa \longrightarrow aABAa + ab$$
$$aABAa + bBb \longrightarrow aABABb + ab \quad etc.$$

The main types of the functional groups which can
enter the polycondensation reaction and the resulting
groupings are:

$$\sim R{-}OH + HOOCR'\sim \xrightarrow{-H_2O} \sim R{-}OOC{-}R'\sim$$
$$\sim R{-}NH_2 + HOOCR'\sim \xrightarrow{-H_2O} \sim R{-}NH{-}CO{-}R'\sim$$
$$\sim R{-}Cl + H{-}Ar\sim \xrightarrow{-HCl} \sim R{-}Ar\sim$$
$$\sim R{-}OH + HO{-}R'\sim \xrightarrow{-H_2O} \sim R{-}O{-}R'\sim$$

$$etc.$$

Thus the polycondensation is a reaction which
converts two simple compounds into a polymer with
simultaneous formation of a simple low-molecular
by-product (most often water). Hence, one difference
between polymerisation and polycondensation
is that the polymerization product (polymer) has the
same elemental composition as the starting substances
(monomers), whereas polycondensations change the
chemical composition. The individual process steps
exhibit the equilibrium characteristic of a reversible
reaction and do not differ mutually in activation
energies or reaction rates. The reaction proceeds
gradually in the whole volume, so that the system
contains the longest polymers along with the shorter
ones and the unreacted monomer. Termination is caused

by decreased reactivity of the functional groups as
a consequence of their lowered concentration in the
high viscosity reaction mass, or by side reactions of
the functional groups which prevent their further
participation in the polycondensation (e.g. splitting
off of ammonia from diamines, dehydratation of glycols,
etc.). The molecular weight of the polycondensate can
be affected by addition of monofunctional compounds which
can stop the growth by reaction with terminal functional
groups. E.g. monocarboxylic acids, monofunctional
alcohols, or amines are used for this purpose.

A typical example of polycondensation is the reaction
of phenol with formaldehyde which gives, according to
the proportions of the two components and the catalyst
used, products with linear or three-dimensional structures.
If phenol is present in excess and the reaction is
acid-catalysed, then it produces a permanently soluble
and fusible polymer (*Novolak*) which can be cross-linked
by addition of further formaldehyde (e.g. in the form
of hexamethylenetetramine) to give a three-dimensional
and infusible product. In the presence of excess
formaldehyde the first polycondensation phase produces
a linear, soluble,and fusible polymer (*Resol*) which can
be transformed by heating into the insoluble and
infusible *Resit*, the transformation going through an
intermediate state (*Resitol*). The final product is known
under the name *Bakelite*.

2.2.4 POLYADDITION

Compounds containing multiple bonds or small rings can
not only be combined mutually, but also can enter
addition reactions with compounds possessing suitable
functional groups /24/. If the starting components
contain a sufficient number of the reactive groups
(two at least), then a macromolecular product can be

formed.

Examples of polyaddition are the formation of
polyurethanes and polyureas from glycols and diamines,
respectively, with diisocyanates:

$$O=C=N-(CH_2)_6-N=C=O + HO-(CH_2)_4-OH \longrightarrow$$

$$\longrightarrow \ ...-\underset{\underset{O}{\|}}{C}-N-(CH_2)_6-NH-\underset{\underset{O}{\|}}{C}-O-(CH_2)_4-O-...H$$

$$O=C=N-(CH_2)_6-N=C=O + NH_2-(CH_2)_6-NH_2 \longrightarrow$$

$$\longrightarrow \ ...-\underset{\underset{O}{\|}}{C}-N-(CH_2)_6-NH-\underset{\underset{O}{\|}}{C}-NH-(CH_2)_6-NH-...H$$

The composition of the product does not differ
from that of the initial mixture. In contrast to
polymerization, however, the structure of the basic
building unit of the polyadduct differs from that of
the initial components.

There are many cases among the individual preparation
methods, however, when the macromolecular compounds
are formed by processes which cannot be classified
unambiguously as belonging to one of the three
above-mentioned methods.

2.3 METHODS OF PRODUCTION OF SYNTHETIC POLYMERS

The polymerization of a single monomer represents the
simplest case. If the polymer is soluble in the monomer,
as polymerization progresses, a more and more viscous
solution is formed until the solution solidifies to give
a block assuming the shape of the vessel (block polymerization). If the polymer is insoluble in monomer, it
separates as fine particles.

If a solvent for the polymer is added to the monomer,
the reaction yields directly a polymer solution (solution
polymerization).

More often used is polymerization in heterogeneous
phases - either in suspension or in emulsion (called
suspension and emulsion polymerizations). Polymerization can also be accomplished in the solid or gaseous
states /31/. The properties of the polymers differ
according to method, so the choice of the polymerization method also depends on the requirements which
must be fulfilled by the product.

Block polymerization has the drawback of difficulty
in removal of the considerable reaction heat. The
non-uniform temperature has its effect on the molecular
weight distribution of the polymer which in turn
influences the properties of the final product. Generally,
polymerization is accompanied by volume decrease
(density increase). Slow reduction of the residual
monomer content in the block results in shrinkage, and
because of the low mobility of macromolecules at normal
temperatures this causes stress in the mass which can
lead to fissures in the case of larger objects. The
mean molecular weight of block polymers (in current
rapid and   economic preparation) reaches the order
of magnitude of 10^4.

In solution polymerization, the reaction heat can
be removed more easily. As every solvent is also to
some degree a radical transfer agent, the molecular
weight of the final product is always lower than that
obtained by block polymerization under comparable
conditions. The polymer solutions are used directly
as adhesives or as varnishes.

Suspension or bead polymerization was also developed
to overcome the poor heat transfer of block polymerization.
The monomer and initiator are dispersed in water as
small globules by stirring. The size or, more correctly,
size distribution of the globules is determined by the
stirring which overcomes the effect of surface tension
tending to coalesce the particles into one phase of

minimum surface area.

The polymerization proceeds with mild heating at constant stirring. With the conversion, the viscosity increases and the globules become sticky, which increases the tendency to form larger aggregates by coalescence. This tendency is reduced by substances which increase the viscosity of the aqueous phase (e.g. poly(vinyl alcohol), gelatine), or by mineral powders (e.g. talc) which cover the sticky surface of the globules. It is not too difficult to remove the mineral substances from the surface of the final polymer beads.

The suspension polymer has a more uniform molecular weight distribution than the block polymer, hence the final articles have usually better mechanical properties. Due to their low content of electrolytes, block and suspension polymers can provide good electroinsulating materials.

The present meaning of the term emulsion polymerization is a process producing the polymer from the monomer particles dispersed in the aqueous phase by both mechanical action and soap (emulsifier) and using a water-soluble initiator (Fig. 16). The process is very complicated and goes through many stages /32/.

2.4 MODIFICATION REACTIONS OF POLYMERS

The modification of plastics has a broad meaning involving numerous ways of chemical and structural transformation. These are intentional transformations of the polymers whose purpose is to obtain new or enhanced properties.

The following basic ways are adopted for these modifications /33/:

Mechanical mixing of two or more polymers /34/.

Chemical reactions of some active compound with reactive groups of the polymeric chain /35/.

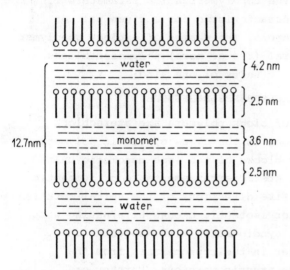

Fig. 16 Representation of solubilization of a monomer in soap solution /24/. The monomer layer is located between non-polar ends of the soap molecules.

Mechano-chemical ways: by mechanical degradation (e.g. calandering, kneading) under suitable conditions in the presence of substances, sometimes polymerisable monomers, which react with the polymer fragments, recombination or chain extension are effected.

In the case of the mechano-chemical modifications /36/ a series of reactions lead to formation of grafted copolymers or block copolymers, which differ by structure and length of the blocks, and to formation of three-dimensional cross-linked fragments. From the structural point of view the resulting product is far from being uniform and well-defined. This copolymer type, which contains various proportions of the above-mentioned structures along with the homopolymers, is called usually an interpolymer.

Statistical copolymers are those with random sequence distribution of monomeric units of two or more types:

....ABAABABBBABABAABAB...

The distribution of the individual monomeric building units is given by the copolymerization parameters r_1 and r_2 which are characteristic for a given pair of copolymerizing monomers. The chains of a block copolymer are formed by several (two or more) polymeric blocks composed of monomeric units of one kind:

...AAAAAAAAAAAAAAAAABBBBBBBBBBBBBBBBBBBB...

A special kind of block polymers are stereoblock polymers which belong to the class of tactic polymers. They are formed by blocks of the same monomeric units but with different steric arrangements, e.g. isotactic sections with opposite directions of the spiral rotation of the structure, or isotactic blocks alternating with atactic blocks, or syndiotactic with isotactic ones. Such polymers differ in their properties from the purely iostactic or atactic versions. Furthermore, stereoblock copolymers are possible in which differing molecular species also possess different forms of stereoregularity.

Grafted copolymers have main chains of one monomer with side chains of other monomer "grafted" thereon:

```
                    B
                    B
                    B
                    B
                    B
AAAAAAAAAAAAAAAAAAAAAAAAAAAAAAAAAAAAAA
      B
      B
      B
      B
```

If the side chains are of the same type as the main chain, the product is called a branched polymer.

The grafted and block copolymers have different properties from those of the statistical copolymers, even if the overall chemical composition is the same.

The individual homopolymeric blocks are large enough
to retain some of the properties of the homopolymers
but by interaction to have also some properties not
exhibited by the original homopolymers.

The increasing number of copolymers posed a nomenclature problem, and the most appropriate classification
seems to be that by Ceresa /36/ which is based on the
nomenclature used in Chemical Abstracts. The individual
types of copolymers are denoted by the prefixes:
co - statistical copolymer, g - grafted copolymer,
b - block copolymer.

The following abbreviations are used by Ceresa
for further denotation of structure of the chains:
cl - cross-linked, br - branched.

The tacticity is denoted by the symbols: a - atactic,
iso - isotactic, syndio - syndiotactic.

The application of these symbols is shown in the
following examples, in which S and B stand for styrene
and butadiene, respectively.

```
      SSSSSSSSSSSSSSSSSS          polystyrene
    SSBSBBBSBSSBSBBSSSBSBS        poly(butadiene-co-styrene)
    SSSSSSSSSSSSSBBBBBBBBB        poly(styrene-b-butadiene)

    SSSSS......SSSSSSSSSSS
       B           B
       B           B
       B           B
       B           B              poly(styrene-g-butadiene)
       B           B
       B           B
       B           B
```

The polymer which forms the main chain or that prepared
in the first step of the synthesis is always given in
the first place in the name.

```
                        S
                        S
                        S
                        S
                        S
                        S
SSSSSSSSSBBBBBBBBBBBBBBBBBBBBBBBBBB
        S
        S
        S                poly[styrene-b-(butadiene-g-styrene)]
        S
        S
        S
        S


SBBSBSSSBSBSBBSBSSBSBBS
    S         S
    S         S
    S         S            poly[(butadiene-co-styrene)-g-styrene]
    S         S
    S         S
    S         S
    S         S


SSSSSSSSS.......SSSSSS
iso  S          S   iso
     S          S
     S a        S a
     S          S      poly[(iso)-styrene-g(a)-styrene]
     S          S
```

REFERENCES

/1/ Treloar L.R.G., Archenhold A.T.,Introduction to
    Polymer Science, Wykeham, London, 1975.

/2/ Flory P.J.,Statistical Mechanics of Chain Molecules,
    Interscience, New York, 1969.

/3/ Flory P.J.,Principles of Polymer Chemistry, Cornell
    University Press, Ithaca, New York, 1953.

/4/ Champetier G. et al., Chimie macromoleculaire,
    Hermann, Paris, 1970.

/5/ Labuna S.S., Chemistry and Properties of Crosslinked
    Polymers, Academic Press, New York, 1977.

/6/ Bassett D.C., Principles of Polymer Morphology,
    Cambridge University Press, Cambridge, 1981.

/7/ Polanyi M.,Naturwissenschaften, 1921, 9, 228.

/8/ Meyer K., Mark H., Ber., 1928, 61, 593.

/9/ Meyer K., Mark H., Ber., 1931, 63, 1999.

/10/ Meyer K., Mark H., Ber., 1931, 63, 2913.

/11/ Baekeland L.H., Ind. Eng. Chem., 1913, 5, 506.

/12/ Staudinger H., Ber., 1920, 53, 1073.

/13/ Staudinger H., Ber., 1924, 57, 1203.

/14/ Staudinger H., Ber., 1929, 62, 2893.

/15/ Staudinger H., Ber., 1930, 63, 2317.

/16/ Staudinger H., Ber., 1930, 63, 2331.

/17/ Staudinger H., Ber., 1930, 63, 2338.

/18/ Staudinger H., Ber., 1931, 64, 1688.

/19/ Carothers W.H., J. Amer. Chem. Soc., 1929, 51, 2548.

/20/ Carothers W.H., Chem.Revs, 1931, 8, 353.

/21/ Flory P.J., Amer. Chem. Soc., 1937, 59, 241.

/22/ Ziegler K., Holzkamp E., Breil H., Martin H.,
Angew. Chem., 1955, 87, 541.

/23/ Natta G., Pino P., Corradini P., Danusso F.,
Mazanti G., J. Amer. Chem. Soc., 1955, 77, 1708.

/24/ Švastal S., Lím D., Kolínský M., Úvod do chemie
a technologie plastických hmot, Práce, Prague,
1954.

/25/ Kovačič L., Bína J., Plasty, vlastnosti,
spracovanie, využitie, Alfa, Bratislava, 1974.

/26/ Štěpek J., Zpracování plastických hmot, SNTL,
Prague, 1966.

/27/ Billmeyer Jr. F.W., Textbook of Polymer Science,
Wiley, New York, 1971.

/28/ Plesch P.H., Chemistry of Cationic Polymerisation,
Pergamon Press, New York, 1964.

/29/ Stevens M.P., Polymer Chemistry. An Introduction
Addison-Wesley, London, 1975.

/30/ Schwarz M., Carbamans, Living Polymers and
Electron-Transfer Processes, Wiley-Interscience,
New York, 1968.

/31/ Schildknecht C.E., Polymer Processes, Wiley-Interscience, New York, 1956.

/32/ Bovey F.A., Kolthoff I.M., Medalia A.I., Mechan E.J.,
Emulsion Polymerisation, Wiley-Interscience,
New York, 1955.

/33/ Heidingsfeld V., Štěpek J., Chem. listy, 1965, 59,
821.

/34/ Olabisi O., Robeson L.M., Shaw M.T., Polymer -
- Polymer Miscibility, Academic Press, New York,
1979.

/35/ Fettes E.N., Ed., Chemical Reactions of Polymers, Wiley-Interscience, New York, 1964.

/36/ Ceresa R.J., Block and Graft Copolymers, Butterworth, Washington, 1962.

Ceresa R.J., Block and Graft Copolymerisation I, Wiley, London, 1973.

3

Structure and Properties of Polymers and Composition of Mixtures of Plastics

J. ŠTĚPEK, V. DUCHÁČEK, M. ŠÍPEK

The development of ideas about the structure of macromolecules passed through several stages, and led to
modern macromolecular theory which is based on the
following presumptions /l/:

   - macromolecules can have varied shapes, linear,
branched, or three-dimensional lattices (networks)
   - macromolecules do not form rigid chains but assume
changing geometry due to continuous motion of the
individual segments
   - macromolecular properties depend on the phase state
(amorphous or crystalline), on molecular weight, and
chemical composition.

So far it has not been possible to develop a general
quantitative theory which would derive the macroscopic
properties of macromolecules from their molecular structure, but a series of structural factors and some of
their effects on physical properties and behaviour of
polymers are known. First of all it must be stated that
plastics include a large number of compounds with
various properties, so that it is very difficult to
classify them strictly according to behaviour. For
example almost every high-molecular organic compound
with linear structure can be obtained in the glassy,

rubber-like, or liquid state. Chemical composition and
molecular weight determine which of these states will
be assumed under normal conditions and the temperature
limits for such state.

The high molecular weights of polymers mean that
their boiling points are far above their decomposition
temperature. So the gaseous state does not exist with
these substances /2/. Polymers can only be in liquid
or solid states. For solid macromolecules it is possible
to differentiate the crystalline state which is highly
ordered and the glassy state which is disordered. The
degree of order of molecules of the glassy state
resembles that of the liquid state, but the freedom for
spatial migration and rearrangement (flow) is suppressed.
All amorphous substances can exist in such a state:
the glassy state is considered very often to be that of
a supercooled liquid with such high viscosity that it
cannot flow and has the properties of a rigid body.
However, the amorphous polymer in the glassy state
differs from liquids so much, that it can be considered
a special, unique state of matter.

In contrast to low-molecular substances, polymers
are characterized by a certain transition state between
the glassy and liquid states, viz. the rubber-like
state. Polymers in this state can be easily and extensively but reversibly deformed, so they cannot be
considered rigid bodies, but they are not liquids
either, because the deformation is not the irreversible
flow which is typical of the plastic or liquid state.

So, polymers can exist in four alternative physical
states, of which three are amorphous (glassy, rubber-like,
and liquid or plastic) and one is crystalline. Each of
these states is characterised by different behaviour
and a number of the physical properties of the polymer.

Smooth and long macromolecular chains (i.e. those
without any marked steric hindrance) allow the formation

of crystalline structure, especially if the intermolecular forces are sufficiently large. But a regular
crystalline structure cannot be formed throughout the
material because different chain lengths inhibit
complete crystallisation. Hence, the X-ray diffraction
patterns of crystalline polymers exhibit both the
diffraction spots characterizing the crystalline regions
and the diffusion rings typical of amorphous substances.

During heating the behaviour of an amorphous polymer
can be characterized by its glass transition temperature
(t_g) and that of a crystalline polymer by its melting
temperature (t_m). If the deformation after a defined
time at a given temperature and tension is chosen to
characterize the behaviour of a macromolecular substance,
and if such measurements are made at temperature
intervals through a broad temperature range, then a
thermomechanical curve may be obtained which is characteristic for that polymer.

In case of an amorphous polymer the thermomechanical
curve will have the shape given schematically in Fig.
17 /2/.

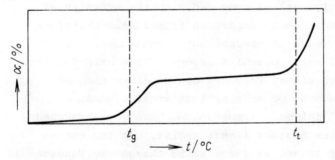

Fig. 17 Thermomechanic curve of an amorphous polymer
/2/ t_g - the glass transition temperature, t_t - the
flow temperature, α- deformation

The t_g thus characterizes the region in which
amorphous polymers - in the course of cooling - solidify abruptly. The temperature interval from t_g to t_t

corresponds to the rubber-like region in which deformation changes very little with temperature and is predominantly reversible. Only in the region of t_t, i.e. the pour point, the heating results in an abrupt and considerable increase of deformation, because the viscous flow becomes significant and is characterized by irreversible deformation. Above this temperature the polymer is in the liquid or plastic state.

Some polymers do not exist in all three amorphous states, because t_g and t_t values can be higher than the decomposition temperature of the polymer, and such polymers then only exhibit the rubber-like and the glassy states or even only the latter as with cellulose and some of its derivatives.

3.1 LINEAR STRUCTURE AND PHASE STATE OF POLYMERS

The basic prerequisite for a material to behave as a plastic is the existence of chain molecules which influence each other by intermolecular forces which are relatively weak as compared with the strength of chemical bonds. Every departure from linear structure leads to decrease of thermoplastic behaviour. Thus the shape of macromolecules is one of the main factors affecting behaviour. Polymers can be classified as linear, branched, or more or less cross-linked.

Technical polymers almost always have some branched macromolecules besides linear chains, but the extent of branching is not so large as to change the fundamental linear character, so both the macromolecular formations will be treated together.

The crystallizability is a feature of some linear polymers and has a significant effect on their properties. From this point of view, linear polymers can be divided according to their tendency to spontaneous crystal-

lization. The crystallizability of polymers is governed by the following factors /1/:

a) polarity of the groups in the chain (cohesive
energy)

b) geometrical regularity of the chain,

c) volume of substituents and, in case of branched
macromolecules, number and length of branches,

d) flexibility of the macromolecular chain.

In an ideal amorphous polymer the molecules should
have the statistically most probable conformation[1]
(from the point of view of entropy, the most probable
conformation is the most disordered one). Real polymers,
however, differ somewhat from this abstract picture.

If a polymer molecule contains strongly polar
groups, random conformation does not occur, the energetically (not entropically) most favourable conformation being predominant. The large cohesion energy
causes mutual activation between chains until they
are closely packed, having conformations of the
lowest possible energies. Thermal motion of individual
segments[2] of macromolecules has the opposite effect.
If at a certain temperature the cohesion energy of the
chains is larger than their kinetic energy, then conditions are suitable for parts of the macromolecules
to exhibit close packaging and incorporation into a
crystalline arrangement. Thermal motion decreases with
decreasing temperature, and when the cohesion energy
exceeds the kinetic energy of the chains, provided
there is not too large steric hindrance, crystallization
takes place (even in polymers which do not crystallize
under normal conditions). For example natural rubber

---

[1] The conformations of macromolecules are different
spatial arrangements of chains resulting from free
rotation around their single bonds.

[2] The term segment denotes a part of the macromolecular
chain which exhibits kinetic independence.

does not crystallize spontaneously at normal temperature because of the geometrical irregularity and thermal mobility of its chains. At low temperature (below 10°C), however, the cohesive forces predominate and crystallization takes place; if left at this temperatures for a very long time, it becomes highly crystalline and even retains this crystallinity during heating up to 32°C (Stark Rubber). In contrast, the styrene butadiene rubber (SBR) does not crystallize at low temperatures. This is particularly true for the "warm" type containing 23% of 1,2-addition, considerable branching, and a mixture of *cis* and *trans* isomers.

It follows that the effect of polarity on the crystallizability cannot be evaluated without taking into account the geometrical regularity of the chain and the bulkiness of side groups.

The effect of geometrical regularity of the chain is exemplified by the behaviour of two polyisoprene isomers - natural rubber (the *cis* isomer) and gutta-percha (the *trans* isomer). The *cis* configuration and non-planar arrangement in natural rubber cause a curling of the chain (Fig. 18) which, at normal temperatures when the

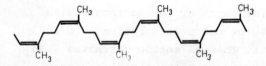

Fig. 18. Structure of the macromolecules of natural rubber (*cis*-1,4-polyisoprene)

kinetic energy is considerable, prevents the mutual approach of the chains essential for crystallization.

In contrast, the *trans* configuration (Fig. 19) allows approach of the chains to each other so that gutta-percha exhibits spontaneous crystallization at room temperature.

Polyethylenes are crystalline at room temperature, because their chains are highly regular, regardless of

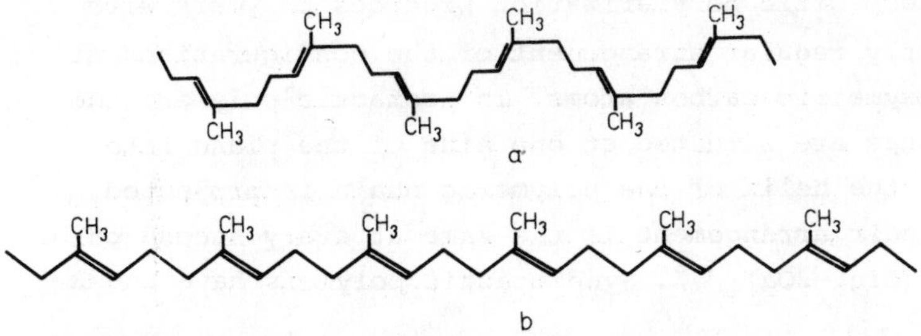

Fig. 19. Structure of the macromolecules of
gutta-percha (*trans*-1,4-polyisoprene)
a - α -modification, b - β-modification

the weak intermolecular forces. Low density polyethylene
is partially branched, which limits crystallinity and
yields a softening temperature of 115-120°C. This
level of crystallinity is favourable for some applications where a tough, flexible polymer is needed.
In extreme contrast, the strictly linear polymethylene
obtained by decomposition of diazomethane is highly
crystalline, has a much higher density, and at high
molecular weight melts at 132°C. The crystallizability
of polymethylene is so strong that it has not been
possible to obtain it in amorphous form /3/, even by
rapid cooling of the melt below the melting point.
Its high crystallinity makes it much more brittle than
low density polyethylene. Of intermediate crystallinity
is high density polyethylene whose low pressure synthesis
creates very little branching, and which has a higher
softening point of about 130°C.

The effect of chain geometry on crystallizability
can also be seen in the tactic polymers. The polymerisation of a monomer of the type CH_2=CHR, i.e. R-sub-
stituted ethylen, produces polymers with asymmetric
carbon atoms, which make possible different configurations, laevo - (L) and dextro - (D) of the substituent
R and the hydrogen atom. In atactic polymers the D and
L configuration alternate statistically (Fig. 20c) /4,5/.

Stereospecific polymerization produces polymers with strictly regular arrangement of the configurations at the asymetric carbon atoms. In isotactic polymers the R groups are situated at one side of the plane into which the helix of the polymeric chain is projected, and their arrangement is the same at every second carbon atom (Fig. 20a) /5/. Syndiotactic polymers have the R

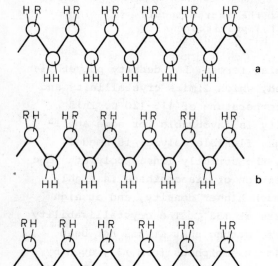

Fig. 20. Three possible configurations of the polymer prepared from CH_2 =CHR monomer /5/
a - isotactic, b - syndiotactic, c - atactic

groups situated alternately at both sides of the plane of the chain, the same arrangement occuring at every fourth carbon atom (Fig. 20b) /5/.

The chain regularity of an isotactic polymer is manifested by a distinct crystallizability as compared with the non-crystallizing atactic polymer. The larger crystalline content in an isotactic polymer influences its physical properties; for example the isotactic

polystyrene has a sharp melting point which is much
higher than the softening temperature of atactic
polystyrene (240°C /5/ vs. 80-90°C), it has lower
solubility in organic solvents (which provides a basis
for separation of the isotactic and atactic species).

The third factor - bulkiness of substituents - in
the case of statistically distributed bulky and
rigid phenyl groups, prevents close approach of the
macromolecules and formation of an ordered crystalline structure. With polyisobutylene, although its
molecules form flexible chains, the short but densely
arranged methyl groups prevent close approach and
crystallization at room temperature:

$$\ldots-CH_2-\underset{\underset{CH_3}{|}}{\overset{\overset{CH_3}{|}}{C}}-CH_2-\underset{\underset{CH_3}{|}}{\overset{\overset{CH_3}{|}}{C}}-\ldots$$

Macromolecules do not form rigid chains but possess
limited flexibility due to conformational change
resulting from thermal motion of the chain segments.
The idea of free rotation around a single bond is
well known in organic chemistry. This rotation changes
the shape of a molecule without changing the individual bond angles therein. In the ideal case, atoms
would be connected by freely rotating bonds and no
steric hindrance would be present, such free rotation
would enable every bond to assume any angle to a
neighbouring bond. There is a strong mutual influence
between macromolecules due to their being closely
packed, or in the condensed state. So free rotation is
restricted by steric hindrance and by electrostatic
fields arising from polar bonds and groups; thus
every rotation (i.e. conformation change) necessitates
some energy input. The chains assume the energetically most probable conformations dictated by the
electrostatic fields and by the substituents in the

chains. The energy needed for a change from the most
favourable conformation is characterized by the height
of the energy barrier. Understanding of the types and
magnitudes of potential barriers which control the
motions of isolated molecules is insufficient to deal
with macromolecules in the condensed phase because of
the steric hindrance and electrostatic fields arising
between adjacent macromolecules. Limited rotation is
thus observed only in some sections of the macromolecules which behave as kinetically independent units.
These sections are called segments. According to their
chain length it is possible to evaluate the flexibility
of the entire molecules. The more rigid the macromolecule, the longer are the segments /2/.

The most flexible chains are those in the polymers
of linear structure. The energy barrier is highest in
macromolecules containing electronegative groups because
of interactions between these groups and hydrogen atoms
in adjacent chains. The cross-linking of macromolecular
substances decreases the mobility of their chains, the
effect becoming distinct when the cross-links occur at
intervals of 40 carbon atoms or less in the main chain.
The rotation is almost impossible in the polymers with
a three-dimensional cross-linked structure. The restriction of rotation is manifested by increased
brittleness, reduced crystallizability, and lower
permeability to gases,

In the last few years, serious criticisms have arisen
about the postulate of segment crystallization which
is explained by a model of crystalline-amorphous
structure, the micellar model. According to this
theory /6/ ordered sections of macromolecules -
"crystallites" - are surrounded by the non-ordered
amorphous polymer, and one linear macromolecule can
extend through several amorphous and crystalline
regions over a distance of about 10 nm (Fig. 21). This

theory faces more and more contradiction with new
findings /7-10/.

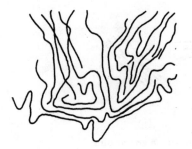

Fig. 21. Schematic
representation of
crystalline-amorphous
structure of a polymer /49/

There are also new ideas about the structure of
amorphous polymers; the density of amorphous polymers
is greater than it should be according to the previous
concepts of freely mobile molecules with random
conformation. In particular V.A. Kargin and coworkers
/2/ succeeded to prove that the amorphous state
represents a system which is organized to a certain
extent.

The lowest degree of organization, which exhibits
amorphous structure under electron diffraction, is
found with spherules comprised of the individual
macromolecules and reaching a diameter of 10 to 20 nm.

With the help of the electron microscope it was
possible to find spherules of up to 200 nm diameter
/11/. The spherules, however, can be arranged further
into macroscopic crystalline formations with well-
developed edges (Fig. 22) /2/. The lattice spacings
of such crystalline structures, however, are much
greater than the corresponding distances in small
molecule crystals, and the spherules themselves (as
structural units of the crystal lattice) are little
ordered systems formed of the randomly coiled
macromolecule, hence normal methods of crystallo-
graphy indicate a non-ordered state.

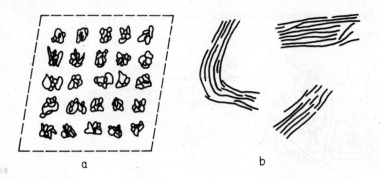

Fig. 22. Schematic representation of amorphous
structure of a polymer /49/
a - globular crystal, b - bundle of macromolecules

A higher degree of organization is seen with the
somewhat disentangled, parallel chains combined in
bundles which show signs of crystallization in the
electron diffractograms /12, 13/. From Fig. 22b it
is seen that these bundles are much longer than the
individual macromolecules, so that in a bundle some
macromolecules can be arranged one after another or
can have their ends in various parts of a bundle,
hence their individual character is not manifested
in a distinct way.

The basic morphological forms of crystalline
polymers are lamellae (a) and fibrillae (b) (see
Fig. 23). The theory of lamellar structure was supported by the discovery of single crystals of various
polymers. The single crystals are plate formations
of constant thickness, which is reproducible under
given conditions of crystallization. The thickness is
independent of the molecular weight of the polymer,
but it depends on the method of preparation and on the
crystallization temperature, varying according to the
temperature from 7 nm $\pm$ 2 nm to 14 nm $\pm$ 2 nm. By
electron diffraction it was found that the axes of the
molecular chains are perpendicular to the plane of the

lamellae. The macromolecules have the lengths of up to
several hundreds of nanometers, and as every individual

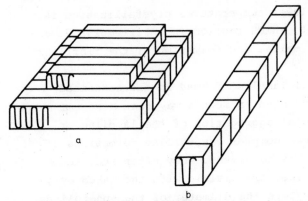

Fig. 23. Schematic representation of basic morphological forms of crystalline polymers /49/
a - lamella, b - fibrilla

macromolecule can only belong to a single crystal
(lamella), it must be folded through an angle of 180°
at intervals corresponding to the lamella thickness,
so that it is alternately folded to form a strip
(Fig. 24a). The folding of the macromolecules is
spontaneous, originating from a tendency to decrease
the surface tension of the crystallizing system. In

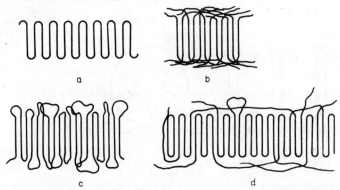

Fig. 24. Schematic representation of development of
crystalline structure of polymers /17/
a - strip, b,c,d - irregularities of the space lattice
at the points of bending of the macromolecules

the places where the molecule is folded the spatial
lattice is disturbed (Fig. 24b,c,d). Such folding
was found even in such rigid linear chains as cellulose
triacetate. At lower temperatures crystallization is
multidirectional, which results in an irregular development of the lamellae and formation of the fibrillar
structure.

The individual lamellae formed in the condensed
state are also twisted in such a way that a spiral shape
is produced. Radial aggregation of the lamellae or
fibrillae produces complex crystalline formations of
spherical shape - spherulites. They often reach such
dimensions that they are visible with the naked eye,
e.g. in polypropylene the diameter of the spherolites
can be as large as 2 nm. The growth of the spherolites
of macromolecular crystalline substances can be
represented by the Bernauer model consisting of several
stages (Fig. 25) /14/. Formerly, spherulites were
considered to be formed by a rearrangement of the
crystallites already formed. The latest investigations,
however, show that they are formed by a direct crystallization of the molecule /15, 16/. The size of spherulites increases with increasing crystallization temperature. To disturb the arrangement of the crystallization centres and to prevent recrystallization of

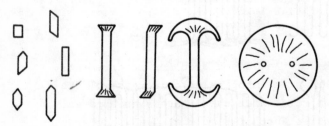

Fig. 25. Development phases of a spherolite of
a macromolecular crystalline compound /14/

the spherulites at their original positions requires
heating to about 50°C above the melting point. It is
obvious that the two-phase model of segment crystallization must be revised, but there remain many problems
to be solved, such as the mechanism of folding, the
degree of organization of "amorphous" polymer, and the
constant thickness of the lamellae.

Besides structural prerequisites, crystallization
also needs suitable kinetic conditions.

Crystallization is temperature-dependent to a considerable extent (Fig. 26). Its rate exhibits a maximum
between t_g and t_m. The crystallization isotherms have
typical shapes (Fig. 27); at the beginning the crystallization rate shows no distinct increase, due to the

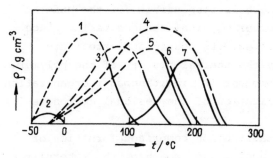

Fig. 26. Dependence of the crystallization rate expressed
as the density change $\Delta\rho$ of some polymers vs. temperature
changes t /92/
1 - polyethylene, 2 - natural rubber, 3 - polypropylene,
4 - poly(hexamethyleneadipamide), 5 - polychlorotrifluoro-
ethylene, 6 - polycaprolactam, 7 - poly(ethylene
terephthalate)

low sensitivity of the method used for measurement;
then there follows an autocatalytic increase of the
crystallization rate which then reaches an apparent
equilibrium; after further time a gradual acceleration
of crystallization occurs. The extent of crystallization
approaches asymptotically to a certain value.

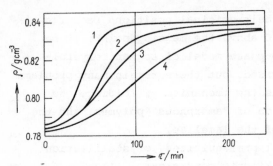

Fig. 27. Crystallization isotherms of polypropylene
(dependence of the polymer density ρ on time τ) /92/
1 - 143°C, 2 - 145°C, 3 - 146°C, 4 - 147°C

The main changes occur in the first part of the
curve - the primary crystallization; the secondary
crystallization has great practical importance. It is
this latter stage which is responsible for the undesirable volume and other physical changes taking place
usually after processing. Its course and extent are
affected markedly by the thermal history of the polymer.
In principle, this process is simultaneous reorganization
and further crystallization of the residual amorphous
phase.

The primary crystallization comprises formation of
the crystallization nuclei and growth of the spherulites.
Its overall rate is often characterized by the half-life,
denoted $\tau_{0.5}$, i.e. the time for a half of the crystallization to take place. This is easily determined from
the crystallization isotherms. Polymers usually
crystallize in less symmetrical crystal systems, some
even in more than one crystal system; in the case of
polyethylene the orthorhombic and pseudomonoclinic
systems were found.

The properties of linear polymers also depend on
their molecular weight. The high molecular weight of
plastics, reaching often 10^{4} to 10^{5}, is the main feature
distinguishing them from low-molecular substances. For

complete characterization of a polymer it is absolutely
necessary to know its molecular weight because this
determines behaviour under various conditions. For
useful mechanical strength a plastic must have a degree
of polymerization above 40 to 80 and beyond this, mechanical strength increases steadily with increasing chain
length.

In practice, polymers contain molecules of various
chain lengths and the distribution of molecular weights
obtained from simple polymerization systems can be
calculated statistically. This distribution can be
expressed by the dependence between the weight fraction
of the polymer of a given molecular weight and the molecular weight (Fig. 28) /17/.

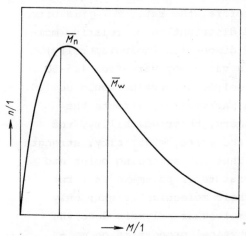

Fig. 28. An example of distribution of relative
molecular weights M in a polymer /17/
$\overline{M}_n$ - the number-average molecular weight, $\overline{M}_w$
- the weight-average molecular weight

As a scatter of molecular weights occurs in every
sample of a polymer, the experimental determination of
molecular weight only gives a mean value. Depending on
the method of the determination the value obtained is
either the weight-average molecular weight $\overline{M}_w$ or the

number-average molecular weight $\bar{M}_n$.

Methods of determination based on measurements of colligative properties, such as ebullioscopy, cryoscopy, and osmometry, or on end groups /18/, give the number of molecules in a known weight of the material. As the Avogadro number is known, the determination leads to the number-average molecular weight $\bar{M}_n$. For typical polymers this number average lies near the top of the distribution curve of molecular weights or at the point of the most probable molecular weight /17/.

In scattering measurements (e.g. light scattering) the contribution of a molecule to the turbidity is a function of its weight /17, 18/. The result is the weight-average molecular weight $\bar{M}_w$ which is equal to or higher than $\bar{M}_n$. Therefore, the ratio $\bar{M}_w/\bar{M}_n$ is often considered a measure of distribution of relative molecular weight degree of dispersion (polydispersity). For typical polymers its value may vary from 1.5-50 /17/. For monodisperse polymers its value would be 1.0.

The molecular weight particularly affects the softening point of polymers, their solubility, the viscosity of solutions and melts, elasticity, strength, and thermal stability. Thus the softening point and the decomposition temperature of polymers at first increase steeply with their molecular weight; only with sufficiently high-molecular polymers does the dependence of various physical properties on molecular weight become insignificant.

In summary, linear polymers may be characterized generally according to their properties:

a) The polymers tending to spontaneous crystallization are usually fibre-forming and film-forming, strong, tough, thermoplastic, elastic, of low solubility or insoluble in common organic solvents, and their temperature range of the rubber-like state is usually narrow, i.e. their transition from solid phase to melt is rapid. Therefore, they are not so suitable for such treatments as calandering or extrusion but are very suitable for the processes requiring rapid melting as in injection moulding, spinning, etc. This group of polymers can be exemplified by polyamides, poly-formaldehyde, polyethylene, and polypropylene.

b) The polymers which do not crystallize spon-taneously but only by tension or temperature change are usually rubber-like at normal temperature, thermo-plastic, elastic, and readily soluble in organic solvents. Typical methods of processing are calandering and extrusion. Polyisobutylene and natural rubber can be given as examples.

c) The polymers which do not crystallize at all are usually brittle, transparent, very readily soluble in organic solvents, and thermoplastic. They are workable by extrusion, injection moulding, and by other methods. Examples are polystyrene, poly(methyl methacrylate), and non-hardened resol resins.

3.2 CROSS-LINKED POLYMERS

The term cross-linked denotes macromolecules in which the main chains are linked by chemical bonds so that they form a three-dimensional network (Fig. 29).

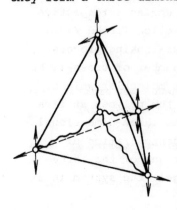

Fig. 29. Schematic representation of an ideal three-dimensional network of a cross-linked polymer /18/

Cross-linked polymers are formed either by cross-linking of a linear polymer or by reaction of one or more monomers with functionality greater than two.

The cross-linking of linear polymers is characterised by rubber, when it is termed vulcanization and produces vulcanized rubber. In recent years cross-linking has become industrially important for other linear polymers, such as polyethylene, polypropylene, and poly(vinyl chloride): these polymers thereby acquire some advantageous properties, especially higher distortion temperature and improved physico-mechanical properties.

A linear polymer containing reactive sites in its chains can react with a cross-linking reagent to form chemical cross-links. The cross-link is able to connect reactive sites in various chains; a bifunctional cross-link connects two reactive sites, which is the most frequent case in practice. A polyfunctional cross-linking reagent can link several chains. In some cases the molecule (atoms) of cross-linking reagent becomes a part of the cross-link (e.g. in the vulcanization of polydienes with sulphur). In other cases the cross-linking reagent causes the reactive sites to form direct links without itself being involved in the cross-link, as with cross-linking by peroxides, or radiation. Therefore, the cross-link is considered to be any connection of two or more chains irrespective of the chemical nature of the junction formed /18/.

In the first stage of the cross-linking process, i.e. on introduction of a small number of cross-links into a linear polymer of a finite molecular weight, the relative molecular weight of the polymer at first increases without formation of the three-dimensional polymeric network. At a certain concentration of cross-links, however, the first signs of the three-dimensional structure appear as gel: the system is at

the gelation point. At any higher concentration of the
cross-links the polymer consists of two phases: the
gel which is insoluble in all solvents (unless they
attack the cross-links or the chain bonds) and the
sol which is soluble and can be extracted. At the
gelation point the properties of the polymer change
distinctly. Thus a mixture of rubber (linear polymer)
and a vulcanization agent can be calandered, extruded,
or shaped in some other ways but soon after the cross-
linking (vulcanization) reaction starts the mixture
reaches its gelation point, gradually stops to be
plastic, and at a certain degree of cross-linking
cannot be shaped any longer.

Cross-linked polymers are mainly infusible, and
organic solvents cause swelling without dissolution.

The concentration of cross-links introduced into
linear polymers during vulcanization is relatively
low. For example, common vulcanizates (cross-linked
rubbers, vulcanized rubber) contain only 10-20
cross-link sites in each original macromolecule; the
section of an original molecule which is between two
subsequent cross-links has a molecular weight of
3000 to 10 000 /18/. In contrast, polymers with
three-dimensional structure formed by polyreactions
of polyfunctional monomers usually have a dense
network with a high concentration of cross-links.
Examples are polymers formed by reaction of glycerol
with dibasic acids (polyesters), phenol with formaldehyde (phenolformaldehyde resins and phenolic plastics),
and glycol maleate polyester resins with styrene.
Diamond (Fig. 30) represents an ideal example of a
three-dimensional structure with a dense network.

The dimensions of linear or branched macromolecules
are finite; in diluted solutions such macromolecules
behave as individual units. A three-dimensional polymer
network, theoretically has both volume and molecular

weight unlimited (infinite), but in practice is limited by the size of the sample.

The physico-mechanical properties of polymers with
three-dimensional structure depend significantly on
the density of the network (concentration of cross-
links); changes in properties depending on the spatial
network can be easily followed and are well understood
in the case of vulcanized rubbers (sparsely cross-
linked linear polymers) /18/, whereas for densely
cross-linked polymers (typified by phennolic plastics),
which differ from the vulcanizates in having low or
almost no elasticity and considerable hardness,
theoretical understanding is less easy because of the
structural complexity.

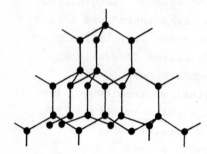

Fig. 30. Schematic
representation of the
diamond structure
(three-dimensional
arrangement of carbon
atoms) /49/

3.3 RESISTANCE AND STABILIZATION OF PLASTICS

In comparison with some other materials, plastics are
very resistant to corrosion and are therefore useful
for the protection of metals, wood, and building
materials. Corrosion can be defined as an unwanted
chemical change in surface composition and consequent
deterioration of properties caused by atmospheric
conditions. The deterioration of plastics can arise
from various sources, such as aggressive chemicals,
increased temperature, weather, ultraviolet radiation,
and microorganisms, but is not classified as corrosion;

it takes place not only during the service life of
articles, but even earlier during polymer production
and during subsequent processing, most often through
thermal or mechanical stress. It is important to know
the conditions under which polymers are degraded and
the methods of protection and studies in this field
have led to general principles which make it possible
to restrict or even prevent degradation.

From the point of view of chemical reactivity,
macromolecular organic compounds behave like the
low-molecular compounds of the same character. However,
the course of reaction is different. The degradation
of polymers involves chemical reactions proceeding
mostly in heterogeneous systems, and the course of
reaction is controlled firstly by diffusion processes.
Diffusion through an organic polymer represents molecular migration of a substance from one region to
another (in the direction of lower concentration or
pressure) made possible by the mobility of the
macromolecules. Continuous vibrational motion of the
segments of polymeric chains produce transient "cavities" which are filled with the migrating substance.
As the location of these cavities is continuously
changed, the substance diffuses through the polymer.
The permeation of the substance through fine pores
and channels which may be present due to imperfect
preparation of the polymer article is not at the molecular level, nor considered to be the true diffusion
process. Such permeation is not specific for the given
material and depends little on temperature, whereas
true diffusion is specific for the material and its
rate is temperature-dependent; the non-specific
permeation usually shows an inverse temperature-
dependence as compared with true diffusion, because
the pores contract on heating due to thermal expansion
of the polymer. In the case of specific diffusion,

thermal energy increases kinetic motion of the chain
segments, and diffusion is accelerated.

3.3.1 PERMEABILITY OF POLYMERS

A package may need to meet various requirements
concerning its permeability. The permeability of package to water vapour or to gases may have considerable
effect on quality of the products packed. The presence
of oxygen can induce undesirable oxidation processes
which can lead ultimately to deterioration of the
product; the loss of aromatic substances or carbon
dioxide is often undesirable. Therefore, in packaging,
considerable attention is paid to the permeability
of packages and to its determination. Ideally permeation
is synonomous with diffusion, but in practice, holes
and pores may also play a part.

Permeation, as mass transfer through a plastics
membrane, is a process which originates from concentration, pressure, or temperature gradients between
the two sides of the membrane. The mechanism of this
process is complex and depends on the nature of both
the membrane and the permeating substance, the latter
being absorbed by the polymer at first, then dissolved,
then it penetrates through the membrane, and finally
is desorbed from the other side of the membrane.

Diffusion through isotropic materials at constant
pressure and temperature obeys the Fick laws /19/.
According to the first Fick law the rate (i) of transfer
of diffusing substance through unit area of a section
is proprotional to the concentration gradient $\partial c/\partial x$.
The equilibrium unidirectional diffusion along the x
coordinate can be expressed as

$$i = -D\frac{\partial c}{\partial x} \tag{1}$$

where i means the mass of the substance which passed
through a unit area in perpendicular direction during
a unit time, D is the diffusion coefficient, and c
stands for concentration of the migrating substance.

Distribution or concentration of the migrating
substance along the coordinate x in dependence on
time τ is expressed by the second Fick law which
follows from Eq. (1) and the continuity equation (2)
for non-reacting systems /20/.

$$\frac{\partial c}{\partial \tau} + \frac{\partial i}{\partial x} = 0 \tag{2}$$

Non-steady state unidirectional diffusion obeys the
equation:

$$\frac{\partial c}{\partial \tau} = \frac{\partial}{\partial x}\left[D\frac{\partial c}{\partial x}\right] \tag{3}$$

At constant diffusion coefficient Eq. (3) is
transformed into the usual form of the second Fick
law:

$$\frac{\partial c}{\partial \tau} = D\frac{\partial^2 c}{\partial x^2} \tag{4}$$

With regard to Eqs (3) and (4) it must be noted
that the diffusion coefficients are, especially in
polymeric systems, significantly dependent on the
concentration.

If a substance at constant surface concentrations
c_1 and c_2 $(c_1 > c_2)$ diffuses through a membrane of
thickness t, and if steady state is reached, i.e. the
concentration of the migrating substance at any point
does not change with time $(\partial c/\partial \tau = 0)$, then Eq. (3)
becomes:

$$\frac{\partial}{\partial x}\left[D\frac{\partial c}{\partial x}\right] = 0 \tag{5}$$

Double integration of Eq. (5) for the initial and limit condition $x = 0, c = c_1$ and $x = t$, $c = c_2$ ($\tau \geqslant 0$) and combination with Eq. (1) lead to Eq. (6) for the steady state diffusion flow.

$$i = -\frac{1}{t}\int_{c_1}^{c_2} D\,dc = \bar{D}\frac{c_1 - c_2}{t} \tag{6}$$

where the integral diffusion coefficient $\bar{D}$ is defined by the relation

$$\bar{D} = \frac{1}{c_1 - c_2}\int_{c_2}^{c_1} D\,dc \tag{7}$$

At constant diffusion coefficient $\bar{D} = D$, the concentration of the substance in the membrane is a linear function of x, and Eq. (6) becomes:

$$i = D\frac{c_1 - c_2}{t} \tag{8}$$

Eq. (8) makes it possible to determine the amount of the substance passed through the membrane of area A during time τ

$$Q = DA\tau\frac{c_1 - c_2}{t} \tag{9}$$

From practical point of view the most important permeability is that to gases and vapours. If the pressures of the gas or of the saturated vapour $(p_1 > p_2)$ at both sides of the membrane are known, and the concentration (c) of the gas in the membrane depends linearly on the external pressure (p)

$$c = Sp \tag{10}$$

where S is the solubility of the gas in the membrane, then Eqs (9) and (10) lead to:

$$Q = PA\tau \frac{p_1 - p_2}{t} \qquad (11)$$

The permeability coefficient P is defined by the equation

$$P = DS \qquad (12)$$

i.e. it is equal to the product of the diffusion coefficient and the solubility.

The diffusion and permeability coefficients and the solubility characterize the process of gas transfer through a plastic membrane.

The diffusion coefficient D is the amount of gas passed through a unit area during unit time at unit concentration gradient, and its dimension is $m^2 s^{-1}$. The permeability coefficient P gives the volume of gas (m^3) passed through a unit area during unit time at unit pressure gradient, and its SI dimension is $m^2 s^{-1} Pa^{-1}$. In the former c.g.s. system, its usual dimension was $cm^2 s^{-1} atm^{-1}$.

In older American and British literature the permeability coefficient is expressed in $cm^2 s^{-1} Torr^{-1}$ or in $cm\ mm\ s^{-1} (cm\ Hg)^{-1}$, because Barrer /21/ defined P as the diffusion flow of the gas (in $cm^3 s^{-1}$, measured at standard conditions: $T = 273.15$ K and pressure $p = 1$ atm $= 101\ 325$ Pa), when Δp (i.e. $p_1 - p_2$) has the value 1 cm Hg, and the thickness of the membrane is $t = 1$ mm.

The gas or vapour solubility S is formally expressed as the amount of the substance in m^3 (measured under standard conditions) dissolved in 1 m^3 of the solvent (polymer) at the partial pressure of 1 Pa at the given temperature. However, usually it is expressed at the given temperature and at the pressure

of 101 325 Pa, which corresponds to the values of the
Bunsen absorption coefficient.

All the relations derived also apply to the solubility of liquid and solid substances, the respective
coefficients being, however, expressed in different
units. The solubility of liquids and solids in polymers
is expressed in g cm^{-3} or in mass fractions.

To understand the proper mechanism of permeability,
it is necessary to know both the diffusion coefficient
and the solubility of the substance in the polymer,
and also the dependences of these quantities on
temperature, structure of the polymer, and character
of the migrating substance.

At low pressures or at enhanced temperatures the
solubility of gases in a polymer is independent of
the pressure, and Eq. (10) is analogous to Henry´s
law /22/.

The solubility of gas (S) is an exponential function
of temperature:

$$S = S_0 \exp\left(-\frac{\Delta H}{RT}\right) \qquad (13)$$

where S_0 is a constant, ΔH is the heat of dissolution
of the gas in the polymer, and R and T stand for the
universal gas constant and absolute temperature,
respectively. The heats of dissolution for most gas-
polymer systems have values from 0 to 12 kJ. The temperature dependence of permeability reads as follows:

$$P = P_0 \exp\left(-\frac{E_P}{RT}\right) \qquad (14)$$

where P_0 is a constant and E_p is the activation energy
of permeability. With respect to Eq. (12) P is mainly
determined by the temperature dependence of the diffusion
coefficient. For a constant diffusion coefficient it
can be written

$$D = D_0 \exp\left(-\frac{E_D}{RT}\right) \tag{15}$$

where D_0 is a constant, and E_D is the activation
energy of diffusion which is usually a linear
function of temperature.

The quantity E_p has no physical meaning. From Eqs
(12)-(15) it follows that

$$E_P = E_D + \Delta H \tag{16}$$

There is no simple relation between permeability
and properties of the migrating substance, because
the whole process is complex. The mass, size, shape
and nature of the migrating molecules determine not
only the diffusion coefficient, but have also a decisive
effect on the solubility value. The nature and structure of the polymer also influence the solubility
coefficient significantly and the diffusion coefficient
decisively. The highest permeability is found with
highly elastic polymers such as rubber, and the lowest
permeability is exhibited by rigid polymers with strong
intermolecular attraction or chemical cross-links.

Considerable attention is paid to these aspects,
and details can be found in numerous monographs /23, 24/.

The permeability of plastics can be measured by both
direct and indirect methods.

In the indirect methods two parameters are determined
experimentally, and the third parameter is calculated
from Eq. (12). E.g. the method by Daynes and Barrer
/25, 26/ determines P and D and calculates S. Other
methods determine D and S and calculate P. Solubility
of gases and vapours is usually determined in sorption
apparatus   by measuring the equilibrium amount of the
gas or vapour absorbed by a known volume or weight of
polymer. In these methods it is presumed that no mol-

ecules of the gas or vapour are adsorbed at the surface of the polymer. Suitable methods for gases are manometry, chromatography, or volumetry, whereas the absorption of vapours is usually determined gravimetrically.

For determination of the diffusion coefficient, it is necessary to solve the second Fick law at the limit and the initial conditions characterizing the given experiment.

Practically the most frequent case is a diffusion between the polymer solution of the initial concentration c_0 and pure solvent in the plane of $x = 0$. Solution of Eq. (4) for these limit and initial conditions

$$\left. \begin{array}{ll} c = c_0 & x < 0 \\ c = 0 & x > 0 \end{array} \right\} \ \tau = 0$$

leads to the expression

$$c(x, \tau) = \frac{c_0}{2} \left[1 - \mathrm{erf} \left(\frac{x}{2\sqrt{D\tau}} \right) \right] \qquad (17)$$

where $\mathrm{erf}\, x = \dfrac{2}{\sqrt{\pi}} \displaystyle\int_0^x e^{-\eta^2}\, d\eta$

is the so called Gauss integral of errors.

The sharp interface between the solution and the solvent at the time $\tau = 0$ is deliquesced symmetrically to the both sides during the diffusion. The diffusion course is represented graphically in Fig. 31.

For $x = 0$ at all times except for $\tau = 0$, the concentration is $c = c_0/2$, and all distribution curves intersect in this point.

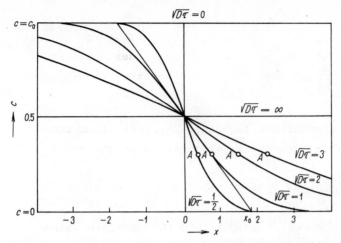

Fig. 31. Graphical representation of the diffusion course between a polymer solution (initial concentration $c = c_0$) and pure solvent at the plane $x = 0$

Differentiation of Eq. (17) according to x gives.

$$-\frac{\partial c}{\partial x} = \frac{c_0}{2\sqrt{\pi D\tau}} \exp\left(-\frac{x^2}{4D\tau}\right) \tag{18}$$

For $x = 0$, Eq. (18) is transformed into Eq. (19),

$$-\left(\frac{\partial c}{\partial x}\right)_{x=0} = \frac{c_0}{2\sqrt{\pi D\tau}} = \operatorname{grad} c \tag{19}$$

where grad c means the concentration gradient.

Plotting of the relation (19) against x gives, for the given time τ, a curve which includes the area

$$A_k = \int_{-\infty}^{+\infty} -\frac{\partial c}{\partial x} = c_0 \tag{20}$$

Introduction of Eq. (20) into Eq. (19) gives the relation

$$D = \frac{A_k^2}{4\pi \operatorname{grad} c^2 \tau} \tag{21}$$

according to which it is possible to calculate value of the diffusion coefficient from the known values A_k and concentration gradient grad c (19).

The diffusion coefficient can also be calculated from the amount of substance which passes through a unit area during the time τ in the plane $x = 0$:

$$\dot{Q} = \int_0^\tau i_{x=0} \, d\tau \tag{22}$$

where $i_{x=0}$ means the diffusion flow of the substance in $x = 0$.

From Eqs (1) and (19) it follows that

$$i_{x=0} = -D\left(\frac{\partial c}{\partial x}\right)_{x=0} = \frac{c_0}{2}\sqrt{\frac{D}{\pi\tau}} \tag{23}$$

Introduction of Eq. (23) into Eq. (22), subsequent integration and modification give the relation

$$D = \frac{\dot{Q}^2 \pi}{c_0^2 \tau}$$
(24)

where C_O is the initial concentration of the solution of polymer.

Besides these methods, the diffusion coefficient can be determined (in homogeneous systems at free diffusion) in such a way that the time dependence of the x position is followed for a chosen concentration ratio $\alpha = c/c_O$ ($\alpha < 1$) or, on the contrary, the time dependence of concentration of the migrating substance is followed for a chosen x. If e.g. for $\alpha = 0.3$ (the points A in Fig. 31) the experimental value x are plotted against $\sqrt{\tau}$, the straight line obtained has the slope k which depends on the diffusion coefficient D as follows

$$k = \sqrt{D}\, \varphi(\alpha)$$
(25)

where the value of function $\varphi(\alpha)$ for a given α can be obtained from tabulated values of Gauss integral and from the relation:

$$\alpha = \frac{1}{2}\left[1 - \operatorname{erf}\left(\frac{\varphi(\alpha)}{2}\right)\right]$$
(26)

For a rough estimate of the diffusion coefficient the tangent is constructed to the distribution curve at the point $x = 0$. This tangent intersects the x axis in the distance x_O (Fig. 31) which is related to the diffusion coefficient as follows:

$$D = \frac{x_0^2}{\pi \tau}$$
(27)

After a suitable modification, the method described can also be used for determination of diffusion coefficients of gases, vapours, or liquids in solid polymers. If a polymer specimen, limited by a plane $x = 0$, has zero concentration ($c = 0$) of the migrating species at the beginning of the experiment, if its thickness t is so large that the migrating species only penetrates to a distance $x < t$ during the experiment, and if the concentration of the migrating species is kept constant ($c = c_0$) outside the polymer throughout the experiment, then the solution of the second Fick law (4) gives the relation:

$$c = c_0 \left(1 - \text{erf}\frac{x}{2\sqrt{D\tau}}\right) \tag{28}$$

where c is the concentration of the migrating substance in the polymer.

The amount $\dot{Q}$ of the substance which diffuses into the polymer through a unit area during a time τ in the plane $x = 0$ is determined by Eq. (22). The diffusion flow at $x = 0$ obeys Eq. (29) which is analogous to Eq. (23).

$$i_{x=0} = c_0 \sqrt{\frac{D}{\pi\tau}} \tag{29}$$

Combination of Eq. (22) with Eq. (29) gives the relation:

$$Q = 2c_0 \sqrt{\frac{D\tau}{\pi}} \tag{30}$$

according to which it is possible to calculate the corresponding diffusion coefficient. Another method /25, 26/ determines the time $\Delta\tau$ (time lag) needed

by a gas in contact with the surface of an evacuated
film of the thickness t to appear also on the other
side of the film and for reaching a steady state. The
solution of the second Fick law for these limiting and
initial conditions at constant D showed that the
diffusion coefficient of a gas or vapour depends on
the time lag $\Delta\tau$ and thickness t according to the
relation:

$$D = \frac{t^2}{6\,\Delta\tau} \tag{31}$$

For a plastic membrane of finite dimensions, limited
by the planes $x = 0$ and $x = t$, the solution of Eq.
(4) for the limiting and initial conditions

$$
\begin{array}{llll}
c = c_0 & 0 < x < t & & \tau = 0 \\
c = 0 & x = 0 & x = t & \tau > 0
\end{array}
$$

gives the relation:

$$c = \frac{4c_0}{\pi} \sum_{v=0}^{\infty} \frac{1}{2v+1} \sin\frac{(2v+1)\pi x}{t} \exp -\left[\frac{(2v+1)^2 \pi^2 D\tau}{t^2}\right] \tag{32}$$

The relation (32) expresses the dependence of concentration of the migrating substance in the membrane as
a function of position and time, if the initial concentration of the migrating substance in the membrane is
$c - c_0$, and in its environment the zero concentration
is maintained for the time $\tau > 0$.

The total amount of the substance which diffuses
from the membrane during the time τ is calculated by
integration of Eq. (32) within the limits from $x = 0$
to $x = t$, and it is

$$Q = \int_0^t c \, dx = \frac{8c_0 t}{\pi^2} \sum_{v=0}^{\infty} \frac{1}{(2v+1)^2} \exp\left[-\frac{(2v+1)^2 \pi^2 D\tau}{t^2}\right] \tag{33}$$

At the following limiting and initial conditions:

$$c = 0 \qquad 0 < x < t \qquad \tau = 0$$
$$c = c_0 \qquad x = 0 \qquad x = t \quad \tau > 0$$

the solution of Eq. (4) leads to the relation:

$$c = c_0 \left\{ 1 - \frac{4}{\pi} \sum_{v=0}^{\infty} \frac{1}{2v+1} \sin\frac{(2v+1)\pi x}{t} \exp\left[-\frac{(2v+1)^2 \pi^2 D\tau}{t^2}\right] \right\} \tag{34}$$

and the amount of substance which diffuses into the plastic membrane during the time τ is given by the relation:

$$Q = c_0 t \left\{ 1 - \frac{8}{\pi^2} \sum_{v=0}^{\infty} \frac{1}{(2v+1)^2} \exp\left[-\frac{(2v+1)^2 \pi^2 D\tau}{t^2}\right] \right\} \tag{35}$$

The relations (33) and (35) enable determination of the respective diffusion coefficients from the known initial concentration c_O and absorption or desorption kinetics of the substance in the polymer.

Determination of the permeability of plastics by the direct method consists in determination of the amount $\dot{Q}$ of gases or vapours which passes through the plastic membrane at the given conditions of state. The permeability coefficient P is calculated from Eq. (11).

$$P = \frac{\dot{Q}t}{A\tau(p_1 - p_2)} \qquad \cdot \qquad (36)$$

The apparatus for determination of permeability
contains a closed cell which is divided by a membrane
of a defined thickness and a defined surface area into
two parts, and one part is filled with the gas; the gas
penetrates into the other part where its amount is
determined from changes of its pressure, volume, weight,
or concentration.

That is why the individual methods of permeability
determination are denoted as manometric, volumetric,
gravimetric, or concentration method, respectively.
Concentrations of the gas or vapour passed are determined
by optical or chemical methods, chromatography, or mass
spectroscopy. The details can be found in the literature
/23, 24/.

Very often the heat conductivity method is used in
which the gas permeating the membrane is determined
either in a flow of carrier gas (differential method)
or in a closed space filled with the carrier gas (integral method).

In these measurements, the permeability coefficients
are determined along with the diffusion coefficients
/27, 28, 29/.

The procedure for the determination of water vapour
permeability through all flat packaging materials inclusive of plastics is prescribed by a Czechoslovak standard
similar to ASTM E - 96. According to this standard the
permeability to water vapour is expressed by the amount
of the water vapour (in grams) passed through the
plastic of surface area 1 m^2 during 24 hours (for thickness
up to 3 mm):

$$P = \frac{\Delta B \cdot 10 \cdot 24}{A\tau} \tag{37}$$

where ΔB is the change (increase) of weight (mg),
A is the sample surface area (cm^2), and τ is the time
(h) corresponding to the weight increase ΔB.

The permeability determination is based on the
following procedure: a test dish with drying agent
(silica gel, calcium chloride) is covered with the
sample and placed in an air-conditioned environment,
and the amount of water vapour passed through the
sample is found from the weight increase of the dish;
the air-conditioning conditions which define the
pressure difference are given in Table 14. For comparison of permeability of various packaging materials the
P values found are usually recalculated for the O.1
mm thickness.

Table 14. The air conditions for determination of
water vapour permeability

Conditions	Temperature (^{o}C)	Relative humidity (%)	The saturated solutions used for maintaining constant air humidity
mild	25 ± 0.5	75 ± 2.0	sodium chloride
tropical	38 ± 0.5	90 ± 2.0	potassium nitrate or sodium tartrate

The procedure for the water vapour permeability
determination of rigid or semi-rigid packages under
tropical conditions is given by a Czechoslovak standard,
in which the package is charged with drying agent,
immediately closed, and placed in the test environment
and the permeability is expressed as grams of water

vapour entering the package during 24 hours. The uptake
of water vapour is determined by weight increase.

The literature gives numerous methods for the
determination of permeability of plastic bottles
/30, 31, 32/. Becker /33/ developed a method for oxygen:
the bottles, filled with oxygen-free water are immersed in an oxygen atmosphere, and the rate of change
of oxygen partial pressure in the bottle is followed
polarographically. It is necessary to know, besides
the experimental da a, the value of the solubility
coefficient of oxygen in water.

Becker /33/ also developed a conductometric method
for carbon dioxide. The bottle filled with a dilute
solution of barium hydroxide, is immersed in a carbon
dioxide atmosphere of constant pressure p = 101 325 Pa.
Carbon dioxide diffuses through the walls of the bottles
and reacts with barium hydroxide as follows:

$$Ba^{2+} + 2OH^- + CO_2 \longrightarrow BaCO_3 + H_2O$$

The formation of sparingly soluble barium carbonate
is accompanied by a decrease of electrical conductivity
of the barium hydroxide solution. From the known
dependence of conductivity on concentration, it is possible
to determine the amount of carbon dioxide reacted.

3.3.2 RESISTANCE TO AGGRESSIVE REAGENTS

The resistance of plastics to attack depends predominantly on their chemical composition. The inertness of
non-polar aliphatic compounds to chemical reactions,
which is well known in organic chemistry, is analogous
to that of non-polar macromolecular compounds. Poly-
olefinic chains, however, undergo substitution and
oxidation reactions relatively easily under certain
conditions. Hence, polyethylene and polyisobutylene are
quite resistant to non-oxidative acids and bases and

their salts as well as to weak oxidation agents, but
they are badly damaged by stronger oxidants and by
oxidizing atmospheres.

The introduction of some polar substituents, such
as hydroxyl, ester or nitrile groups, in the polyethylene
chain, decreases the chemical resistance of the polymer
drastically, as seen in the low resistance of poly(vinyl
acetate) or polyacrylonitrile to acids and bases. Halogen
substituents are exceptions; thus polytetrafluoroethylene,
polytrifluorochloroethylene, poly(vinyl chloride), etc.
resist acids, bases, salts, and to a certain degree
oxidants. The most resistant polymer is polytetrafluoroethylene whose carbon skeleton is effectively
protected by strongly electronegative fluorine groups.
It resists practically all chemical agents, and only
melted alkali metals (sodium, potassium) or complexes
(sodium naphthenate) cause superficial etching.

The resistance of polymers especially to oxidizing
agents is strongly decreased by the presence of double
bonds in their chains (as e.g. in polybutadiene and
natural rubber).

Heteroatoms, e.g. S, N, O, present in individual
structural units of the polymer undergo degradation by
chemical reagents. Especially facile are hydrolytic
reactions caused by acids and bases, as with polyamides,
polyesters, polyurethanes, and cellulose. The hydrolysis
of a polymer containing a heteroatom in its backbone
is manifested by a striking decrease of molecular weight
(e.g. polyamides), whereas the hydrolytic splitting of
side groups leads predominantly only to property changes
(e.g. poly(vinyl acetate)).

Resistance of plastics to organic solvent depends
on the polarities of the polymer and solvent and on the
phase state of the polymer. Generally, a polymer containing polar groups is attacked by polar solvents, the
mutual action being the larger (the polymer being

dissolved ultimately), the more similar are the
intermolecular forces in the polymer and in solvent.
So the non-polar polymers like polyethylene, polyisobutylene, or polystyrene swell or are dissolved in
non-polar solvents (e.g. petrol, benzene, tetrachloromethane) but resist polar solvents (water, alcohols).
On the contrary, polar polymers like poly(vinyl alcohol), poly(vinyl acetate), carboxymethylcellulose,
polyamides, etc. resist non-polar solvents but swell
or are dissolved in polar solvents (water, alcohols,
methyl chloride, phenol, etc.). As already mentioned,
however, an important condition for the dissolution
is approximate equality of magnitudes of intermolecular
forces of the polymer-solvent pair. So e.g. poly(vinyl
alcohol), whose molecules exhibit large polarity, is
dissolved easily in strongly polar water, but is not
dissolved in less polar alcohols. On the contrary,
poly(vinyl acetate) which is less polar than poly(vinyl
alcohol), is soluble in methanol, but insoluble in
water in which there are much stronger intermolecular
forces.

The chemical resistance of plastics is also significantly affected by their supermolecular structure or
phase state, i.e. whether they are crystalline or
amorphous. Typical examples are the non-polar polymers
polyethylene and polyisobutylene. They both can be
expected to be dissolved well in non-polar solvents.
However, the crystalline structure of polyethylene
increases its resistance to non-polar solvents so that
they dissolve it only at elevated temperatures, e.g. at
the boiling point of tetrachloromethane. Amorphous
polyisobutylene is dissolved in these solvents at normal
temperature. The different solubility of crystalline
and amorphous polymers is a basis for the fractional
separation of stereoregular (tactic) polymers from the
atactic form.

Table 15. Qualitative comparison of chemical resistance of the most common plastics /34/

Polymer	Water	Bases	Acids	Oxidation reagents	Organic solvents
polytetrafluoroethylene	1	1	1	1	1
polytrifluorochloroethylene	1	1	1	1	1-2
poly(vinyl chloride)	1	1	1	1-2	2
poly(vinylidene chloride)	1	1	1	1-2	2
polyethylene	1	1	1	1	2-3
polypropylene	1	1	1	1	2-3
polyisobutylene	1	1	1	1	3
polystyrene	1	1	1	3	3
vinyl chloride-vinyl acetate copolymer	1	1	2	2	2-3
poly(vinyl acetate)	2	3	3	3	3
poly(vinyl alcohol)	3	3	3	3	1-2
poly(methyl methacrylate)	1	1	1	3	2-3
poly(vinyl butyral)	1	1	2	3	3
polycaprolactam	1 a)	2		3	1-2
polyurethanes	1	3	3	3	2
linear polyesters	1	3	1	3	2
three-dimensional polyesters	1	3	1	2	2
phenolic resins	1	3	1	1	1

a) It absorbs up to 12% and about 3% water at the boiling point of water and at normal relative humidity, respectively.

1 - it resists, 2 - it is attacked, 3 - it does not resist

Table 15 /34/ compare qualitatively the chemical resistance of common plastics. Table 16 /35/ gives some properties and resistance to chemical reagents of the

Table 16. Density, water absorption, and chemical resistance of some plastics /35/

Density (g cm^{-3})	Water absorption[a] (mg mm^{-2})	Polymer	Weak acids	Concentrated acids	Weak bases	Concentrated bases	Alcohols	Esters	Ketones	Ethers	Chlorinated hydrocarbons	Benzene	Petrol	Propellants	Mineral oils	Animal and plant oils
1.3-1.35	1000	cellulose acetate (plasticized)	-	-	-	-	0	-	-	-	-	0	0	0	+	0
1.2	300	cellulose acetobutyrate (plasticized)	+	-	+	-	0	-			-	-	+	-	+	+
1.45-1.8	200-300	melamine resins	0	-	+	-	+	+	+	+	+	+	+	+	+	+
1.4-1.8	250-500	phenolic resins	+	-	+	-	+	+	+	+	+	+	+	+	+	+
0.92-0.96	0	polyethylene	+	+	+	+	+	0	0	0	-	-	-	-	0	0
1.13	300-1000	polyamide	-	-	+	-	+	+	-	+	+	+	+	+	+	+
1.2	0	polycarbonates	+	+	0	-	+	-	-	-	-	-	+		+	+
1.21	100	polyurethane U	+	-	+	+	0	+	0	+	0	+	+	+	+	+
0.93	0	polyisobutylene (Oppanol B 200)	+	+	+	+	+	-	+	-	-	-	-	-	-	-
0.9	0	polypropylene	+	-	+	-	0	-	-	-	-	+	-	-	+	+
1.18	75-100	poly(methyl methacrylate)	+	-	+	-	0	-	-	-	-	+	-	-	+	+
1.05	0	polystyrene	+	+	+	+	+	-	-	-	-	-	-	-	+	0
1.38	20	poly(vinyl chloride)	+	+	+	+	+	-	-	-	-	-	+	-	+	+
1.38	200	celluloid	+	-	+	-	0	-	-	-	-	0	0	0	+	0

[a] Measured at usual conditions (20°C, 101 kPa) after 7 days, + - stable, 0 - partially stable, - - unstable

most frequently used polymers.

3.3.3 THERMAL STABILITY AND THERMAL-OXIDATION RESISTANCE

The effect of heat on polymers can manifest itself in
two ways:
    a) the polymer softens or melts. The kinetic energy
of the chains becomes large enough to overcome the intermolecular forces, and the plastic flows easily;
    b) the structure is changed. Some macromolecular
compounds undergo scission to lower molecular weight
products or even to monomer without changing chemical
composition - i.e. are depolymerized, others release
low-molecular weight fragments with simultaneous change
in chemical composition - i.e. are decomposed. The both
processes are called degradation.
    The course of thermal degradation depends on chemical
composition. The thermal stability of pure polymers
(i.e. those which do not contain any substances facilitating degradation, such as residual initiators or, in
some cases, compounds of multivalent metals) is limited
by the dissociation energies of their chemical bonds.
    The thermal stability of C—C bonds of the main chain
was found experimentally to decrease in the series /36/:

$$...C—C—C... > ...\underset{\displaystyle |}{\overset{\displaystyle |}{C}}—C—C... > ...C—\underset{\displaystyle \underset{\textstyle C}{|}}{\overset{\displaystyle \overset{\textstyle C}{|}}{C}}—C...$$

exemplified by polymethylene (polyethylene), polypropylene,
and polyisobutylene respectively. Weak points also occur
at the C—C bonds which have a double bond in the β
position. Polymers with quaternary carbon atoms or with
carbon atoms without hydrogen atoms, such as (see p.175),
undergo depolymerization. Aliphatic polyhydrocarbons with
no tertiary hydrogen atoms virtually do not depolymerize.
Depolymerization takes place in macromolecular compounds
which contain groups which are chemically unreactive at
the depolymerization temperature or have high bond energy.

$$-CH_2-\underset{\underset{COOCH_3}{|}}{\overset{\overset{CH_3}{|}}{C}}- \qquad -CH_2-\underset{\underset{C_6H_5}{|}}{\overset{\overset{CH_3}{|}}{C}}- \qquad -CH_2-\underset{\underset{CH_3}{|}}{\overset{\overset{CH_3}{|}}{C}}-$$

$$-CF_2-CF_2- \qquad\qquad -CF_2-\underset{\underset{C_6H_5}{|}}{CF}-$$

The rate of formation of monomer in the pyrolysis of polymers such as poly(methyl methacrylate), poly-tetrafluoroethylene, or poly—α—methylstyrene is explained by the chain reaction in which monomer is released from the ends of the chains or at scission points followed by an unzipping mechanism along the whole chain without any transfer /37/. Some polymers, however, give lower yields of monomer due to transfer made possible by the presence of tertiary hydrogen atoms. This is the case with polystyrene and poly(ethyl acrylate). The transfer facilitated by these tertiary hydrogen atoms also determines the molecular weight of the residual chain. The depolymerization mechanism can be represented as follows:

Thermal initiation:

$$\sim CH_2-CHX-CH_2-CHX\sim \longrightarrow \sim CH_2-\dot{C}HX + \dot{\,}CH_2-CHX\sim$$

Propagation:

$$\sim CH_2-CHX-CH_2-\dot{C}HX \longrightarrow \sim CH_2-\dot{C}HX + CH_2{=}CHX$$

Transfer:

$$\sim CHX-CH_2-CHX-CH_2\sim + \dot{\,}CHX-CH_2\sim \longrightarrow$$
$$\longrightarrow \sim CHX-CH_2-\dot{C}X-CH_2\sim + CH_2X-CH_2\sim$$

The termination can go by disproportionation of two radicals to give saturated and unsaturated ends of shorter chains:

$$\begin{array}{l}\sim CH_2-\dot{C}HX \\ \sim CH_2-\dot{C}HX\end{array} \longrightarrow \sim CH_2-CH_2X + \sim CH{=}CHX$$

Polymers containing reactive substituents where they can be easily split off release low-molecular products (e.g. water, hydrogen chloride, or alcohol), and are decomposed prior to depolymerization. Thus poly(vinyl chloride), poly(vinyl acetate), poly-chloroprene, are decomposed with simultaneous formation of hydrogen chloride or acetic acid:

$$\sim CH_2-\underset{\underset{Cl}{|}}{CH}-CH_2-\underset{\underset{Cl}{|}}{CH}\sim \xrightarrow{-HCl} \sim CH=CH-CH=CH\sim$$

$$\sim CH_2-\underset{\underset{OCOCH_3}{|}}{CH}-CH_2-\underset{\underset{OCOCH_3}{|}}{CH}\sim \xrightarrow{-CH_3COOH} \sim CH=CH-CH=CH\sim$$

The chlorinated polyether Penton

$$\left[-OCH_2-\underset{\underset{CH_2Cl}{|}}{\overset{\overset{CH_2Cl}{|}}{C}}-CH_2-\right]_n$$

however, has high thermal stability, because chlorine is bound to the carbon atom adjacent to quaternary carbon, so that there is no β-hydrogen atom, and elimination of HCl is impossible. A similar situation is encountered in polytetrafluoroethylene: here also hydrogen atoms are absent (elimination of HF is impossible), and elimination of F_2 has a poor energy balance. In addition, even the C—C bond in the fluorinated polymers is substantially stronger than that in similar hydrocarbon polymers. That is why polytetrafluoroethylene is much more resistant to heat than polyethylene. Although the degradation of polytetrafluoroethylene begins above 360°C, its rapid depolymerization only proceeds at 600-700°C (especially at reduced pressure); decomposition at 100 kPa gave 15.9% C_2F_4, 25.7% C_3F_6, and 58.4% C_4F_8. The composition of the depolymerization products, however, depends strongly on pressure: e.g. 97% C_2F_4 was formed at 0.7 kPa /34/.

Thermal oxidative degradation including that of
hydrocarbons, was dealt with in numerous publications
/38-42/. Engler and Bach (1897) formulated the theory
of oxidation to peroxides. Important results were obtained
for gas phase oxidations which were shown to be chain
reactions, and confirmed recently by electron para-
magnetic resonance.

The course of oxidation reactions in the liquid phase
is complex; still more complicated are solid-phase oxidations, especially those of polymers containing additional groups or heteroatoms. Thus the oxidation of
poly(vinyl chloride) is complicated by the accompanying
dehydrochlorination and formation of double bonds.
Epoxides and polyacrylates, are only oxidized at high
temperatures, and no inhibitors are known so far as most
conceivable organic antioxidants are decomposed at these
temperatures.

Thermal oxidation is important with polyolefins which
are susceptible to oxidation. From some reports it follows
that their oxidation follows the mechanism for low-molecular
liquid hydrocarbons.

The oxidation reactions of polyolefins have rather
high activation energies:

polypropylene (PP) $\qquad$ 91 kJ mol^{-1}
ethylene-propylene copolymer (EPM) $\quad$ 129 kJ mol^{-1}
high density polyethylene (HDPE) $\quad$ 134 kJ mol^{-1}
low density polyethylene (LDPE) $\quad$ 137 kJ mol^{-1}

hence their rates depend very much on temperature.

3.3.4   RADIATION RESISTANCE

Solar radiation initiates the decomposition of polymers
(by oxidation, scission, cross-linking, etc.) within
wavelengths from 300 to 400 nm. Though only 5% of total
solar radiation impinging on the surface of the Earth falls
in this range, the degradation of polymers by light is

one of the most significant factors in natural ageing.
If an organic macromolecular compound absorbs a radiation quantum of e.g. 300 nm wavelength, its energy
increases by almost 377 kJ mol^{-1}, sufficient to split
C—C bonds, whose energy in macromolecular organic
compounds is about 335 kJ mol^{-1} /34/.

The absorbed light energy causes strong activation
of the molecule, or its electronic excitation, and this
energy can be either transferred to another molecule
by impact or emitted in the form of longer-wave radiation.
The energy can be released as fluorescence, phosphor-
escence, or heat, and this fact forms the basis of qualitative differentiation of plastics by means of ultraviolet
absorption. In many cases, the activated molecule can
initiate a photochemical reaction. If ultraviolet radiation only is taken into account, the only absorbing
group in polymers is the carbonyl group. Thus non-degraded
pure poly(vinyl chloride) absorbs practically no ultraviolet radiation and should be stable. In reality, however,
a certain amount of hydrogen chloride is always released
during processing, and oxidation at the double bonds so
formed produces carbonyl groups which then accelerate
the degradation processes induced by ultraviolet radiation. The activation of molecules and the degradation
reactions depend on the radiation intensity, wavelength,
and composition of the polymer, but not on the total
amount of the radiation energy absorbed.

The basic difference between thermal and radiation
effects on polymers lies in the relative rates of these
reactions which cause discolouration. So during the
thermal degradation of poly(vinyl chloride), very rapid
brown colouration is observed compared with the colouration due to light which is masked by oxidation of the
conjugated double bonds responsible for colour intensity.
Also it is known that poly(vinyl chloride) discoloured
by heat can be decolourized by photooxidation. The dif-

ferent effects of light and heat are also obvious from
the changes of mechanical properties: these properties
do not change during thermal degradation, but if
poly(vinyl chloride) is exposed to ultraviolet radiation, then the physico-mechanical properties such as
tensile strength and flexural strength deteriorate
very rapidly, and brittleness increases sharply.

It will be shown later how the light effect can
be moderated or prevented by the application of UV
absorbers or suitable pigments.

3.3.5 WEATHER RESISTANCE

If a plastic is exposed to weather, its appearance as
well as its mechanical properties are impaired through
photochemical degradation, photochemically activated
oxidation or/and hydrolysis of macromolecules /34, 42/.
The results depend very much on the climate, on inclination to the sun, and also for experimental specimens
on the colour and material of the support. It is a
problem to decide on the unit of time for comparison
of exposure results. The most common unit used to be
one day, but some workers prefer one hour of sunshine
or the whole year as more objective, because differences
were found between testing in Arizona, Florida, and
New Jersey. Now it is usual to determine solar radiation
impinging on the sample and express it in $J \ cm^2$. Several
laboratory methods were developed for accelerated testing
of the effect of climatic conditions, e.g. the Weathero-
meter uses a carbon arc lamp providing periods of thermal
and light exposure and intermittent water spray.

Evaluation of the effects of weather conditions on
selected plastics are given in ref. /36/. According to
their resistance, these materials are divided into four
groups: The first group comprises the materials which
show no or only negligible signs of damage after 2.5

to 3 years under various climatic conditions (mild zone, subarctic climate, semi-desert, and tropical climate). The second group of materials endured 6 to 7 months exposure without substantial change. The third group of materials were markedly affected after a short exposure. The materials whose resistance varies according to composition or additives fall into the fourth group.

The first group - long-term durability: poly-tetrafluoroethylene, glass fibre reinforced polyester laminates, poly(methyl methacrylate), and allylic casting resins (in the last two cases the optical and electrical properties only were evaluated).

The second group - short-term durability: cellulose propionate, laminated phenolic materials (without glass fibers), melamine-formaldehyde and urea-formaldehyde resins, and allylic casting resins (in the last case mechanical, optical, and electrical properties were evaluated).

The third group - poor durability: polyamides, polystyrene, ethylcellulose, cellulose nitrate, poly-(vinyl alcohol), silicone rubber, and phenolic casting resins.

The fourth group - additive-dependent: cellulose acetate and acetobutyrate, phenolic moulding materials, polyethylene, poly(vinyl chloride), and poly(vinyl acetate).

3.3.6   RESISTANCE TO HIGH-ENERGY RADIATION

High energy ionizing radiation such as the X-rays and gamma radiation, induces structural changes in plastics. Both cross-linking and chain scission take place. The two processes occur simultaneously, and the polymer structure decides which will predominate.

Predominantly cross-linking polymers are polyacrylic

esters, high molecular weight polystyrene, polyesters, polyamides, polyethylene, natural rubber, styrene-butadiene and butadiene-acrylonitrile copolymers, polychloroprene, polydimethylsiloxanes, polypropylene, chlorinated polyethylene, poly(vinyl butyral), and under certain conditions, poly(vinyl chloride).

Predominantly degrading polymers are poly(methyl methacrylate), poly(vinyl chloride), polytetrafluoroethylene, polytrifluorochloroethylene, polyisobutylene, cellulose, poly(vinyl acetate), poly—α—methylstyrene, and low-molecular weight polystyrene /36/.

The cross-linking proceeds more easily with higher molecular weight polymers, the degree being dependent on the radiation dose. A dependence between chemical structure and resistance to ionizing radiation is shown in the following series:

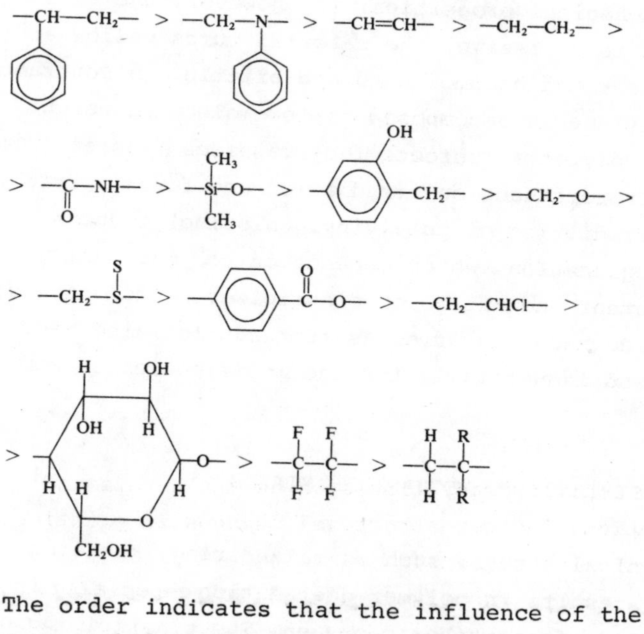

The order indicates that the influence of the benzene ring on stability depends on whether it forms part of the polymeric chain, or a substituent, as in the cases of very stable polystyrene and anilineformaldehyde. The results depend on the properties tested, but it can be stated generally that polymers containing phenyl substituents are resistant, whereas polytetrafluoroethylene or polyisobutylene are not.

Recent research in highly heat- and ionization-resistant polymers has identified polysulphones, polybenzimidazoles, or polyimides /38/ containing many aromatic rings in the backbone of the macromolecules. The degradation of polymers initiated by ionizing radiation is accompanied by the formation of double bonds, evolution of gaseous products (hydrogen, low-molecular weight hydrocarbons, hydrogen chloride, tetrafluoroethylene in the case of fluoroplastics), and, if oxygen is present, also by the formation of peroxidic, carbonyl, or carboxyl groups, and is accompanied by yellowing. Polyethylene forms mostly network structures which enhance mechanical properties; if, however, the radiation dose is excessive, the material turns yellow and finally black and becomes hard and brittle. In contrast, polyisobutylene is decomposed to low-molecular weight products. Polytetrafluoroethylene releases fluorine /34/ or tetrafluoromethane on irradiation and forms a friable powder. Irradiation of poly(vinyl chloride) induces complex degradation, which depends on polymer purity and on the content and nature of plasticizer, which consists in either degradation or cross-linking. Ionizing radiation is used industrially for the modification of some plastics.

3.3.7 RESISTANCE TO MECHANICAL STRESS

Any mechanical stress, such as calandering, kneading, or grinding, results in polymer degradation especially in the presence of atmospheric oxygen. The plasticizing of rubber on a two-roll mill utilizes these effects intentionally to decrease the average molecular weight.

This mechanical degradation, which is often combined with weathering or chemical attack, can proceed
by either radical or ionic mechanisms, the latter being
the case in macromolecules containing weak ionic bonds
or in media of high dielectric constant which lower the
strength of ionic bonds. Rapid degradation during kneading, mixing, etc. depends on shear rate, temperature,
the presence of oxygen, on the amount, character, and
grain size of additives, on the composition and structure of the polymer.

Mechanical stress also decreases the chemical and
ageing resistance of some plastics. This is true for
polyethylene, polystyrene, poly(methyl methacrylate),
unplasticized poly(vinyl chloride), and polyamides.
This phenomenon is generally called environmental stress
cracking and was previously observed with metals /34/.
It is characterized by the formation of fine fissures
which are responsible for a decrease in mechanical properties; in the presence of liquids, the effect is accen-
tuated, especially in stressed bent or shaped sections.
This behaviour is explained by the mechanical energy
causing deformation of bond angles or even chain
scission, so that the activation energy for the chemical
reaction of the polymer exposed to an aggressive medium
is lowered. This is the case with the accelerated action
of oxygen or ozone on rubber under stress.

The effect of solvents probably consists in breaking
the side bonds which results in decreased coherence of
chains: in the stress cracking of polyethylene, the most
active compounds are those which lower the viscosity of
polyethylene solutions, by decreasing the strength of
intermolecular bonds.

The stress cracking of some polymers is also affected
by the method and quality of processing, which is
especially true for those soluble in the corresponding

monomers or those depolymerized by heat. Processing
temperature is important; thus polystyrene extrusion at
too high a temperature can lead to immediate stress
cracking; this behaviour is due to styrene vapour formed
by depolymerization acting simultaneously with the stress
due to extrusion.

All processing parameters which increase internal
stress (e.g. incorrect cooling of the article, overfilling of the mould, or action of external forces)
facilitate stress cracking.

3.3.8 BIOLOGICAL RESISTANCE

The biological deterioration of plastics by microorganisms (especially mildew), is very wide spread
(cf. Section 1.2.3). This attack is difficult to control
and is encountered not only in tropical climates, but
also in temperate regions, especially with plasticized
poly(vinyl chloride), poly(vinyl alcohol) films plasticized with glycerol, and phenolformaldehyde moulding
materials, particularly on machined surfaces where filler
is exposed /34/.

Whether or not a plastic will be attacked by microorganisms depends first of all on whether its chemical
composition can provide nutrients, and also on the environment and conditions of use. Deterioration is induced
by the metabolites or enzymes produced by microorganisms;
the surface may be etched, and thin films may be perforated. Some microorganisms only attack plasticizers,
residual emulsifying agents, or fillers present in polymers. Of the emulsifying agents the least resistant to
microorganisms are derivatives of higher fatty acids,
their resistance decreasing with an increasing number
of carbon atoms. As a counter-effect, the alcohols formed
by hydrolysis of esters tend to inhibit the growth of
microorganisms. Microbial resistance is also increased

by ether linkages in the carbon chain. Maleate,
phthalate, and phosphate esters and salts remain
intact. Stabilizers (especially those based on stearic
and lauric acids) can also be substrates for microorganisms.

3.3.9 LOSS OF PLASTICIZERS

Polymers plasticized with low-molecular weight plasticizers become hard and brittle after a time, with
accompanying weight loss and changes in the properties
of articles.

The loss of a plasticizer can be due to:

a) Volatility. Here the partial pressure of the saturated vapours of the plasticizer, diffusion through the
polymer, and rate of air change at the surface are
important; plasticizer evaporates from the surface of
the polymer, and is replenished by migration from within
the polymer. The volatility data of some well-known
plasticizers for poly(vinyl chloride) are given in Table
17.

b) Extraction. This may occur through contact with
organic solvents, water, oils, etc.

c) Migration. This arises through contact with materials in which the plasticizer is soluble.

d) Reaction. This can occur with reactive additives.

It is difficult to prevent deterioration of this type,
but in some cases it is possible to replace a low-molecular weight plasticizer by one of higher molecular
weight (e.g. polyesters, poly(ethylene glycol) derivatives,
epoxides). The effective alternative is copolymerization
with monomers containing bulky side groups, providing
internal plasticization.

Table 17. Volatility of some usual plasticizers

Plasticizer	Loss of the plasticizer (%)	
	after 10 days at $60^{o}C$	after 3 days at $85^{o}C$
dicapryl phthalate	1.4	1.7
dioctyl phthalate	0.9	1.3
dimethyl cyclohexyl-phthalate	1.2	-
dibutyl sebacate	1.27	-
dioctyl sebacate	5.2	12.8
dibenzyl sebacate	0.46	0.8

3.4. EFFECTS OF ADDITIVES ON PROPERTIES OF PLASTICS

The requirements which must be fulfilled by plastics are so varied that pure uniform polymers are practically never used, they are modified by auxiliary materials /43-47/.

3.4.1 PLASTICIZERS

Plasticizers are organic substances of low volatility which impart flexibility, formability, and lower melt viscosity, glass transition temperature and modulus of elasticity. They are used with a limited number of polymers, especially those which exist in the glassy state at normal temperature, such as homopolymers and copolymers of vinyl chloride, cellulose derivatives, poly(vinyl acetals), and poly(vinyl alcohol). The liquid tarry substances used sometimes in rubber for improvement of workability and/or reduction of price are usually called extenders.

From the kinetic point of view, the efficiency of
plasticizers is evaluated according to reduction of
the glass transition temperature or brittle temperature.
The theoretical and mathematical treatment of plasticization is complicated by the existence of crystalline
regions in polymers, the size of spherolites, and the
heterogeneity of the plasticized system. In a plasticized polymer the brittle temperature decreases with
increasing concentration of the plasticizer, the effect
being greater at low concentrations of the plasticizer
as illustrated in Fig. 32 for poly(vinyl chloride)
plasticized with tricresyl phosphate /49/.

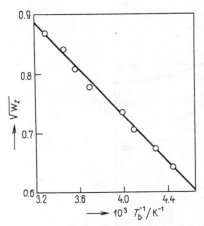

Fig. 32. Effect of tricresyl
phosphate on the brittle
temperature of poly(vinyl
chloride) /48/
T_b - brittle temperature,
w_z - weight fraction of the
plasticizer in the polymeric
mixture

The glass transition temperature of the polymer /50/
(Fig. 33) can be defined as the temperature below which
there is no more reduction of the free volume (Fig. 34)
/45/. A liquid plasticizer, having a greater free volume,
increases the free volume of the whole system and makes
possible Brownian motion at lower temperatures than the
glass transition temperature of the polymer alone. Indeed,
the glass transition temperature of the plasticized polymer is lower than would correspond to additivity of the
glass transition or melting temperatures of the individual

components. The glass transition temperature of a polymer does not change with crystalline content, and the
plasticizer has effect only on the amorphous portion.

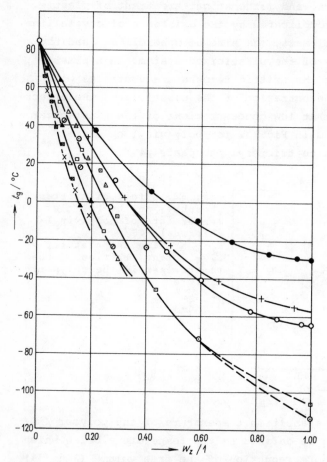

Fig. 33. Effect of various plasticizers on the glass
transition temperature of polystyrene /48/

t_g - the glass transition temperature, w_z - the weight
fraction of the plasticizer in the polymeric mixture,
● - 2-naphtyl salicylate, + - phenyl salicylate,
O - tricresyl phosphate, ☉ - methyl salicylate, ⊡ - nitro-
benzene, ▲ - chloroform, x - methyl acetate, ☒ - ethyl
acetate, ▲ - carbon disulphide, ☐ - benzene, ⊘ - toluene,
△ - amyl butyrate

If the polymer is completely soluble in the plasticizer, or if it is soluble at least in a useful concentration range, the plasticizer is regarded as
compatible. If, however, the polymer solubility in the
plasticizer is limited and results in two phases, one
polymer-rich and one plasticizer-rich, then the plasticizer is regarded as incompatible. Although the polymer is plasticized in both cases, the deformation
of the incompatible system is not as homogeneous as
for a compatible system, and takes place predominantly
in the plasticizer-rich phase.

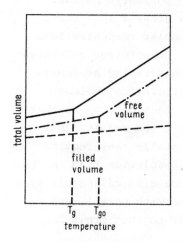

Fig. 34. Effect of a plasticizer on the variation of the
total volume with the temperature

Full line - plasticized polymer.
Broken line - pure polymer

The plasticizers for poly(vinyl chloride) fall into
two groups according to the nature of the non-polar
parts of their molecules.

The first group is the polar-aromatic type /51/ with
molecules having polar groups attached to a non-polar
but polarizable component exemplified by phthalate
esters or tricresyl phosphate. These contain polarizable
benzene rings with relatively loose π electrons and
their molecules behave as dipoles modifying the attractive forces between the polymer chains. This class exhibit
easy incorporation, good gelling properties and good

solubility at low temperature. They are also called
solvent type plasticizers, cannot easily be extracted with petrol and have little effect on brittle temperature.

The second group of plasticizers have molecules
with polar and non-polar parts, the non-polar part,
however, being non-polarizable; they are the polar-
aliphatic type. This class includes esters of aliphatic
alcohols with aliphatic carboxylic acids or phosphoric
acid. The dipoles of the ester groups penetrate between
the polymeric chains, but the aliphatic groups attenuate
the attractive forces between the polymeric chains.
Thus their plasticizing effect is greater than that
of the previous type at the same molar concentration,
and the aliphatic sections of the plasticizer molecules
retain their mobility and confer elasticity at lower
temperatures. However, their incorporation is less
facile than that of the group of plasticizers, so that
a high content of the plasticizer often results in its
migration out of the plastic. Generally they require
higher temperatures for polymer dissolution than in the
case of the previous type. They are also called oil type
plasticizers, are easily extracted with petrol, and
give the plastic better resistance to both frost and
high temperatures.

No plasticizer is known so far which would combine
the good properties of the above two groups and lack
their drawbacks. Therefore, in practice a compromise
must be chosen, often by using mixtures of plasticizers.

Polymeric plasticizers have become important because
of their permanence. The main difficulties in their
use stem from their imperfect miscibility and lower
efficiency when compared with a weight basis with monomeric plasticizers.

With respect to properties and practical applications,
plasticizers can be classified into several groups:

phthalates, phosphates, polymeric plasticizers, hydrocarbon plasticizers, halogenated hydrocarbons, etc.

Esters of phthalic acid: the most widely used are
di(2-ethylhexyl) phthalate (DOP), diisooctyl phthalate
(DIOP), and dibutyl phthalate (DBP). The first two
esters are the most frequently used phthalate plasticizers for poly(vinyl chloride) and vinyl chloride-
vinyl acetate copolymers, having good compatibility,
low volatility, colour fastness and water resistance,
and conferring good flexibility at low temperatures.
Dibutyl phthalate is suitable primarily for coating
compositions from cellulose derivatives and for poly-
(vinyl acetate) adhesives, it is unsuitable alone for
plasticizing poly(vinyl chloride) due to its volatility.

Esters of phosphoric acid: tricresyl phosphate
(TCP), and tri(2-ethyl hexyl) phosphate (TOP) exhibit
low volatility and are only slightly extracted with
oils. They are mostly used in combinations with other
plasticizers. Tricresyl phosphate is used for plasticizing poly(vinyl chloride) especially for low flammability or good dielectric properties; it is unsuitable
for low temperature performance and is classified as
hazardous to health. Tri(2-ethyl hexyl) phosphate is
better at low temperatures but worse for extraction
with petrol and oils.

Esters of aliphatic di- and polycarboxylic acids:
di(2-ethyl hexyl) sebacate (DOS), dibutyl sebacate
(DBS), diisobutyl adipate (DIBA), and monoisopropyl
citrate. The best-known representative of this class
is di(2-ethyl hexyl) sebacate. It possesses high plasticizing efficiency, low volatility, and low extraction
with water and soap solutions. It is suitable for
poly(vinyl chloride) articles designed for very low
temperatures. The remaining three examples are recommended for use in articles which may come into contact
with foodstuffs. Economically advantageous are other

adipate esters which exhibit relatively low dissolving
power and are therefore useful for preparing stable
plastisols, but they are more volatile than the sebacic
acid esters.

Polymeric plasticizers usually have molecular weights
from 600 to about 8000, and include poly(ethylene glycol),
polyesters, and macromolecular epoxides. Poly(ethylene
glycol) is used with proteins, casein, gelatine, and
poly(vinyl alcohol) /17/.

The polyester plasticizers are produced by condensation
of dicarboxylic acids with diols /52/. These products
can be represented schematically as follows:

$$K-G-(DK-G-)_n K \qquad\qquad A-DK-(G-DK-)_n A \; ,$$

where K and DK are mono- and dicarboxylic acids,
respectively, A and G monoalcohol and diol, and $n = 2$
to 40 for the adipates and 3 to 35 for the sebacates.

The acids used include succinic, glutaric, pimelic,
suberic, azelaic, and thiodipropionic, but the common
polyester plasticizers are based on sebacic, adipic,
and phthalic acids.

These plasticizers are rarely used alone, but in
combination with monomeric plasticizers, impart low
volatility and resistance to migration and to extraction with petrol, oils, water, and detergents. They
are employed especially with poly(vinyl chloride) and
nitrocellulose lacquers.

3.4.2 LUBRICANTS AND RELEASE AGENTS

3.4.2.1 L u b r i c a n t s

Lubricants facilitate the processing of plastics. For
example un-plasticized poly(vinyl chloride) is rather
difficult to extrude or injection mould without additives which have either lubricating or plasticizing

effects. Lubricants are used in processing other polymers such as polystyrene and polytetrafluoroethylene. If well selected, they improve not only workability but also properties like surface appearance, resistance to heat and light, and durability. Common lubricants are stearic acid and its salts with the cations Ca^{2+}, Li^+, Ba^{2+}, Al^{3+}, Sr^{2+}, and Mg^{2+}, natural waxes, microcrystalline paraffins with high softening point (in order to avoid dull and greasy surfaces, mineral and vegetable oils, and animal fats).

Formerly it was usual to improve the workability of un-plasticized poly(vinyl chloride) by addition of "small amounts" (up to 10%) of primary plasticizers, which, however, impaired some properties, particularly impact strength and notched impact strength, numerically by as much as 50%. This effect was especially significant with poly(vinyl chloride) which un-plasticized has notched impact strength at normal temperature only of 3 to 4 kJ m^{-2}. Lubricants have none of these disadvantages, but improve some properties.

The main effects of lubricants, however, are to prevent sticking of the hot polymer mass to metal parts in extruders and moulds and to decrease the internal friction in the material and so avoid overheating, which together enhance material flow and increase the effective capacity of processing equipment.

A lubricant requires /43, 53/ at least partial solubility in the polymer, or sufficient absorption to prevent exudation, so it must be a relatively long chain hydrocarbon terminated by a polar group, and have a melting temperature which will ensure coverage of the surface of the particles formed of macromolecular agglomerates with a strong film. It is presumed that too low a melting temperature gives films of insufficient strength, and too high a melting temperature precludes the required uniform film layer.

Lubricants can be divided into two groups /54/:

a) lubricants with external action are negligibly soluble in the polymer and form an interlayer between the polymer and the working surface of processing equipment;

b) lubricants with internal action are compatible with the polymer, decrease the viscosity of the melt and reduce heat evolution due to friction in the polymer being worked.

In practice lubricants function to some degree in both ways. The solubility limit of lubricant in polymer can be used as a criterion for classification. Above this limit the lubricant begins to act externally. With poly(vinyl chloride), lubricants with external action increase the time required for gelation, whereas those with internal action reduce it.

Lubricants may be classified according to their chemical composition /55, 56/:

1. Hydrocarbons show lubricating effect increasing with chain length, so solids are more efficient than oils (with 5 to 15 carbon atoms in the molecule). Examples: paraffin wax, and polyethylene with molecular weight about 2000.

2. Fatty acids act as external lubricants. Their compatibility with the polymer increases with increasing molecular weight. Examples: technical stearic acid (stearin), and 12-hydroxystearic acid.

3. Higher alcohols differ from fatty acids and are more compatible with the polymer. At the concentrations used technically (about 2%) they do not exude to its surface and function as internal lubricants. Example: cetylstearyl alcohol.

4. Organic soaps facilitate the release of articles from moulds, because they are external lubricants. In some cases they also improve the heat and light resistance of the polymer. Examples: stearates of Li, Ca, Ba, Pb.

5. Waxes, the most efficient being esters of fatty acids and higher monofunctional alcohols, have long chains responsible for good lubrication, and the ester group provides good compatibility with the polymer. Examples: partially saponified montan wax, the ester with two C_{28}–C_{32} chains, stearyl stearate.

6. Simple esters of fatty acids are food "separators" of articles from moulds but also act as secondary plasticizers, thus affecting the mechanical properties of the polymer. Examples: butyl and octyl stearates.

7. Polyfunctional alcohols partially esterified with fatty acids act as external lubricants. Example: glycerol monoricinoleate.

8. Amides of fatty acids have pronounced external lubricating action. Example: ethylene bis(stearamide).

The amounts used vary from 0.5% to 3%, and sometimes lubricants are combined to obtain required properties.

3.4.2.2. R e l e a s e A g e n t s

Release agents facilitate the removal of articles from moulds and can thus increase productivity /57/.

Besides the external lubricants, waxes, soaps, and silicone or polytetrafluoroethylene coatings or sprays function as release agents. Silicone coatings (e.g. polydimethylsiloxane) have high thermal stability (about 300°C) and perform well, but some difficulties can be encountered during subsequent printing or metallizing. Films of cellophane or poly(vinyl alcohol) may also be used.

3.4.3 THERMAL STABILIZERS

Thermal stabilizers are used with poly(vinyl chloride) and a wide range of other halogenated polymers.

Thermal degradation does not cause substantial changes

in mechanical properties, but discoloration can be
severe and thermal stabilizers for poly(vinyl chloride)
are assess primarily according to their ability to
prevent darkening.

The choice of suitable stabilizers must take account
of other additives present in the polymer, especially
plasticizers and fillers which often accelerate degradation and discolouration. Efficient stabilizers:

prevent dehydrochlorination or bind hydrogen chloride
to prevent corrosion of equipment,

prevent the formation of chromophoric groups (carbonyl,
carboxyl, etc.) responsible for discolouration,

they should be compatible with the polymer, resist
water extraction, cause no turbidity in the product nor
spoil dielectric properties, and be non-toxic.

During the last 25 years many compounds have been
patented which fulfil these requirements /43, 58-61/
and it is now practicable to process polymers at temperatures above their decomposition point.

For simplicity the heat stabilizers can be divided
into two main classes: metallic and organic.

The first comprises metal (Pb, Cd, Sr, Ba, Zn, Mg,
Li, Ca, Na) salts of organic or inorganic acids, or
mixtures of such salts, and particularly a large group
of organotin stabilizers.

The second class includes epoxide compounds and
chelatation agents (organic phosphites) applied mostly
in combination with the other stabilizers, α-phenylindole and stabilizers containing ures and β-aminocrotonic acid.

The metallic stabilizers were the first stabilizers
used successfully for poly(vinyl chloride) on a large
scale, although calcium stearate alone has little effect,
and found broad application especially in combination
with secondary stabilizers (epoxides, chelation agents).
Although a great number of results have been published,
the mechanism of their action is not fully established yet.

The metallic stabilizers can be further classified according to their effect: lead stabilizers, barium, calcium and strontium stabilizers, cadmium and zinc stabilizers, and organotin stabilizers.

The mechanism of lead stabilizers has been known longest. First, these substances bind the hydrogen chloride split off. Except for basic lead phosphate and phosphites, none of the lead stabilizers is an antioxidant. They are used especially for extrusion, because in calandering the degradation mechanism is radical oxidation. The salts of Ba^{2+}, Ca^{2+}, and Sr^{2+} are slow stabilizers; a pale yellow colouration appears in the polymer initially and turns black slowly. Therefore, these stabilizers are presumed primarily to bind hydrogen chloride. Mostly they are not used alone (in contrast to the lead stabilizers), but in combination with cadmium and zinc stabilizers. The latter initially produce a high colour, quickly followed by an abrupt blackening of the polymer. Besides the catalytic effect of the reaction products ($HZnCl_3$) on dehydrochlorination, the rapid colouration is also explained by the cadmium and zinc compounds allowing oxidation; at the beginning, they catalyze scission of the double bonds which constitute chromophores when conjugated and thereby retard colouration; later on, the oxygen incorporated into the polymer forms chromophores which deepen the colouration /62-64/.

The stabilizing effect of metal salts is determined not only by the cations but also by the anions /65/, because the anions affect compatibility of the stabilizer with the polymer and thereby also the degree of dispersion of the stabilizer.

Increased stabilization may be achieved by combinations of metal stabilizers; synergism /58/ was found in many cases. This means that the combined effect of the individual components of the mixture is higher than

the effect of the same amount of the most efficient
component alone (Fig. 35).

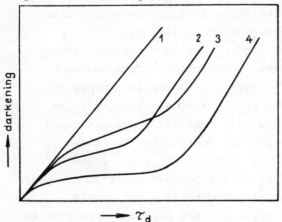

Fig. 35. Schematic representation of the synergism
of stabilizers of poly(vinyl chloride)

1 - the non-stabilized polymer, 2 - the polymer
containing a stabilizer A, 3 - the polymer containing
a stabilizer B, 4 - the polymer containing the A + B
mixture of the same total concentration, τ_d - time of
degradation

Examples are combinations of fatty acid salts of lead,
cadmium, barium, zinc, or calcium with epoxide stabilizers, which at a certain ratio are more efficient
than the individual components. The reason for this
effect has not been elucidated yet, although a number
of theories have been developed. The synergistic mixtures containing zinc stabilizers, particularly those
combined with the systems Ba/Cd or Ca, are noteworthy.
To be effective, zinc stabilizers must be well dispersed
because stabilization is limited to a low concentration
(0.1 to 0.4 weight parts per 100 weight parts of the
polymer), and beyond a critical concentration, which
varies with the polymer, stability is reduced. Zinc
2-ethylhexanoate, which is soluble in poly(vinyl chloride)
at low temperatures, is important. Decrease in the
stability of poly(vinyl chloride) stabilized with zinc

compound may be avoided by combination with epoxide
stabilizers. It is important to bear in mind that
there exists also an opposite effect to synergism, in
that some combinations can exhibit a worse effect
than the individual components alone, and this is called
antagonism.

Of the organotin compounds /43, 59/ the only useful
stabilizers are some tin(IV) compounds, most frequently
dialkyltin(IV) or trialkyltin(IV) derivatives, whereas
tin(II) compounds as well as tetrasubstituted tin(IV)
compounds (e.g. tetrabutyltin) show only mild stabilizing effects in exceptional cases. In practice,
dibutyltin(IV) and di-2-ethyl hexyltin(IV) compounds
are efficient. Substitution of the remaining two
valencies of four-valent tin is possible; the most
efficient examples are the dialkyltin(IV) salts of
organic carboxylic or thiocarboxylic acids. The large
number of organotin(IV) stabilizers produced can be
divided into two main groups: stabilizers with Sn—O
bonds in their molecule and stabilizers with Sn—S bonds.

Stabilizers containing the Sn—O bond include
alkyltin maleates, alkoxides, and polymeric alkyltin
oxides, especially alkyltin salts of carboxylic acids,
the original organotin stabilizers. From the large
number produced, the oldest and most frequently used
is dibutyltin di-laurate (DBTDL) in spite of its
inefficiency at higher temperatures, limited compatibility with poly(vinyl chloride), and difficulties in
subsequent welding. Higher stability was achieved with
organotin maleates, especially the dialkyl derivatives,
which react as dienophiles with conjugated double
bonds in the degraded polymer, thus reducing colouration.

Attempts to remove some technological drawbacks
(e.g. volatility) of the earlier organotin stabilizers
led to polymeric stabilizers.

Analogous to the organotin salts of carboxylic acids, are the stabilizers containing Sn—S bonds. The simplest examples are the organotin sulphides which structurally can be divided into four groups: organotin thiolates, organotin mercaptoalcohols, organotin thioglycolates, and polymeric organotin stabilizers containing sulphur.

A supplementary class of organotin stabilizers contain Sn—N bonds, is exemplified by sulphonamides and sulphimides.

Although the organotin stabilizers can bind hydrogen chloride, it is presumed that they prevent degradation by blocking weak positions in the polymer molecule or by addition to double bonds whereby they prevent oxidation.

The group of organic stabilizers comprises epoxides, chelating agents (organic phosphites), α-phenylindole, compounds based on urea, and derivatives of β-aminocrotonic acid.

Stabilizers of the epoxide type are able to bind hydrogen chloride

$$-CH\!-\!\!-\!CH- + HCl \longrightarrow -\underset{\underset{OH}{|}}{C}H\!-\!\underset{\underset{Cl}{|}}{C}H-$$

Used alone they do not prevent colouration at the beginning of degradation, but this is overcome by combinations with the more efficient primary stabilizers, e.g. organic soaps based on Zn^{2+} and Cd^{2+}.

Organic phosphites are used also to improve the effect of metallic stabilizers, and by chelatation, they prevent the turbidity which would arise with the metal chlorides arising from metallic stabilizers. Some phosphites increase the efficiency of metallic stabilizers due to their antioxidant properties. Alkyl aryl phosphites, e.g. isooctyl diphenyl phosphite, are the

most efficient, but triphenyl phosphite is also
frequently used. These chelating agents are usually
added in plasticizer solutions for good dispersion
in the polymer.

α-Phenylindole and the stabilizers based on urea
are mainly used for European emulsion-type poly(vinyl
chloride) pre-stabilized with sodium carbonate.
α-Phenylindole is used predominantly for the unplasticized polymer /58/.

Of the urea types most common are urea itself or
aryl derivatives of urea or thiourea. Stabilizers
based on β-aminocrotonic acid esters were developed
primarily for emulsion-type poly(vinyl chlorides) and
are widely approved for food contact, at levels up to
3% (1.5 % in Czechoslovakia).

Stabilizers generally are mixed with plasticizers
and pigments for incorporation at levels of 0.5 to 5%
in the polymer.

3.4.4 ANTIOXIDANTS

The main intermediate in thermal oxidation of polymers
is hydrogen peroxide. Therefore, compounds which can
react with peroxidic radicals or hydrogen peroxide
itself are used for protection of the polymers easily
degraded by thermal oxidation (polyolefins, polystyrene,
polyamides, etc.). Such antioxidants include substituted
phenols, aromatic amines, organic sulphides and phosphites
/39-41, 66/.

For correct application of antioxidants the following
facts must be appreciated /58/:

- most antioxidants can only act as oxidation
inhibitors above a critical concentration,

- an antioxidant not only terminates active radicals
but can also initiate oxidation under certain conditions,

- combinations of antioxidants can exhibit either

synergism or antagonism.

Fig. 36 shows the dependence of the inhibition period (I_p) for polypropylene on the antioxidant concentration at various temperatures. The length of the inhibition period, during which no oxidation occurs, is a measure of the inhibiton due to the antioxidant. The critical concentration increases with temperature and reaches 2.10^{-3} mol kg^{-1} of polymer at $210°C$. The effect of various antioxidants are shown in Fig. 37; the rate of oxidation is the same after the antioxidants are consumed.

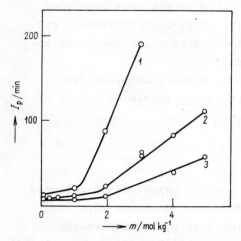

Fig. 36. Dependence of the inhibition period I_p of oxidation of polypropylene on molality m of 2,2´-methylene-bis(4-methyl-6-tert.-butylphenol) at various temperatures and at partial pressure of oxygen equal to 39.5 kPa /67/

1 - $190°C$, 2 - $200°C$, 3 - $210°C$

The inhibition period (I_p) does not represent constancy of physical properties of the stabilized polymer, because in many cases changes occur prior to oxidation. Therefore, full evaluation of an antioxidant must follow its effects on both oxygen consumption and physical properties.

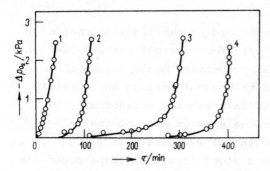

Fig. 37. Effect of various antioxidants at the concentration of 0.003 mol kg^{-1} on oxidation kinetics of polypropylene at 140°C and at the starting partial pressure of oxygen p_{O_2} = 39.5 kPa /67/

The polymer: 1 - without antioxidant, 2 - with benzidine, 3 - with diphenylamine, 4 - with phenyl-2-naphthylamine (Neozon D)

As seen from Fig. 37, the inhibition period depends on both the concentration and the nature of the antioxidant; in most cases the inhibition period hardly changes until the critical concentration of the antioxidant is reached after which it increases sharply (Fig.38). Sometimes the inhibition period reaches a maximum then decreases with more antioxidant (curve 2

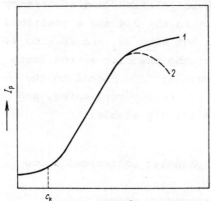

Fig. 38. Schematic representation of a dependence of inhibition period I_p of oxidation of a polymer on concentration of antioxidant AH, c_k - critical concentration

in Fig. 38) because of a sudden initiation effect of the radicals formed by the reaction of antioxidant with peroxidic radicals.

This explanation is supported by the findings that a decrease in molecular weight of the polymer and deterioration of physical properties occur during the inhibition period, and by the observation that certain combinations of antioxidants exhibit antagonism, i.e. have lower effect than the same amounts of the individual components used in the mixture.

Antioxidants can be divided into two groups according to reaction mechanism /67/:

Compounds which terminate (inhibit) the peroxidic or growing radicals (e.g. substituted phenols, amines, and thiols).

Compounds which decompose hydrogen peroxides (e.g. organic sulphides and phosphites).

Phenol derivatives exhibit food compatibility with polyolefins, quite high boiling points, high efficiency at temperatures up to $220^{o}C$, and prevent changes in mechanical properties of the polymer. Their efficiency depends on the positions of substituents, and their chemical nature determines compatibility with the polymer. Thus the non-polar character of an alkyl group increases compatibility with polyolefins, and efficiency is enhanced with substitution in the 2,4 and 6 positions of not only the alkyl groups but also NH_2 and SH groups.

The mechanism of action of the phenolic antioxidants (AH) consists in hydrogen transfer from phenol to the growing radical (RO˙) which is thereby terminated, and the new radical formed is relatively stable:

$$RO˙ + AH \longrightarrow ROH + A˙$$

The most frequently used phenolic antioxidants are given in Table 18.

Amines, most frequently secondary amines and p-phenylenediamine derivatives, do not usually affect mechanical properties of polymers processed below $250^{o}C$,

but are only suitable in coloured products. The
antioxidant efficiency of amines is increased by
the presence of —OH group.

Table 18. The most frequently used phenolic
antioxidants /67/

---

p-hydroxyphenylcyclohexane
bis(p-hydroxyphenyl)cyclohexane
discresylpropane
mixtures of alkylphenols
2,6-ditert.butyl-4-methylphenol
2,4,6-tritert.butylphenol
condensation products from dialkylphenols and formaldehyde
reaction products from phenol and styrene
1,1´-methylenebis(4-hydroxy-3,5-ditert.butylphenol)
2,2´-methylenebis(4-methyl-6-tert.butylphenol)
2,6-(2-tert.butyl-4-methyl-6-methylphenol-p-cresol)
phenylethylcatechol and phenylisopropylcatechol
2,2´-thiobis(4-methyl-6-tert.butylphenol)
4,4´-thiobis(3-methyl-6-tert.butylphenol)

The concentrations used are low (0.003 to 0.5 mol kg^{-1}
of polymer). Table 19 gives some amine-type
antioxidants and their recommended doses for polyolefins.

Of the thiols (formerly called mercaptans), which
possess quite high inhibition efficiency, especially
recommended are mercaptobenzothiazole, mercapto-
benzimidazole, 2-naphthalenethiol, and some aliphatic
thiols, e.g. dodecanethiol. They are used in relatively
high concentrations (about 1%), which can often affect
physical properties of the final article, because their
compatibility with polyolefins is poor.

The most frequently used organic sulphides are
diphenyl sulphide, di-2-naphthyl sulphide, thiophene

types, and disulphides.

Table 19. Amine-type antioxidants and their recommended
doses added to polyolefins /67/

Amine	Dose (mol per 1 kg mixture)
diphenylamine	0.01
phenyl-1-naphthylamine	0.005
phenyl-2-naphthylamine	0.005
mixture of phenyl-1-naphthylamine and diphenyl-p-phenylenediamine	0.05 % by wt.
diphenyl-p-phenylenediamine	0.003 to 0.005
N,N´-phenylcyclohexyl-p-phenylene-diamine	0.005 to 0.008
N,N´-bis-2-naphthyl-p-phenylenedia-mine	0.008

They belong to the antioxidants which decompose
peroxides.

The group also includes antioxidants based on
phosphorous acid, such a simple esters of phosphorous
acid or trinonyl-phenylphosphite (*Polygard*), tri-p-tert.-
butylphenylphosphite, etc. Even more efficient are
esters of pyrocatecholphosphorous acid, e.g. its
2,6-di-tert.butyl-4-methylphenyl ester. The other
esters used are derived from 1-naphthol, pyrocatechol,
etc.

The phosphites react with hydroperoxides to yield
phosphates.

Some phosphites are very efficient, non-toxic, and
cause no colouration. Some combinations of antioxidants
are synergystic, e.g. 2,6-di-tert.-octylcresol (a
phenolic-type antioxidant which has a very low efficiency) with didecyl sulphide (also of low efficiency)
exhibits a very much increased efficiency. The effects
of individual antioxidants and their combinations on

the inhibition period of oxidation is presented in
Fig. 39 /34/. Curve 3 representing the most efficient

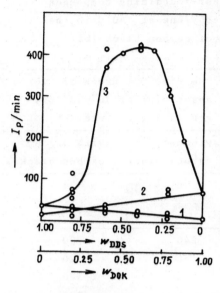

Fig. 39. Effect of didecyl
sulphide (DDS) and of
2,6-di-tert.octylcresol
(DOK) on the inhibition
period I_p of oxidation of
polypropylene stabilized
either with each of the
substances alone or with
their combinations of
various molar ratios and
the total dose of 0.2 mol
kg^{-1}. The oxidation temperature 200°C. The par-
tial pressure of oxygen
39.5 kPa /34/

1 - DDS, 2 - DOK, 3 - combined DDS + DOK, w - weight
fraction

synergistic mixture of these antioxidants shows a maximum
at the 1:1 ratio. In other cases, however, the maximum
effect can be achieved at quite different ratios, e.g.
1:20.

Active carbon black is often used for stabilizing
polyolefins, providing absorption of ultraviolet radiation as well as considerable antioxidant action. The
material with larger surface area is more efficient, the
effect is attributable to hydroxyl, carbonyl, and
quinoid groups at the surface, and comparable with
macromolecular antioxidants. 3% active carbon black in
polyethylene is more efficient at 140°C than
2,6-di-tert.butyl-4-methylphenol. Application of the
active carbon black is somewhat complicated by the fact
that it exhibits antagonism in combinations with some
phenolic or amine-type antioxidants but synergism with
sulphur-containing antioxidants in most cases (Table 20).

Table 20. Influence of sulphur-containing compounds
(0.1%) on oxidation of polyethylene at 140°C in the
presence and in the absence of carbon black /62/

Compound	The time (h) necessary for absorption of 10^4 mm^3 oxygen in 1 g polyethylene	
	without carbon black	with 3% carbon black
polyethylene (reference sample)	6	35
polyethylene + 1% sulphur	-	120
thiobis(2-naphthol)	240	730
diphenyldisulphide	6	120
bis(2-tolyl)disulphide	6	520
2-naphthalenethiol	6	900
2-mercaptobenzothiazol	35	380
1-dodecanethiol	6	160

3.4.5 ABSORBERS OF ULTRAVIOLET RADIATION

The energy of ultraviolet radiation is large enough to
initiate degradation reactions in the polymer either
by formation of primary radicals or by decomposition
of hydrogen peroxide /68/. UV absorbers must meet the
following basic requirements /69-71/:

- they must not transmit light of the wavelength
300-400 nm

- no chemical reaction must take place between the
absorber and the polymer

- the absorber must transform the energy absorbed
into some harmless form, i.e. long-wave radiation or
heat, that energy being insufficient to cause decomposition

of the polymer

- the absorber must be stable, i.e. the absorption and transformation of energy must proceed without its decomposition or any change.

The substances which meet most of the above requirements are colourless and exhibit little or no absorption in the long-wave end of the ultraviolet region. According to their structure they can be classified into following groups: derivatives of hydroxybenzophenone aromatic acid, hydroxyphenylbenzotriazole, hindered amines (HALS), metallic chelates, substituted acrylonitriles, and inorganic pigments. The first four groups are of the main importance for polymeric materials of packaging.

The first group is characterized by the presence of at least one ortho-hydroxyl group in the benzene ring: if one hydroxyl group only is present, the substances are practically colourless (they absorb the ultraviolet radiation below 390 nm), if there is one ortho-hydroxyl group in each of the two benzene nuclei, the substances are yellowish (their absorption extends to the visible spectral region). The following substituted hydroxy-benzophenones are therefore useful absorbers:

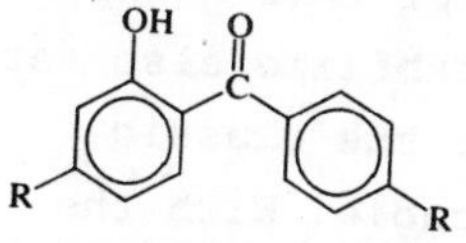

Benzophenone itself exhibits practically no UV absorption, but introduction of —OH or —OCH$_3$ groups in 2 or 4 positions imparts the ability to absorb the longer-wave radiation in the ultraviolet region. The 2-substitution is more important as 4-hydroxy-benzophenone transmits 90% of ultraviolet radiation. Any further —OH group introduced extends the absorption region by 10 nm (5 nm in the case of —OCH$_3$ group); the chain length of alkoxy groups on the benzene ring does not affect the extent of UV absorption but affects the

compatibility of the absorber with the polymer /72/.
The benzophenone derivatives most frequently used
/73/ are: 2,2´-dihydroxy-4-methoxybenzophenone
(*Cyasorb 24*) and 2,2´-dihydroxy-4,4´-dimethoxybenzophenone
(*Uvinul D 49*).

Derivatives of salicylic acid are only used to a
limited extent, because their absorption in the wavelength range 290-320 nm is not satisfactory, they change
their structure on longer irradiation. This group includes phenyl salicylate (*Salol*) and its derivatives
like tert.-butylphenyl salicylate.

Derivatives of hydroxyphenylbenzotriazoles represent
a newer type of UV absorber, the most important representative being the alkyl derivative of 2-hydroxyphenyl-
benzotriazole (*Tinuvin P*) which has high absorptivity
in the ultraviolet region and importantly exhibits a
very sharp absorption edge at about 400 nm. It is also
highly stable to light.

Hindered amine light stabilizers (HALS) represent
the newest type of UV absorbers used primarily for
polyolefins, but at the same time they have very good
perspectives for the stabilization of other polymers
including PVC. An attractive property of this group
of UV absorbers is their ability to stabilize also very
thin films and fibres where the use of the classic
types of UV absorbers is hardly applicable. With the
aim of achieving maximum resistance to elutriation and
minimum volatility, the development of new types of HALS
is oriented towards oligomeric and polymeric structures
/74/. On the other hand, the limitation of the mobility
of a UV absorber in a polymer matrix leads practically
in all cases to the reduction of its stabilization
efficiency /75, 76/. These facts are connected with
the mechanism of the HALS action based on the formation
of hydroxylamine /77/ which can have basic importance
for the next stabilization. This compound, however,
is one of the most efficient donors of hydrogen /78/,

in addition to its powerful antioxidizing effects /77/.
Hydroxylamine ether as a next product of the stabilization reaction of HALS is also a compound with the
properties of a primary antioxidant.

The original mechanism /79/ of HALS action could
contribute to the development of production methods.
HALS were first commercialized in the 1970s. Products
with such trade names as *Tinuvin 144* and *770* (low molecular products), *Tinuvin 622* and *Chimasorb 944*
(oligomeric products), to *Ciba-Geigy* as well as *Hostavin
N 20* (Hoechst) are very well known.

Tinuvin 622 was certified by the Food and Drug
Administration in 1982 for use in food packaging, the
first HALS ever to achieve that distinction /80/. It
is reported to be compatible with all polyolefins and
to show the following performance properties: it is
thermally stable at temperatures in excess of $300^{o}C$,
imparts long-term heat stability to the polymer, has
no effect on substrate colour, and has low volatility,
low extractibility and a low migration rate. It is
cost-effective despite a price double of most benzophenones, for it can be used at one half to one fifth
normal concentration with equal protection.

B.F. Goodrich's HALS, *Good-rite UV 3034*, is effective
in polyolefins as well as such other materials as acry-
lics, polycarbonate and polyesters. It is synergistic
with benzophenones and benzotriazoles. It is also
effective at low loads (as little as 0.05 weight parts
per 100 parts of a polymer) /81/.

Spinuvex A-36, a high performance HALS manufactured
by Montefluos, is claimed to be effective in polyolefin
moulding, and extrusion materials, acrylonitrile-butadiene-styrene copolymers, and impact polystyrene /81/.
It can be blended with other polymer additives for
synergistic effect.

There is hardly a doubt that the years to come will
see an extensive broadening in the supply of hindered

amines. But that does not mean less availability of
the more conventional and lower-cost types, for which
there will always be ready market - the more so in view
of recent product upgradings /80, 81/.

All the UV absorbers mentioned are characterized by
the ability to form a chelate ring by hydrogen bond
between hydroxyl and carbonyl groups or nitrogen-
containing groups.

This characteristic is considered to be fundamental
to UV absorption in these compounds /82/; the energy
absorbed is sufficient to open the hydrogen bond. This
conclusion is supported by the fact that compounds
unable to form a hydrogen bond (benzophenone, 2-methoxy-
benzophenone) exhibit no absorption effect in the
ultraviolet region. The use of UV absorbers is steadily
increasing, because they prolong the service life of
plastics, particularly outdoors.

3.4.6 FILLERS AND REINFORCING MATERIALS

Fillers are additives which modify characteristics of
the polymer (mechanical properties, surface appearance,
density, melt-flow properties, etc.) or decrease price.
With thermosetting resins, particularly phenoplastics
and aminoplastics, fillers improve mechanical and
dielectric properties, abrasion resistance, heat
resistance, ignition resistance and chemical
resistance. However, the filler must be carefully chosen
and its amount properly adjusted. The mechanical properties depend usually on the concentration of the filler.
They are used extensively with thermosetting resins
and rubbers, but their use with thermoplastics is limited.

In the case of phenolic plastics, fillers improve
most properties. Fillers of organic origin include
wood flour (for consumer products and electroinsulating
products), cellulose fibres (textile or paper) which

improve impact strength, flour of various fruit shells
which improve fluidity of the moulding material and
both dielectric properties and appearance of final
products. Also widely used are protein flours (e.g.
casein, soya) for aminoplastics. Of inorganic materials,
asbestos is widely used (anthophyllite particularly to
improve heat and acid resistance of polyolefins) and
also the fibrous chrysotil to improve mechanical properties. Mica imparts food dielectric properties and heat
resistance to plastics. Lead and barium compounds serve
as X-ray shielding substances. Glass fibres are extremely
important, but they are rather included among reinforcing
additives. Further inorganic fillers for thermosetting
resins include: slate, cement, graphite, quartz sand,
talc.

The fillers used in elastomers are primarily those
with reinforcing qualities connected with their large
specific surface, their chemical character, the ability
to produce chains and aglomerates, and give good dispersion in the polymer. Various substances are used, e.g.
carbon black, hydrated precipitated silicates, pyrogenic
or precipitated silica, kaolin, chalk, baryta.

Fillers are of limited use with glassy plastics like
poly(vinyl chloride), polystyrene, and poly(methyl
methacrylate). In the case of poly(vinyl chloride),
a small addition of a fine, polar filler does not
seriously detract from its basic properties, but coarser
particles worsen the physical properties substantially.
Emulsifiers, being surface-active substances, facilitate
the dispersion of fillers in the polymer and improve surface compatibility essential for application of the adsorption
forces. Therefore the interaction between a silicate
filler and emulsion poly(vinyl chloride) is more efficient than that with suspension polymer /83/.

In order to improve interaction, fillers can be
subjected to surface modification. Addition of bentonite

(below 9%) increases the hardness of polystyrene, and the surface modification of the bentonite particles with octadecylamine increases its affinity and, hence, its reinforcing effect. However, it was also found that fillers added to polystyrene and poly(methyl methacrylate) always cause a decrease in flexural and impact strength, more significantly with coarser particles /84/.

When using fillers it must be remembered that working surfaces (extrusion screws, barrels and dies, injection channels) are sensitive to mechanical and chemical wear. Also to attain good appearance of the articles, the filler must contain no moisture and its particles must have adequate hardness.

Reinforcements are substances which, due to their shape and structure, reinforce the resins used, particularly thermosetting resins like phenol-formaldehyde and urea-formaldehyde resins and cross-linked polyesters. They are fibers and fabrics based on cotton, paper, glass, metals, or minerals which are impregnated with the resins and moulded into the required shape. The laminated or reinforced plastics (or laminates) produced have excellent mechanical properties. Maximum reinforcement is achieved with glass fibres which are the usual reinforcement for polyester styrene laminates. The most frequently used glass fibres are made of special alkalifree borosilicate glass with less than 1% soluble alkali metals and with good hydrolytic resistance /85/.

Glass fibre reinforcement is used increasingly for thermoplastics (polyethylene, ABS terpolymers, poly(vinyl chloride), polyamides, etc.). The glass fibres must be delubricated, because lubricants impair adhesion to the resin. This adhesion is further improved by application of a thin layer of a substance which can bind the glass with the resin. One coupling agent used is methacrylatochromium(III)   chloride (*Volan A*) /85/

which is also suitable for glass fibres in epoxide
and phenolic laminates. However, some organosilicon
compounds are more efficient such as vinyltrichlor-
osilane (*Garan*). Further agents are isocyanates and
polyvinylpyrrolidones /86/.

3.4.7 CHEMICAL BLOWING AGENTS

Blowing agents are organic or inorganic substances
which are decomposed by heat to produce gas. Their
characteristic is the temperature necessary for gas
evolution which determines applicability to a given
plastic and its processing conditions. According to
chemical composition, blowing agents can be classified
as follows /87, 88/:

   Inorganic blowing agents: carbonates, hydrogen
carbonates. They are cheap but insoluble in plasticizers and organic solvents, their dispersion is
difficult and they decompose during long storage.

   Organic blowing agents: azo-bis(formamide)
(azodicarbonamide) represents the most frequently used
azo compound. It is non-toxic, odourless, and incom-
bustible. Although it is insoluble in usual solvents
and plasticizers, it is easily dispersed. It is manufactured under various trade names as e.g. *Vinylfor-AC,
Unifoam AZ, Celogen AZ, Genitron AC, Porofor AC,* etc.
Economically it is very advantageous, because it
releases a considerable volume of gas (220-240 cm^3g^{-1}
at $200^{\circ}C$).

   The second representative of this group is azo-bis-
(isobutyronitrile) whose applications are limited by
toxicity of its decomposition products. Some trade
names are: *Genitron AZDN, Porofor N.* It releases 130
cm^3g^{-1} gas at $115^{\circ}C$.

   Diazoaminobenzene (*Porofor DB*) evolves 115 cm^3g^{-1}
gas at $103^{\circ}C$.

A typical representative of nitroso compounds is
N,N'— dinitrosopentamethylenetetramine known under
the trade names *Cellmic A, Porofor DNO/N*. It is a very
widely used blowing agent, mainly for rubbers. It has
a disagreeable smell which is especially distinctive
in its applications to poly(vinyl chloride), but may
be suppressed by addition of urea. This blowing agent
releases about 240 cm^3g^{-1} gas at about 200°C.

The main sulphonyl hydrazide derivatives used are
benzenesulphonyl hydrazide, commercial name *Celogen
BSH, Genitron BSH,* and *Porofor BSH,* which is decomposed
at 95°C to give 130 cm^3g^{-1} gas, p-toluenesulphonyl
hydrazide (*Celogen TSH, Cellmic-H*) which gives 120
cm^3g^{-1} gas at 110°C, and p-toluenesulphonyl semicarbazide
(*Celogen RA*) whose decomposition begins at about 235°C
to give 140–150 cm^3g^{-1} gas. A wide spread representative
of this group of compounds is p, p '-oxy-bis(benzenesulphonyl hydrazide) (*Celogen OT, Porofor DO 44, Genitron OB*). It is decomposed at 150°C and the amount of
the gas released is 125 cm^3g^{-1}.

A number of other compounds serving as blowing agents
are reported in literature, e.g. cyanamide or derivatives
of cyclopentanetetracarboxylic acid containing at least
two hydrazide groups, 2,4-dioxy-1,2-dihydro-4-benzoxazine
and its derivatives which release no corrossive gases
and colouring or toxic products, various nitrogen-
containing compounds as urea and its derivatives, biuret,
or acetylurea.

The decomposition of blowing agents can be controlled by the presence of so-called kickers, i.e. substances acting as activators. There exist slow kickers,
e.g. barium of barium-zinc salts (which raise the
decomposition temperature), fast kickers, e.g. cadmium
or cadmium-zinc salts (which lower the decomposition
temperature), and medium kickers, e.g. lead or lead-zinc
salts. It is, however, difficult to classify a given

kicker as slow or fast, because of other criteria,
like the expansion rate of the plastic and the structure of the foam formed.

Blowing agents should have the following characteristics:

1. stable during storage,

2. easily dispersed in the polymer,

3. release gas within a narrow temperature range
appropriate to the processing temperature,

4. the decomposition rate should be independent of
moulding pressure,

5. the gas released must be neither corrosive nor
toxic,

6. the decomposition products must be non-toxic,
odourless, and colourless,

7. the decomposition products should not affect
physical or chemical properties of the foam,

8. the decomposition products must be compatible
with the polymer to avoid blooming,

9. the decomposition should not be significantly
exothermic,

10. inexpensive.

3.4.8 MACROMOLECULAR MODIFIERS

The aim of adding such modifiers is to improve or
supplement the properties of the polymer. The most
common additives are those which improve the main drawback of glassy polymers, their poor impact strength.
The low impact and/or notched impact strengths restrict
the use of hard, i.e. unplasticized poly(vinyl chloride)
and account for its slower development of production
and application compared with that of other tough polymers, e.g. tough polystyrene (tPS), even though it is
very slightly better than tPS.

Improvement of the notch impact strength also aims to retain high overall strength and improve, or at least maintain, the workability, heat stability, resistance to ageing, and possibly also maintain transparency, which is difficult with polymeric mixtures. In the case of PVC, the tough materials are obtained by mixing the powdered poly(vinyl chloride) with additive which for ease and economy should also be in powder form. If the modifying additive is only available in granulated form or as a compact material, mixing is technologically complicated.

As the modifying agents do not usually meet all the aims given above, each application necessitates careful selection. As toughened poly(vinyl chloride) is increasingly used in packaging, the suitable polymeric modifiers can serve as examples:

Chlorinated polyethylene (CPE) produced by chlorination of polyethylene /89/ still holds the leading position and satisfactory grades depend on the molecular weight of the starting polymer, and its branching, and on the method, temperature, and extent of chlorination /90/. The grades used with poly(vinyl chloride) contain about 35-45% chlorine, have glass transition temperatures in the range $-30^{\circ}C$ to $-10^{\circ}C$, and molecular weight above 100,000. Poly(vinyl chloride) and chlorinated polyethylene can be mixed in practically all ratios, but at more than 50% modifier, difficulties are encountered because chlorinated polyethylene has a much higher melt viscosity. The first producer of tough poly(vinyl chloride) based on chlorinated polyethylene was Farbwerke Hoechst (FRG), and the trade name *Hostalit Z.*

The range of ethylene-vinyl acetate (E/VAC) copolymers includes low-molecular waxes, thermoplastics, and rubbers. One of the first industrial products was *Elvax 150* (Du Pont de Nemours, U.S.A.). It contains 37% vinyl acetate, is available in granulated form and

is used for both direct processing and modification
of other polymers, e.g. poly(vinyl chloride) and
polyethylene. It improves workability, but it is not.
suitable for transparent products nor the optimum
additive for poly(vinyl chloride). A better additive
is *Levapren 450* (Bayer, FRG). However, ethylene-vinyl
acetate copolymers have limited application with
poly(vinyl chloride) because they cannot be supplied
in powder form. Powder materials are made by grafting
(vinyl chloride) on to ethylene-vinyl acetate copolymer.
These graft copolymers are produced either for direct
processing as toughened material (containing 5-10%
ethylene-vinyl acetate) or as additives (with as much
as 50% ethylene-vinyl acetate).
Commercial products are: *Levapren VC45/55* (Bayer, FRG),
Vinnol 315/65 (Wacker Chemie, FRG), and *Hi-Blend*
(Japanese Geon, Japan).

Tough poly(vinyl chloride) based on E/VAC sometimes
shows over-milling, where reduction of particle size
during kneading due to mechanical stress and heat,
below the optimum value affects the notch impact strength.
Over-milling is also influenced by some lubricants and
stabilizers.

Rubber-like polymers of acrylic acid esters, particularly butyl acrylate and isooctyl acrylate /91/,
have sufficiently low glass transition temperature
(below $-50^{\circ}C$). BASF produces *Vinoflex 719* - a tough
grafted poly(vinyl chloride) based on alkyl acrylate
and designed for external application. The K value of
the poly(vinyl chloride) portion is 61 and the chlorine
content is 50.5%. *Vinoflex 719* is used with suspension
poly(vinyl chloride).

Modifying grafted terpolymers MBS (methyl metha-
crylate-butadiene-styrene) have been developed for
packaging /92, 93/ and are expected to be efficient with
respect to toughness but also impart surface gloss,
and are suitable for food contact, inert to the goods

packed, transparent, and colourless. Transparency
necessitates that the modifying component has a refractive index approximating to that of the matrix
(e.g. 1.535 for poly(vinyl chloride)) /94/. The commercial grades for transparent tough poly(vinyl chloride) include *Paraloid K M 607* (Röhm and Haas, FRG),
Kureha BTA III N (Kureha Chemical Industry Co., Japan),
and *Kane Ace B 11, 12, 16, 18 A* (Kanegafuchi Chemical
Industry Co. Ltd., Japan).

The last manufacturer also markets types for special
technologies, e.g. *Kane Ace B 18* (in powder form) for
transparent tough poly(vinyl chloride) bottles made
by blow moulding. Also Pechiney-Saint Gobain (France)
offers a modifier for poly(vinyl chloride) based on
an alkyl acrylate copolymer containing about 50% vinyl
chloride and also designed for transparent articles
(*Lucovyl H 4010*).

The additives for general use are acrylonitrile-
butadiene-styrene terpolymers (ABS) which are usually
based on polybutadiene rubber, the glass phase containing styrene and acrylonitrile. Commercial types are
Marbon Resin 301 (in Europe or *Blendex 301* elsewhere)
produced by Marbon Chemical (Borg Warner Corp., USA)
and *Sicoflex S 160* produced by Mazucheli Celluloide
S.p.A. in Italy. The Czechoslovak analogue is
Forsan 940 M produced by Chemopetrol. Another rubber
recommended for tough poly(vinyl chloride) is lithium
polybutadiene *Acadene NF 35 A* (Asahi Chemical Ind. Co.,
Japan) which in combination with E/VAC copolymer, ABS,
MBS, or Hypalon (i.e. chlorosulphonated polyethylene)
enhances their efficiency.

U.S.I. Chemicals (National Distillers and Chemical
Corp.) recommends the polyethylene powder *Microthene F*
as a modifying additive.

3.4.9 FLAME RETARDANTS

Agents reducing flammability or flame propagation are
used primarily in lacquer technology to produce non-
flammable coatings, and in building technology for
the production of expanded polymers. Non-flammable
plasticizers (Table 21) can play a similar role.

Table 21. The plasticizers preventing combustion of
plastics /97/

Plasticizer	PVC	PET	PO	PUR	CA	CN	PVAC
tricresylphosphate	X	X	X	O	O	X	X
diphenyl cresyl phosphate	X	O	O	O	X	X	O
diphenyl octyl phosphate	X	O	O	O	X	X	O
tris(2-chloroethyl) phosphate	X	X	X	X	X	X	X
tributyl phosphate	X	O	O	O	O	X	X
tris(dichloropropyl) phosphate	X	X	X	X	O	O	O
triphenyl phosphate	O	O	O	X	X	X	O

X - used, O - not used, PVC - poly(vinyl chloride),
PET - linear polyesters, PO - polyolefins, PUR - polyurethanes, CA - cellulose acetate, CN - cellulose nitrate, PVAC - poly(vinyl acetate).

Special plasticizers are the chloroparaffins, chlorinated biphenyls, 2,3-dibromopropyl phosphate, etc.
/95, 96/. The selection of other additives is much
broader, the most used being antimony trioxide, chromic
oxide, kaolin, chalk, various silicates, asbestos,
finely divided mica, graphite, barium and magnesium
sulphates, aluminium(III) hydroxide-oxide, zinc
phosphate, antimony(III) phosphate and zinc borate.

Less important materials include those which decompose
at high temperatures to release gases which either
cause formation of a heat-insulating foam or directly
extinguish the flame /96/. This group includes starch,
paraformaldehyde, dicyandiamide, aminoacetic and
salicylic acids, ammonium phosphate, pentaerythritol,
benzenesulphonyl hydrazide, urea, and carbonates.
Special flame suppressants are also known, e.g. tetra-
bromobiphenyl, pentabromophenol, tetrabromophthalic
esters, bromoalkyl phosphates, and certain blowing
agents /98, 99/.   Some combinations exhibit synergism
/100/, e.g. antimony trioxide with zinc borate or with
microfine mica.

However, non-flammability is determined also by
other components, particularly the polymer.

3.4.10 PIGMENTS AND BRIGHTENING AGENTS

3.4.10.1 P i g m e n t s

Pigments are coloured insoluble powders which impart
to coatings colour and covering power.

According to their origin, pigments are classified
as inorganic, organic, and bronzes (which are powdered
metals). Natural inorganic pigments include: chalk,
gypsum, siennes, ochres, and graphite. Manufactured
ones are lithopone, titanium white, chrome yellow and
orange, cadmium red, blue pigments based on hexa-
cyanoferrate(II), green aluminium hydroxide-oxide,
ferric black, green chromium(III) oxide, ultramarine,
vermilion, zinc white, red lead, white lead, etc. The
organic pigments most frequently used are non-migrating
pigments stable above $140^{o}C$, such as the indanthrene
vat dyes, phthalocyanine pigments and some insoluble
salts of soluble acid dyestuffs.

The features of pigments and dyestuffs to consider

for plastics are:

- density. The density difference between pigment
and binder affects the stability of coating compositions
- covering power. The ability to cover variations
in substrate colour with uniform coating
- shape, size and size distribution. These determine
the stability of suspensions used and affect colour
hue. A suitable diameter is from 0.5 to 3 μm.
- texture. Soft pigments are anatase titanium white,
zinc white, and talc. Rutile titanium white and
ultramarine are hard
- tinting power. This is the ability of a certain
amount of the pigment to modify a colour to another hue.
Sometimes an addition of titanium white enhances tinting
power
- wettability. This relates to the displacement of
surface moisture and air by the binder. If the pigment
is easily wettable by polar liquids, e.g. water, it
is called hydrophilic, whereas the pigments with good
wettability by non-polar liquids are called hydrophobic.
Carbon black and most organic pigments are hydrophobic.
Metal oxides, hydroxides, and salts of oxygen-contain-
ing acids are examples of hydrophilic pigments
- fastness. This is stability of shade on exposure
to light and is determined primarily by chemical
composition. Most changes in colour are induced by
photochemical oxidations of reactive groups in the
pigment. So inorganic pigments are relatively more
stable
- chemical reactivity. The pigment may react with
plastics, other additives, the atmosphere, or with
other materials which may contact the article. Therefore,
lead stabilizers or lead pigments cannot be combined
with lithopone or with some polymers pre-stabilized
with sulphur compounds
- heat stability. This is important for plastics
processing. Generally, organic pigments are less stable

to high temperatures, but stability depends not only
on the temperature but also on the time of exposure.

3.4.10.2 O p t i c a l   B r i g h t e n i n g   A g e n t s

The optical brightening agents absorb some ultraviolet
rather than visible radiation (in the wavelength region
340-400 nm) /101/. The energy absorbed is re-emitted
during irradiation in the form of visible light of
wavelength 430-460 nm (fluorescence).

Although most optical brightening agents are used
to improve whiteness with paper and textiles, they
also find application in plastics. They are used to
counteract a yellowish hue of some polymers which spoils
bright colouration, especially in blue, pink, or violet
hues. Their best effect is in daylight. The amounts
used vary from 0.001% to 0.05% and they are either
added to the monomer before polymerization, ground
with the polymer, or dissolved with the polymer in a
solvent. They must not be toxic, nor form coloured
products or react with the resin.

With rutile titanium white, which absorbs ultraviolet
radiation, and has a high refractive index, the efficiency of optical brighteners is reduced considerably.
UV absorbers have a similar effect in reducing or
nullifying the effects of optical brightening agents.

3.4.11 ANTISTATIC AGENTS

The electrically non-conducting character of most
plastics has the consequence that friction (e.g. during
winding or unwinding films) produces electrostatic
charges which have several undesirable effects such as
difficult feeding of the film through packing or printing machines, pick-up of loose materials and dirt on the
surface of packaging materials and packages, and

disagreeable electric shocks to operators. The magnitude of these charges is determined by the construction and arrangement of the processing or packaging equipment and by the surface resistivity of the polymer. Materials having surface resistivity above 10^{11} Ω (e.g. polyolefins or poly(vinyl chloride)) exhibit a high tendency to become charged (Table 22). With surface resistivity below $10^{11} \Omega$ electrostatic charge rapidly decays so that there is no practical problem (e.g. cellophane).

Table 22. Order of magnitude of surface resistivity of some packaging materials /102/

Material	Surface resistivity(Ω)
non-lacquered cellophane	10^8
lacquered cellophane	10^9
polyamide	10^{10}
cellulose acetate	10^{13}
cellulose acetate with antistatic treatment	10^9
polyethylene	10^{13}
polyethylene with antistatic treatment	10^9
polypropylene with weldable coating	10^{10}
biaxially oriented polypropylene	10^{14}
oriented polyester	10^{14}

The precautions taken against electrostatic charge consist either in rendering air conductive in the region of highest charge density by means of high-voltage or radioactive pointed ionizors /102/, or in antistatic modification of the plastic. In the latter case the conductivity of the material is increased by surface treatment or by incorporating antistatic agents with the polymer.

These agents are either strongly hydrophilic or possess an electrically conductive structure. As seen from the following examples, they contain nitrogen, phosphorus, sulphur, etc. Out of a large number of compounds the following products of American Cyanamid Co. /103/ have found broad application:

Catanac 477, N-(3-dodecyloxy-2-hydroxypropyl)ethanolamine $C_{12}H_{25}OCH_2CH(OH)CH_2NHCH_2CH_2OH$, melting point 59-60°C, is recommended for poly(vinyl chloride) and polyolefins, especially high density polyethylene, at a dosage above 0.15%. Higher doses (1.5%) are recommended for polystyrene and polypropylene. The substance is stable up to high processing temperatures (250°C).

Catanac 609, N,N-bis(2-hydroxyethyl)-N-(3´-dodecyloxy-2´-hydroxypropyl)methylammonium methyl sulphate, $(C_{12}H_{25}OCH_2CH(OH)CH_2N(CH_3)(CH_2CH_2OH)_2^{(+)} CH_3SO_4^{(-)}$, is used in the form of a 50% solution in a water-isopropyl alcohol mixture primarily for the surface modification of vinyl chloride homopolymers and copolymers (e.g. for gramophone records).

Catanac SN, stearamidopropyldimethyl-2-hydroxyethyl-ammonium nitrate, $(C_{17}H_{35}CONHCH_2CH_2CH_2N(CH_3)_2CH_2CH_2OH^{(+)} NO_3^{(-)}$, is used in the form of a 50% solution in water-isopropyl alcohol added to the polymer during the milling or kneading process. It has almost no corrosive effects, but decomposition begins above 180°C and becomes rapid above 250°C. It is miscible with water, acetone, and alcohols, but resistant to acids and bases. It is applied to unplasticized poly(vinyl chloride), acrylates, toughened polystyrene and acrylonitrile-butadiene-styrene terpolymers, paper, textiles, and a number of other materials. Table 23 gives its effects in some polymers.

The films and plates made of poly(vinyl chloride), polyethylene, and polypropylene   food use are preferably modified with esters of monocarboxylic acids and

glycerol of general formula $HOCH_2CH(OH)CH_2OCOR$ or
with ethoxylated fatty acids of general formula
$RCOOCH_2CH_2(OCH_2CH_2)OH$, which also possess anti-slip
properties. They are marketed as *Antistat-Slip Agents-
Drewplast* (Drew Chemical Corp., U.S.A).

Table 23. Effect of an antistatic agent on surface
resistivity of some polymers /103/

Polymer	Antistatic agent Catanac SN (%)	Surface resistivity ($M\Omega$)
unplasticized poly(vinyl chloride)	0	5.0×10^8
	1.5	1.1×10^4
	2.0	4.2×10^2
acrylonitrile-butadiene-styrene terpolymer	0	1.6×10^6
	1.5	8.1×10^4
polystyrene	0	3.5×10^7
	2.0	1.6×10^3

REFERENCES

/1/ Losev I.P., Petrov G.S., Chemie umělých prysky-
    řic, SNTL, Prague, 1955.

/2/ Karghin V.A., Slonimskiy G.L., Úvod do fyzikální
    chemie polymerů, SNTL, Prague, 1963.

/3/ Tobolsky A.V., Vlastnosti a struktura polymerů,
    SNTL, Prague, 1963.

/4/ Stuart H.A., Die Physik der Hochpolymeren, 3. Band,
    Springer, Berlin, 1955.

/5/ Stevens M.P., Polymer Chemistry - An Introduction,
    Addison-Wesley, London, 1975.

/6/ Herrmann K., Gerngross O., Abitz W., Z. Phys. Chem.,
    B 10, 1930, 371.

/7/ Keller A., Growth and Perfection of Crystals,
    Wiley, New York, 1958.

/8/ Jenckel E. et al., Kristallisation der Hochpoly-
    meren, Westdeutscher Verlag, Köln, 1958.

/9/ Geil P.H., Polymer Single Crystals, Interscience,
    New York, 1963.

/10/ Mandelkern L., Crystallization of Polymers,
     McGraw-Hill, New York, 1964.

/11/ Moore L.D., Peck V.G., J. Polym. Sci., 1959,
     36, 141.

/12/ Karghin V.A., Kitaygorodskiy A.I., Slonimskiy G.L.,
     Kolloid. Zh., 1957, 19, 131.

/13/ Kitaygorodskiy A.I., Cvankin D.Ya., Vysokomol.
     Soedin., 1959, 1, 269.

/14/ Bernauer F., Gedrillte Kristalle, Band 2,
     Forschungen zur Kristallkunde, Borntrager, Berlin,
     1929.

/15/ Kahle B., Elektrochem., 1957, 61, 1318.

/16/ Buchdahl R., Miller R.L., Newmans S., J. Polym.
     Sci., 1959, 36, 215.

/17/ Billmeyer F.W.Jr., Textbook of Polymer Science,
     Wiley, New York, 1971.

/18/ Meissner B., Fyzikální vlastnosti polymerů, Part 1,
     Textbook of Prague Institute of Chemical Tech-
     nology, SNTL, Prague, 1971.

/19/ Jost W., Diffusion in Solids, Liquids, Gases,
     Academic Press, New York, 1960.

/20/ Bird R.B., Stewart W.E., Lightfoot E.N., Přeno-
     sové jevy, Academia, Prague, 1968.

/21/ Barrer R.M., Trans. Farady Soc. 1939, 35, 628.

/22/ Hála E., Reiser A., Fyzikální chemie 1, Academia,
     Prague, 1971.

/23/ Reytlinger S.A., Pronikaemost polimernych materia-
     lov, Khimiya, Moscow, 1974.

/24/ Crank J., Park G.S., Diffusion in Polymers,
     Academic Press, London and New York, 1968.

/25/ Daynes H.A., Proc. Roy. Soc., London, A 97,
     1920, 286.

/26/ Barrer R.M., Skirrow G., J. Polym. Sci., 1948,
     3, 549.

/27/ Šípek M., Jehlička V., Nguyen X.Q., Chem. listy,
     1982, 76, 273.

/28/ Pasternak R.A., Schimschmeier J.F., Heller J.,
     J. Polym. Sci., A 2, 1970, 8, 467.

/29/ Isuda H., Rosmgren K.J., J. Appl. Polym. Sci.,
     1970, 14, 2839.

/30/ Wucherpfennig K., Bretthauer G., Ratzka D.,
     Brauwissenschaft, 1967, 20 (7), 275.

/31/ Calvano H.J., Brummer J.G., Speas C.A., Mod. Packag.,
     1968, 41 (11), 143.

/32/ Müller B., Verpack. Rdsch., 1967, 18 (10), 1186.

/33/ Becker K., Verpack. Rdsch., 1971, 22 (7), 57.
     ibid. 1969, 20 (6), 51.

/34/ Doležel B., Koroze plastů a pryží, SNTL, Prague,
     1981.

/35/ Seachtling H., Zebrowski W., Kunststoff-Taschen-
     buch, Hanser, Munich, 1979.

/36/ Achhammer B.G., Troyon M., Kline G.H., Kunststoffe,
     1959, 49, 600.

/37/ Grassie N., Developments in Polymer Degradation 2,
     Applied' Science Publishers, London, 1979.

/38/ Stevens M.P., Polymer Chemistry - An Introduction,
     Adison-Wesley, Massachusetts, 1975.

/39/ Scott G., Atmospheric Oxidation and Antioxidants,
     Elsevier, Amsterdam, 1965.

/40/ Scott G., Developments in Polymer Stabilization,
     Applied Science Publishers, Berking, 1979.

/41/ Pospíšil J., Antioxidanty, Nakladatelství ČSAV,
     Prague, 1968.

/42/ Conley R.T., Thermal Stability of Polymers,
     M. Dekker, New York, 1970.

/43/ Štěpek J., Daoust H., Additives for Plastics,
     Springer Verlag, New York-Heidelberg-Berlin, 1983.

/44/ Gächter R., Müller H., Taschenbuch der Kunststoff-
     Additive, Carl Hanser Verlag, München, 1979.

/45/ Seymour R.B., Additives for Plastics, Academic
     Press, New York, 1978.

/46/ Mascia J.B., The Role of Additives in Plastics,
     J. Wiley, New York, 1975.

/47/ Ritchie P.D., Critchley S.W., Hill A., Plasticizers,
     Stabilizers and Fillers, Iliffe Books, London,
     1972.

/48/ Štěpek J., Zpracování plastických hmot, SNTL,
     Prague, 1966.

/49/ Mark H., Whitby C.S., Advances in Colloid Science.
     Vol. 2., Scientific Progress in the Field of
     Rubber and Synthetic Elastomers, Interscience,
     New York, 1964.

/50/ Bueche F., Physical Properties of Polymers,
     Interscience, New York, 1962.

/51/ Leuchs P., Kunststoffe, 1956, 46, 547.

/52/ Thinius K., Chemie, Physik und Technologie der
     Weichmacher, Verlag Technik, Berlin, 1960.

/53/ Jacobsen B., Brit. Plast., 1961, 34 (6), 328.

/54/ Oakes V., Hughes B., Plastics, 1966, 31, 1132.

/55/ Gentner A.W., Kunststoffe, 1961, 51, 8.

/56/ Reithmayer S., Gummi, Asbest, Kunst., 1965,
18, 435.

/57/ Borovik M.C., Modern Plastics Encyclopedia,
Van Nostrand-Reinhold, New York 1960.

/58/ Chevassus F., De Broutelles R., La stabilisation
des chlorures de polyvinyle, Amphora, Paris,
1957.

/59/ Thinius K., Stabilisierung und Alterung von Plast-
werkstoffen, Band 1, Stabilisierung und Stabili-
satoren, Academie-Verlag, Berlin, 1969.

/60/ Hawkins W.L., Polymer Stabilization, J. Wiley,
New York, 1972.

/61/ Jellinek H.H.G., Degradation and Stabilization
of Polymers, Vol. 1, Elsevier, Amsterdam-Oxford-
New York, 1983.

/62/ Perry G., Rubber Age, 1959, 85, 449.

/63/ Fuchsmann J., SPE J., 1959, 5, 787.

/64/ Fuchsmann J., SPE J., 1961, 7, 590.

/65/ Štěpek J., Franta I., Doležal B., Mod. Plast.,
1963, 40 (12), 145.

/66/ Bayern G., Antioxidants 1971, Noyes Developments,
Pearl River, 1971.

/67/ Neiman M.B., Starenie i stabilizaciya polimerov,
Nauka, Moscow, 1964.

/68/ Kirchhof P., Gummi Asbest, 1958, 11, 614.

/69/ Coste J.B., Hansen R.H., SPE J., 1962, 8, 431.

/70/ Heller J., Kunststoffe, 1961, 51, 13.

/71/ Penn W., Rubber Plast. Weekly, 1961, 141 (7,8), 43.

/72/ Larin N., Teze dokladov o konferencii o starenii
i stabilizacii plastmass, Goskhimizdat, Moscow,
1962.

/73/ Zussman H.W., Modern Plastics Encyclopedia,
Van Nostrand-Reinhold, New York, 1960.

/74/ Klemchuk P.P., Advances in the Stabilization and
Controlled Degradation of Polymers, 2nd Intern.
Conf., Lucern, 1980.

/75/ Citovický P., Sedlář J., Mejzlík J., Chrástová V.,
25th Microsymposium IUPAC, Prague, 1983.

/76/ Gumunus F., Res. Disclosure, 1981, 357.

/77/ Sedlář J., Přísady do polymérnych materiálov,
Dom techniky ČSVTS, Bratislava, 1983.

/78/ Brownlie I.I., Ingold K.U., Can. J. Chem., 1967,
45, 2427.

/79/ Shilov Yu. B., Denisov E.T., Vysokomol. Soed.,
A 16, 1974, 2313.

/80/ Anonym, Mod. Plast. Intern., 1982, 12 (9), 60.

/81/ Anonym, Mod. Plast. Intern., 1983, 13 (10), 50.

/82/ Chaudet J., SPE J., 1961, 7, 26.

/83/ Heidingsfeld V., Zelinger J., Sborník VŠCHT
v Praze, C 12, 1967, 71.

/84/ Uskov I.A., Tarasenko J.G., Solomko V.P., Vysokomol. Soedin., 1964, 6, 1768.

/85/ Sternschuss A., Zvonař V., Slezák O., Kučera M.,
Polyesterové skelné lamináty, SNTL, Prague, 1961.

/86/ Jelínek Z.K., Zilvar V., Chem. prům., 1956, 6,
332.

/87/ Scheurlen H.A., Kunststoffe, 1957, 47, 446.

/88/ Lasman H.R., Modern Plastics Encyclopedia, Van
Nostrand-Reinhold, New York, 1963.

/89/ Fuchs W., Lonis D., Makromol. Chem., 1957, 22, 1.

/90/ Burnell C.N., Parry R.H., Rubber Age, 1968, 100, 47.

/91/ Reiff G., Kunststoffe, 1968, 58, 277.

/92/ Shargodskii A.M., Fedoseeva G.T., Savelev A.P., Sareda E.A., Plast. Massy, 1972, (9), 42.

/93/ Bochkarova G.G., Ovchinikov Yu. V., Plast. Massy, 1972, (11), 3.

/94/ Sato K., Japan Plast. Age, 1969, 7 (5), 52.

/95/ Jawwein W., Plaste, Kaut., 1958, 5, 81.

/96/ Ulbrich K., Farbe, Lack, 1957, 63, 154.

/97/ Mack G.P., Modern Plastics Encyclopedia, Van Nostrand-Reinhold, New York, 1963.

/98/ Anonym, Rubber World, 1958, 137, 754.

/99/ Anonym, Deutsche Farben-Z, 1958, 12, 132.

/100/ Vít J., Nehořlavé povlaky na bázi chloroprenového kaučuku. Thesis, Prague Institute of Chemical Technology, Prague, 1961.

/101/ Villaume F.G., Modern Plastics Encyclopedia, Van Nostrand-Reinhold, New York, 1964.

/102/ Čurda D., Obalová technika v potravinářství, Prague Institute of Chemical Technology, Prague, 1971.

/103/ Anonym, Polymer Additives, Noyes Data Corp., New Jersey.

4

Production of Basic Types of Plastics, Their Properties and Application to Packaging Technology

J. ŠTĚPEK, V. DUCHÁČEK

4.1 PLASTICS BASED ON NATURAL MACROMOLECULAR MATERIALS

Research into the structure and composition of natural
macromolecular materials led to understanding of the
chemistry of polymers. Natural polymers like cellulose
and its derivatives, natural rubber, shellac, casein,
and gelatine were the first macromolecular compounds
which found technical application.

Usually, natural polymers are divided into three
classes: polysaccharides, proteins, and polynucleotides
/1/. The synthesis and structure of proteins and polynucleotides is currently one of the most significant
regions of polymer chemistry. In this book we shall
deal with those polymers best known in packaging applications.

4.1.1 CELLULOSE DERIVATIVES

Cellulose is a polysaccharide and attracted the attention
of early polymer chemists, being easily accessible in
cotton and wood. Dry cotton contains 90% of cellulose,

while wood contains 50% along with 30% lignin (a natural phenolic polymer reinforcing the wood structure), saccharides, and salts.

Historically, the production of cellulose and its derivatives is connected with the origin of the plastic industry in 1865 (Parkers, Great Britain) or 1868 (Hyatt, USA) when the plasticization of cellulose nitrate (nitrocellulose) with camphor made the shaping of simple objects possible /2/. In the first part of this century cellulose nitrate was gradually replaced (except for lacquers and explosives) by regenerated cellulose and cellulose esters.

Cellulose itself does not behave as a plastic as it cannot be softened by heat and is insoluble. First it must be transformed into a water-soluble derivative which can be easily shaped, after which the cellulose is chemically regenerated - hence the name "regenerated cellulose". Later the cellulose esters and ethers having thermoplastic character were produced.

They were used for production of a wide variety of goods /3/. These included amyloid, a strong translucent parchment produced by the action of dilute sulphuric acid on cellulose and subsequent washing with water, cuprate silk (Bemberg silk) whose basic component is the cuprammonium complex of cellulose $\frac{1}{2}\left[Cu(NH_3)_4\right]^{2(+)}$ $\left[C_6H_7O_5Cu.2H_2O\right]^{(-)}$, formed by dissolving cellulose in ammoniacal cupric hydroxide (Schweizer reagent), and alkali-cellulose, the raw material for viscose which is used to make viscose rayon and cellophane, a common packaging material.

In the production of viscose, cellulose sheets like blotting paper are treated with aqueous sodium hydroxide solution for 2-4 h. The sodium-cellulose formed is left to stand for 2-3 days (ripening), which results in reduction of its molecular weight. The sodium-cellulose is transformed to cellulose xanthate by carbon disulphide;

the aqueous solution of this ester is called viscose, which is ripened for a further 4-5 days to secure uniform distribution of the xanthate groups. The viscose is then transformed to fibers or films by extrusion or casting and treated with acid to effect cellulose regeneration. The cellulose macromolecule (Fig.40) is composed of glucose structural units each of which contains three hydroxyl groups (two secondary and one primary), which can undergo reactions characteristic of

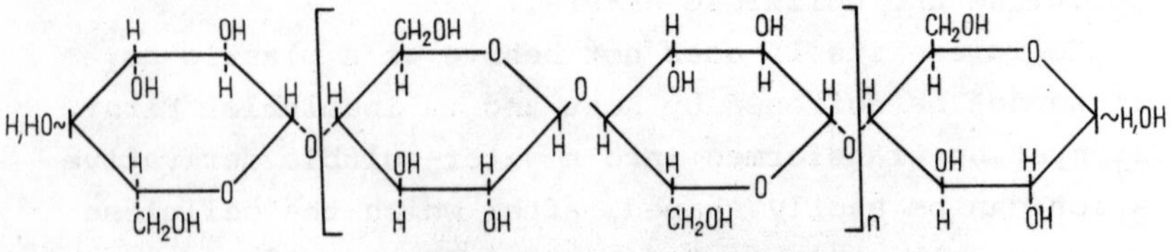

Fig. 40. Chemical structure of cellulose /1/

alcohols. The process can be represented by the following simplified scheme:

$$R(OH)_3 \xrightarrow[H_2O]{m \text{ NaOH}} R(ONa)_m (OH)_{3-m} \xrightarrow{m \text{ CS}_2} R(O-CS-S^{(-)}Na^{(+)})_m (OH)_{3-m}$$

cellulose   alkali-cellulose   viscose ($m \approx 0.5$)

$$\xrightarrow{H_2SO_4} R(OH)_3$$

regenerated
cellulose

The high flammability of cellulose nitrate caused its early withdrawal for the production of solid articles except for table-tennis balls where it can hardly be replaced by another polymer. The nitration of cellulose proceeds according to the scheme:

$$R(OH)_3 \xrightarrow[H_2O]{HNO_3, \ H_2SO_4} R(ONO_2)_3$$

cellulose trinitrate, $t_g = 53°C$

The cellulose nitrate containing 10.5 to 12.5 % N (collodion cotton) was used in the production of celluloid and photographic films, and is still employed in coatings for wood, metals and cellophane films.

Organic acids react with cellulose in the presence of condensation agents to produce esters. Acetylation is carried out in three steps /2/: Cellulose is first treated with glacial acetic acid for 1-2 h, then with acetic anhydride catalyzed with sulphuric acid at about 50°C, by refluxing in the presence of methylene dichloride. The reaction mixture is typically:

cellulose	100 weight parts	preparatory phase
glacial acetic acid	35 weight parts	
acetic anhydride	300 weight parts	acetylation
methylene dichloride	400 weight parts	
sulphuric acid	1 weight part	

The second reaction step produces cellulose triacetate which is only rarely used directly as a plastic (for films and fibres). The third step of the process is partial hydrolysis to the diacetate with water which takes several days. The reaction scheme is as follows:

$$R(OH)_3 \xrightarrow[H_2SO_4, \ CH_2Cl_2]{\substack{\text{acetic acid} \\ \text{acetic anhydride}}} R(OCOCH_3)_3 \xrightarrow[\text{heating}]{H_2O} R(OCOCH_3)_2OH$$

cellulose   cellulose   cellulose
                           triacetate   diacetate
                           $t_g = 105°C$ $t_g = 120°C$

Cellulose diacetate is soluble in acetone whereas the triacetate only dissolves in some more expensive solvents, e.g., chloroform. The diacetate is used for films, fibres, and coating compositions. Dimethyl phthalate is a suitable plasticizer for this material.

Similar procedures are used for cellulose butyrate, propionate, and (mixed) acetobutyrate.

Well known commercial names of cellulose esters are *Tenite* (Eastman, USA) and *Cellidor* (Bayer, FRG). The latter are more common in Europe and are denoted as follows:

Cellidor A (2,5-acetate) is characterized by high mechanical strength, and resistance to stress cracking, to light, to oils and petrol, and is used for automobile components, etc.

Cellidor S has a higher ester content and decreased wettability.

Cellidor U has the same composition as type S, but is manufactured as a self-extinguishing grade.

Cellidor CP is cellulose propionate; it resists boiling water and has better mechanical properties than the acetate at low temperatures.

Cellidor B/B is the acetobutyrate characterized by exceptional dimensional stability, weldability, and weathering resistance, making it the most useful type.

Ethylcellulose and benzylcellulose are ethers, similar to cellulose acetate both in their properties (stiffness, transparency, chemical resistance) and applications (films, coating compositions). Other cellulose ethers are water-soluble polymers used as protective colloids in pharmaceutical and cosmetic preparations, as additives to water-soluble coating compositions, as impregnants for paper and textiles, and as water-soluble adhesives. Hydroxypropylcellulose as a sandwich material with water-soluble films has been used for degradable bottles /1/.

Table 24. Characteristic properties of packaging materials based on cellulose derivatives /46/

Property	Ethyl-cellulose	Cellulose acetate	Cellulose propionate	Cellulose acetobutyrate	Cellulose nitrate
density (g cm^{-3})	1.09–1.17	1.28–1.32	1.17–1.24	1.15–1.22	1.35–1.40
workability	good	excellent	excellent	excellent	excellent
tensile strength (MPa)	14–56	31–56	14–55	18–49	49–56
elongation at break (%)	5–40	20–50	30–100	50–100	40–45
thermal conductivity (W m^{-1} K^{-1})	0.15–0.29	0.16–0.32	0.16–0.32	0.16–0.32	0.23
thermal expansion $\alpha \times 10^5$ (K^{-1})	10–20	10–15	11–17	11–17	8–12
relative permitivity at the frequency of 60 Hz	3.0–4.2	4.0–5.0	3.7–4.3	4.0–5.0	7.0–7.5
10^3 Hz	3.0–4.1	4.0–5.0	3.6–4.3	4.0–5.0	7.0
10^6 Hz	2.8–3.9	4.0–5.0	3.3–3.8	4.0–5.0	6.4
refractive index	1.47	1.49–1.50	1.46–1.49	1.46–1.49	1.49–1.51
water absorption (%) for a 3 mm film after 24 h	0.8–1.8	2.0–7.0	1.2–2.8	0.9–2.2	1.0–2.0
effect of sunlight	small	strong	small	small	very strong

Except for hydroxyethylcellulose, the cellulose ethers are obtained by condensing sodium-cellulose with the corresponding alkyl chlorides or sulphates /2/ as follows:

$$\text{sodium-cellulose} \xrightarrow[50-100^\circ C]{m\ CH_3Cl\ or\ (CH_3)_2SO_4} R(OCH_3)_m(OH)_{3-m}$$

$$\text{methylcellulose}$$
$$(m = 1.7 \text{ to } 1.9)$$

$$\text{sodium-cellulose} \xrightarrow[90-150^\circ C]{m\ C_2H_5Cl\ or\ (C_2H_5)_2SO_4} R(OC_2H_5)_m(OH)_{3-m}$$

$$\text{ethylcellulose}$$
$$(m = 2.4 \text{ to } 2.5)$$

$$\text{sodium-cellulose} \xrightarrow[40-50^\circ C]{m\ ClCH_2COONa} R(OCH_2COONa)_m(OH)_{3-m}$$

$$\text{carboxymethylcellulose}$$
$$(m = 0.7 \text{ to } 1.4)$$

$$R(OH)_3 + mn\ CH_2\!\!-\!\!CH_2 \xrightarrow[\substack{\text{solvent} \\ \text{(alcohol, ketone)}}]{30^\circ C} R(O(CH_2CH_2O)_n H)_m (OH)_{3-m}$$

$$\text{hydroxyethylcellulose}$$
$$(n = 1-3,\ m = 2.0-2.5)$$

4.1.2 PROTEIN DERIVATIVES

Proteins are components of living cells and essential to life /4/. The international term "protein" was used first by the German chemist Mulder at the suggestion of Berzelius in 1838.

The authors of textbooks and monographs dealing with proteins tried to classify these substances in some way /2,4,5/, following the purpose of the publication, so it is difficult to adopt a uniform system.

For our purpose it is suitable to classify proteins according to molecular shape, thus in two groups, fibrous and globular.

The fibrous proteins form the structural material of animals just as cellulose serves in the plant kingdom. Examples are keratin (in hair, nails, hooves and horny scales, feathers), fibroin (in natural silk), collagen (in connective tissues), and myosin (in muscular tissues). A characteristic feature of all fibriform proteins is insolubility in water. On boiling in water, collagen forms soluble gelatine.

The globular proteins, in contrast, are soluble both in water and in aqueous solutions of acids and bases. Examples are albumin (in egg-white), casein (in milk), and zein (in maize).

In packaging the most important are the globular proteins (cf. Section 1.1.6).

Casein is a classical adhesive for paper, wood, and textiles, and a basis for plastics processed by compression moulding or fibre spinning. It is obtained industrially by precipitation from buttermilk /6/, commonly by organic or inorganic acids (e.g. lactic, hydrochloric, or sulphuric). Precipitation at 35-37°C is followed by coagulation at 10°C. After decantation, the casein is washed with warm water (45°C),then with cold water, and finally dried to give a powdery product soluble in aqueous alkalis (as the caseinate). Casein can be insolubilized by heat treatment (150°C).

For compression moulding casein is wetted with water (20-25%) and mixed with plasticizers, fillers, and pigments, and cured with formaldehyde. The final plastic is odourless and non-flammable. It resists alcohols, diethyl ether, benzene, oils, and fats, but swells in water.

For the production of fibres, casein was dissolved in aqueous sodium carbonate at 24°C and left to stand for 48 h (ripening). The solution was then processed by wet spinning, similar to that for viscose, in which

it extruded through jets into a precipitation bath of
aqueous sulphuric acid and aluminium sulphate or sodium
chloride. The resulting fibres, whose commercial name
was *Lanital*, were usually mixed with wool (25-50%).
Lanital absorbs as much as 15% water.

Analogous processes were used for fibres from another
globular protein, arachin, obtained from peanuts after
removal of the oil. The commercial name of this textile
was *Ardil*. Similarly zein yielded *Vicara*, and soya
proteins *Silkoon* /6/.

4.1.3 DERIVATIVES OF NATURAL RUBBER

Natural rubber is found in several plants in the form
of milky liquid called latex, but the only plant of
economic importance is the tree *Hevea brasiliensis* of
the *Spurge genus* (*O.N. Euphrobiaceae*) /7/.

4.1.3.1 R u b b e r   H y d r o c h l o r i d e

The hydrochloride of natural rubber is produced by
introducing dry hydrogen chloride into a solution of
plasticized rubber at about $10^{\circ}C$ /7/. The required
degree of addition is reached after 6 hours with continuous agitation, the product is allowed to stand 12 h
(ripening), and cast on a continuous belt or paper.
Both the films and layers on paper show high flexibility
and impermeability to moisture and are suitable for
packing food products. The films of natural rubber
hydrochloride contain 30% chlorine and are known under
the commercial name *Pliofilm*.

Rubber hydrochloride is resistant to dilute bases
and acids. It is soluble in aromatic and chlorinated
hydrocarbons but insoluble in ethanol, petrol, diethyl
ether, and acetone. It is used also in bonding rubber
to metals.

4.1.3.2 T h e r m o p r e n e s (C y c l i z e d R u b b e r)

Rubber treated with sulphuric or chlorosulphuric acid, p-toluenesulphonyl chloride, or chlorocarbonic acid yields materials similar to balata[1], shellac[2], or ebonite[3]. These substances, generally called thermoprenes, have the same empirical formula as rubber, but their unsaturation is lower. They are presumed to have cyclic structure.

Depending on the reagent and the time reaction, various derivatives are formed ranging from the materials used in coating composition, printing inks or impregnants, to thermoplastics for hot forming. Cyclized natural rubber also forms the basis of some rubber-to-metal bonding agents.

4.2. SYNTHETIC POLYMERS

4.2.1 POLYOLEFINS AND FLUORINATED PLASTICS

Crude oil and natural gas are raw materials whose processing provides the cheapest chemicals for polymer production, viz. ethylene and propylene. These are monomers

---

[1] Balata is a natural polymer of isoprene structurally similar to gutta-percha, i.e. trans-1,4-polyisoprene.

[2] A natural resin produced as a secretion by a tropical (lac)insect. It is a complex mixture of cross-linked polyesters based on hydroxycarboxylic acids, predominantly 9,10,16-trihydroxyhexadecanoic acid /1/.

[3] Ebonite or hard rubber is a hard horny material of low ductility, flexibility, and elasticity resembling phenoplasts. It is produced by vulcanization of rubber with a large amount of sulphur (usually above 30 %).

which also serve as starting materials for a wide
range of further monomers whose polymerization and
copolymerization produce a variety of polymeric materials. As monomers they yield polyolefins whose span
of properties is considerable, ranging from linear
polymers of high crystallinity and high melting point
to completely amorphous rubberlike materials. In this
chapter we will only deal with those which have found
applications in packaging and do not fall into the
group of synthetic rubbers which are dealt with later
(e.g. butyl rubber, Section 4.2.8).

4.2.1.1. Polyethylene

Polyethylene, already important in packaging, is
strictly the homopolymer of ethylene, but by common
usage is applied to copolymers in which ethylene
predominates. The physico-mechanical properties of
these materials depend considerably on molecular
weight, on morphology, and crystallinity. These
characteristics for polyethylene depend primarily on
the method of production.

It was generally accepted that ethylene can only
be polymerized to highmolecular weight polymer under
high pressure /8/ until Ziegler's discovery of low
pressure polymerization in 1955 /9/. The low pressure
polymerization of ethylene using catalysts prepared
from alkylaluminium derivatives and titanium(IV)
chloride and its industrial application, due especially
to Natta /10/, are undoubtedly the most significant
successes of modern polymer chemistry. The two scientists
were awarded the Nobel Prize (Karel Ziegler in 1964 and
Giulio Natta in 1965).

High activity Ziegler catalysts and various polymerization conditions permit polyethylene production by
both batch and continuous methods /1,8/.

Almost simultaneous with Ziegler's discovery, a
number of patents were published by the American firms
Phillips Petroleum Co. and Standard Oil Co. describing
new ways of producing polyethylene under very mild
conditions - pressure below 7 MPa and temperature below
300°C in a hydrocarbon solvent medium with the use
of new catalysts /11-15/. The main representatives of
these new catalysts are chromium(VI) oxides on alumina
or silica-alumina carriers and compared with Ziegler's
catalysts, they have two special advantages: easy
preparation and revivability.

Both low-pressure polymerization with Ziegler
catalysts and medium-pressure polymerization with metal
oxide catalysts give linear polymers. Both use a
hydrocarbon solvent and require the absence of water
and polar compounds and removal of catalyst residues
from the polymer (for good dielectric properties), and
in the both cases the polymerization lasts several hours.
For these reasons and because each of the two methods
has a number of variants, it is not possible to state
which is more favourable technologically or with regard
to the polymer produced. Table 25 compares the properties
of polyethylenes produced by different methods.

Two basic types of polymers can be differentiated
in the case of polyethylene by various criteria. At
first the two types were distinguished as the high-
pressure and the low-pressure polyethylenes. Later it
was possible to prepare at high pressure products
exhibiting the properties of the original low-pressure
polymers, and vice versa. Therefore, other criteria
were sought and the polymers began to be classified
according to density, hardness, branching etc. It was,
however, impossible to determine the boundary between
the basic types with sufficient accuracy, because the
selected properties change continuously and not
abruptly. Therefore, the determination of the boundary

Table 25. Some properties of polyethylene (PE) produced by various ways /8/: (A) — DTNH high-pressure PE, (B) — Super Dylan low-pressure PE (Ziegler), (C) — Marlex 50 medium-pressure PE (Philips)

Properties	A	B	C
density (g cm⁻³)	0.91–0.92	0.93–0.95	0.96
number of double bonds per 1000 carbon atoms	0.6	0.7	1.5
content of double bonds of various types (%): RCH=CH₂	15	43	1
RR´C=CH₂	68	32	5
RCH=CHR´	17	25	<1.5
number of —CH₃ groups per 1000 carbon atoms: total	21.5	3	<1.5
—CH₃ end groups	4.6	2	
CH₃ groups at branching positions	2.5		
—CH₃ groups at branching positions	14.4	1	<1
—C₂H₅ groups at branching positions	64	87	93
crystalline portion (%): from X-ray analysis	65	84	93
from NMR analysis	19	36	39
diameter of the crystallites (nm)			

was always a matter of convention. As all the properties
can be derived from the structure of the polymer, it
appeared most reasonable to consider the branching of
macromolecules to be the basic criterion. So, according
to this criterion we can differentiate the linear and
the branched types of polyethylene. The linear type
(lPE) is still denoted more frequently as high-density
polyethylene (HDPE). The branched type (bPE) as low-
density polyethylene (LDPE).

Besides morphology and crystallinity, and the density connected therewith, properties of the polymer
are influenced by molecular weight. In Fig. 41 it can
be seen that polyethylene can exhibit a number of
different properties.

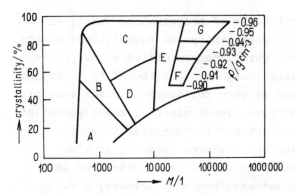

Fig. 41. Effect of relative molecular weight on
crystallinity of polyethylene /1,6/

A - vaseline, B - soft polyolefins, C - brittle
polyolefins, D - tough polyolefins, E - high molecular
polyolefins, F - most usual polymers, G - linear polymers

The structure of polyethylene is determined by the
chain segments which form crystalline regions whose
number and size decrease with increasing branching of
the chains. The polymethylene prepared from diazomethane
is a strictly linear polymer, its regular structure
only disturbed by end groups. Table 26 compares the
properties of polymethylene and branched polyethylene.

Table 26. Some properties of polymethylene and of branched polyethylene /8/

Properties	Poly-methylene	Branched polyethylene
degree of branching, number of $-CH_3$ groups per 100 $-CH_2-$ groups	0	3
density (g cm^{-3})	0.98	0.92
crystalline portion at $20^{o}C$ (%)	95	60
melting point (^{o}C)	138	110
elongation at break (%)	10	up to 400
cloud point (5% solution in xylene) (^{o}C)	90	70

The chemical resistance of polyethylene increases with increasing of crystallinity. At normal temperatures, polyethylene resists water, non-oxidizing acids, bases, salts, and their solutions, but is attacked by oxidizing agents. Polyethylene is very resistant to polar solvents like alcohols, aldehydes, lower glycols, and glycerol at normal temperatures, but not to non-polar solvents especially at higher temperatures, when it is also attacked by some polar solvents; it is dissolved in boiling tetrachloromethane, benzene, toluene, etc.

Polyethylene possesses excellent resistance to low temperatures and becomes brittle only at $-120^{o}C$. The articles produced from high density polyethylene retain shape up to $100^{o}C$, and can be sterilized with boiling water. Low density polyethylene softens at somewhat lower temperatures.

Some commercial names of polyethylene in Europe are: *Hostalen* (Hoechst, FRG), *Vestolen A* (Hüls, FRG), *Baylen* (Bayer, FRG), *Alkathene* (I.C.I., Great Britain), *Bralen* (Slovnaft, ČSSR), *Liten* (Chemopetrol, ČSSR).

Because of their relatively economic production, performance characteristics and low density (Table 26), polyethylene films play an important role in packaging. Polyethylene containers for milk are now widely accepted /16, 17/.

The combination of high permeability to oxygen and carbon dioxide and low permeability to water vapour makes polyethylene suitable for packing goods which require "breathing" without dehydration, e.g., fresh fruit and vegetables. On the other hand, permeability to aromatics and oils prevents wider application of polyethylene for packing toilet articles and perfumes.

In thick sections, polyethylene is white and opaque, but may be transparent in thin films. The transparency depends on crystallinity, i.e. on branching of the macromolecules and rate of cooling of the melt. It is improved with more branching or faster cooling, and slightly with increasing molecular weight.

4.2.1.1.1 *Copolymers of Ethylene with Other Monomers*

Radical polymerization at high pressures and temperatures is used for the production of copolymers of ethylene with polar monomers, as e.g. vinyl acetate and ethyl acrylate /2/. With decreasing ethylene content the crystallinity of the material decreases and its flexibility at room temperature increases.

Two types of ethylene copolymer find application in the field of packaging.

Ethylene-vinyl acetate copolymers are used for food packaging especially as shrinkable films. They are marketed commercially as *Elvax* (Du Pont, USA), *Ultrathene UE* (National Distillers and Chemical Corp., USA), *Montothene G* (Monsanto, USA), *Alkathene VJF 502* (I.C.I., Great Britain), and *Levaprene* (Bayer, FRG) /18-20/.

Blends of ethylene-vinyl acetate copolymers with
low density polyethylene or with polypropylene provide
intermediate properties.

Copolymers of ethylene with vinylcarboxylic acids,
e.g. methacrylic acid, were introduced by Du Pont
under the generic name *Surlyn*, and called ionomers.
The introduction of mono- or di-valent metals (which
form intermolecularionic bonds) inhibits crystallization and produces a transparent polymer. This ionic
cross-linking imparts toughness, elasticity, and low
solubility, but, in contrast to chemical cross-linking,
it preserves thermoplasticity, so that extrusion and
injection moulding processes can be used. Three basic
types of Surlyn differ in their workability and are
suitable for films and coatings, for injection moulding,
and for electrotechnical articles (Table 27).

Ionomers also exhibit better adhesion to certain
other materials, and are useful in lamination or
printing (in contrast to polyethylene) or in combining
with metal foils. The ionomer melt contains labile
cross-links, so ionomer films are of elastomeric character and can be vacuum-shaped more easily than polyethylene films.

4.2.1.2 P o l y p r o p y l e n e

The polymerization of propylene can lead to the atactic,
syndiotactic, or isotactic versions of polymer /21/.
Atactic polypropylene is a soft material which can be
used alone only with difficulties, but the isotactic
polymer obtained by stereospecific polymerization with
the catalysts of the Ziegler-Natta type is a highly
crystalline plastic with melting point about 170°C.

Polypropylene resists aggressive chemicals better
than polyethylene especially at increased temperatures
/22/, but is also dissolved in aromatic and chlorinated

Table 27. Properties of the ionomer Surlyn A [a]/3/ of various types

Properties	A	B	C
density (g cm^{-3})	0.94	0.94	0.96
melting index according to ASTM-D-1238	0.5-1.2	2-10	0.3
water absorption (%)	1.4	1.4	0.1
tensile strength (MPa)	35	28	35
elongation at break (%)	250/540[b]	450	400
impact strength (kJ m^{-2})	740	850	360
inner insulating resistance (m)	10^{18}	10^{18}	10^{19}
relative permitivity at 60 Hz frequency	2.4	2.4	2.5
factor of insulating material at 60 Hz frequency	0.003	0.003	0.001
dielectric strength (kV m^{-1})	25.5	25.5	28.0

A - film, B - moulding types, C - electrotechnical types
[a] The properties depend on the molecular weight, crystallinity degree, and type of the metal ion.
[b] Longitudinal/transversal

hydrocarbons above 80°C. It absorbs mineral and vegetable oils only slightly and without impairment of its mechanical properties. Its main physical properties are given in Table 28.

Polypropylene is processed in similar ways as polyethylene (blow-moulding, extrusion, injection, etc.). Some commercial names are *Vestolen* (Hüls, FRG), *Moplen* (Montedison, Italy), *Daplen* (Danubia, RSR), *Tatren* (Slovnaft, ČSSR), *Mosten* (Chemopetrol, ČSSR).

Polypropylene is used in packaging as films, binding tapes, containers, and caps.

Table 28. Physical properties of polypropylene /3/

density	$0.90 - 0.91 \text{ g cm}^{-3}$
flexural strength	43 MPa
tensile strength	30 MPa
elongation at break	700%
modulus of elasticity	1500 MPa
impact strength	$10 - 15 \text{ kJ m}^{-2}$
compressive strength	110 MPa
heat resistance (Martens)	$30 - 40 \text{ }^{\circ}\text{C}$
heat resistance (Vicat)	$30 - 90 \text{ }^{\circ}\text{C}$
linear thermal expansion in the interval 30-120°C	$(1.6 - 1.8) \times 10^{-4} \text{ K}^{-1}$
volume expansion in the interval 30-120°C	$(4.8 - 6.0) \times 10^{-4} \text{ K}^{-1}$
thermal conductivity at 20°C	$0.30 - 0.33 \text{ kW m}^{-1}\text{K}^{-1}$
specific heat at 20°C	$1.7 \text{ kJ kg}^{-1} \text{ K}^{-1}$
relative permitivity at 10^6 Hz frequency	2.0
factor of insulating material at 10^6 Hz frequency	0.0005
dielectric strength (thickness of 0.1 mm)	$70 - 80 \text{ kV mm}^{-1}$
surface insulation resistance	$5.10^{11} \text{ } \Omega$
specific volume resistance	$10^{14} - 10^{16} \text{ } \Omega\text{m}$
water absorption after 7 days	0.1%
permanent thermal stability	$100 ^{\circ}\text{C}$

Special polypropylenes in which isotactic and atactic blocks alternate in the macromolecular chains have been made and exhibit elastomeric properties.

4.2.1.3 Other Stereoregular Polyolefins

With Ziegler catalysts it is possible to prepare a variety of stereoregular polymers /2/. Industrial

products include poly(l-butene) /21/ and isotactic
poly(4-methyl-l-pentene).

4.2.2 POLYSTYRENE AND STYRENE COPOLYMERS

4.2.2.1 P o l y s t y r e n e

Styrene polymerization was described in 1839 /23/, but
its industrialization came a hundred years later.
Nowadays, polystyrene ranks with polyolefins and
poly(vinyl chloride) in usage.

The monomer may be produced by alkylation of benzene
and subsequent dehydrogenation, by oxidation of ethyl-
benzene /3/ or by pyrolysis of natural gases or crude
oil condensation of acetylene with benzene /24/.

Styrene can be polymerized in four ways /21/:

a) Bulk polymerization, for which continuous operation
is most advantageous; the pre-polymerized styrene passes
through a high tower divided into several zones of
different temperatures from $100^{o}C$ to $220^{o}C$ /25/. The
polystyrene is obtained in the form of a granulate.

b) Solution polymerization which produces polymers
of lower molecular weight which are more easily dissolved.

c) Emulsion polymerization which is energetically
favourable.

d) Suspension polymerization, which gives a high-
quality product.

Polystyrene has a density of 1.05 g cm^{-3}, is hard,
brittle, and transparent: 90% of visible light is
transmitted. It is resistant to water, solutions of
non-oxidizing salts, acids, and bases, to alcohols,
and mineral oils. It is dissolved in aromatic hydrocarbons, petrol, chlorinated hydrocarbons, and esters.
It is easily dyed and printed, and has excellent
dielectric properties. Commercial European types
include: *Edistir* (Montedison, Italy), *Polystyrol*

(BASF, FRG), *Trolitul* (Dynamit Nobel, FRG), *Vestyron* (Hüls, FRG), *Distrene* (Distrene Ltd., Great Britain), *Krasten* (Chemopetrol, ČSSR).

Due to its easy processability (injection moulding) and brittleness, polystyrene is mainly used for non-exacting consumer goods like toys, cups, various decorative articles, etc. Foamed polystyrene has found wide application in the building industry and in packaging. Polystyrene films excel in lustre, transparency, water resistance, and dielectric properties, and the oriented films compete with polyethylene and poly(vinyl chloride) films.

4.2.2.2 S t y r e n e   C o p o l y m e r s

A significant group are binary and ternary copolymers of styrene with acrylonitrile, butadiene, or methyl methacrylate, developed to solve the problem of poly-styrene brittleness which limited its applications. The most widely used copolymer of styrene grafted on butadiene-styrene copolymer is superior to a blend in processability and quality (cf. Section 2.4). The toughened types lack the transparency of polystyrene, but the toughness is several times higher, the other physical properties being maintained. Table 29 lists the physical properties of Czechoslovak types of standard and the toughened polystyrenes. Injection moulding, extrusion, and vacuum forming are used to produce various vessels, containers, dishes, drawers, etc.

The binary copolymer of styrene with acrylonitrile (SAN) is harder than the toughened polystyrenes, has excellent lustre and transparency, but does not match the toughened polystyrenes in notched impact strength. On the other hand, it has the best chemical resistance of all polystyrene materials. It is known as *Kostil*

Table 29. Physical properties of Czechoslovak standard polystyrenes Krasten 1,2 and Krasten 1,4 and tough polystyrenes Krasten 2,2 and Krasten 2,3 /3/

Properties	1,2	1,4	2,2	2,3
density (g cm^{-3})	1.05	1.05	1.05	1.05
tensile strength (MPa)	40	38	30	35
flexural strength (MPa)	90-120	85-110	70-90	80-100
impact strength (kJ m^{-2})	20	16	25-55	40-60
compressive strength (MPa)	120	120	100	100
heat resistance (Vicat) ($^\circ$C)	86	86	85	90
the minimum temperature ($^\circ$C) of permanent use	-10	-10	-30	-30
the maximum temperature ($^\circ$C) of permanent use	50	50	60	65
surface insulation resistance (Ω)	10^{14}	10^{14}	10^{14}	10^{14}
specific volume resistance (Ωm)	10^{13}	10^{13}	10^{13}	10^{13}
relative permitivity at 10^6 Hz frequency	2.3	2.3	2.3	2.3
water absorption (%) after 7 days	0.1	0.1	0.2	0.2

(Montedison, Italy), *Lacqsan* (Aquitaine-Total Organico, France), *Luran* (BASF, FRG), and *Plastik SNP*
(U.S.S.R.).

The copolymer of styrene with methyl methacrylate
has better resistance than polystyrene to weathering,
higher temperatures and solvents. It is mainly
processed by injection moulding and is produced in
U.S.S.R. as *Plastik MS*.

The terpolymers of styrene with butadiene and
acrylonitrile (ABS) and with butadiene and methacrylate
(MBS) are commercially important.

ABS resembles the toughened polystyrene types in
being a two-phase system composed of rubber particles
dispersed in a glassy matrix /21/. It has excellent
/3/ toughness and resistance to cold flow, and to
elevated temperatures. It is easily electroplasted
and its applications are very broad. In packaging it
is used for containers and pallets. As a composite
material, reinforced with glass fibres, it has even
better mechanical properties. Commercial names include:
Editer (Montedison, Italy), *Cryolac* (Marbon Chemical,
USA),   *Lacqran* (Aquitaine-Total Organico, France),
Novodur (Bayer, FRG), *Sicoflex* (Mazzuchelli, Italy),
Terluran (BASF, FRG), *Forsan* (Chemopetrol, ČSSR).

Compared with ABS, the MBS types are transparent,
of good lustre and higher surface hardness, other
properties being somewhat inferior. In packaging, MBS
can be used for bottles and transparent vessels; it
can also substitute glass for pipelines in the food
industry.

4.2.3 VINYL POLYMERS

The vinyl polymers are important plastics obtained by
polymerization of compounds of general formula
$CH_2{=}CH{-}R$. Therefore, their macromolecular chains have

similar structures:

$$\ldots\ -CH_2-\underset{R}{CH}-CH_2-\underset{R}{CH}-CH_2-\underset{R}{CH}-\ \ldots$$

where R is Cl (poly(vinyl chloride)), OH (poly(vinyl alcohol)), $OCOCH_3$ (poly(vinyl acetate)), etc.

4.2.3.1 Poly(Vinyl Chloride) (PVC)

Poly(vinyl chloride) is the most significant vinyl polymer and widespread due to relatively cheap production of vinyl chloride monomer and the useful properties of the polymer. It is easy to process by calendering, extrusion, injection moulding, blowing, vacuum forming, etc., can be gelled with various organic and inorganic esters and plasticized to different degrees, and demonstrates considerable chemical resistance.

Suspension, emulsion, and bulk polymerization techniques are used industrially (see Section 2.3). As the polymer is insoluble in monomer, all the polymerization methods give more or less porous powdery products.

Emulsion polymerization gives a latex of 40-50% solids content which may be spray-dried to obtain the polymer. The paste-forming polymer powder obtained in this way is mainly used in combination with plasticizers for pastes which are applied by dipping, casting, or spreading (cf. Sections 5.2.1 to 5.2.3).

The major part of world production of poly(vinyl chloride) is based on suspension polymerization (cf. Section 2.2.1) and gives products which contain much smaller amounts of auxiliary materials.

During the last twenty years bulk polymerization has become industrially established, because it produces very pure porous poly(vinyl chloride) powder.

The molecular weight of individual types of poly(vinyl chloride) varies within the limits 10^4 to 10^5. Commercial types of PVC are characterized by the K values[1] /3/ from 55 to 85 and are indicative of molecular weight.

With increasing molecular weight the mechanical properties of the polymer and its thermal resistance are improved, but its processability is impaired. The higher-molecular products can be plasticized to a high degree. Highly plasticized products, like dipped articles, are manufactured from polymers with K above 70, whereas polymers with K below 60 are used for unplasticized goods, e.g. bottles.

Freedom from residues of polymerization additives is greatest with bulk polymerization, being followed by suspension polymerization, and the largest amounts of impurities (especially emulsifiers) are present in emulsion polymer. The bulk and suspension types have thus lower water absorption, better electrochemical properties, and can be used with suitable stabilizers for manufacturing transparent articles. On the other hand, emulsifier residues act as lubricants in processing /26/.

1/ *The K value is a conventionally accepted quantity which is calculated from the dependence of viscosity of the solution of a linear polymer on its concentration and is thus closely related to its viscosity-average molecular weight. The K constant is related to the Fikentscher k constant (1000k = K) which is defined by the relation*

$$\log \frac{\eta_c}{\eta_0} = \left(\frac{75k^2}{1 + 1{,}5kc} + k \right)c$$

where η_c *and* η_0 *stand for the viscosities of the polymer solution of concentration c(given in g* dl^{-1}*) and of the solvent, respectively.*

Poly(vinyl chloride) is used either without plasticizers (but with stabilizers, lubricants, and
modifiers), to give hard articles, or with plasticizers to give semi-rigid to highly elastic products.

As plasticizers can modify the mechanical properties
of poly(vinyl chloride) within broad limits without
decreasing its excellent chemical resistance, this
polymer is one of the most versatile plastics. This
chapter will mention briefly its degradation and
stabilization /27/ (see also Sections 3.4.3-3.4.5).

Poly(vinyl chloride) is not depolymerized by heat,
but changed chemically, predominantly by loss of
hydrogen chloride; degradation by light is also
accompanied by loss of water, carbon dioxide, carbon
monoxide, and hydrogen. During thermal degradation
polymer turns yellow, red, brown, and finally black.
The intensity of the initial colouration depends on
traces of polymerization initiator or other additives;
later the colour results from the molecular structures
formed by degradation. Degradation is accelerated by
oxygen, polymerization initiators, and various multivalent metal ions. Photochemical degradation follows
a different course from that of thermal degradation.
The latter is accompanied by a distinct change in
colour but small changes in mechanical properties,
whereas the former predominantly decreases the mechanical values. PVC discoloured thermally is decolourized
by ultraviolet radiation. It is clear from these brief
comments that poly(vinyl chloride) degradation is a
complex set of processes affected by many factors.

Poly(vinyl chloride) exhibits high chemical resistance
especially to non-oxidizing acids. Thus it embrittles in
sulphuric acid at concentrations above 50%. Plasticization
(above 30%) decreases the resistance markedly. Concentrated organic acids (e.g. formic or acetic) cause
swelling of PVC above 40°C but alkali hydroxides have

little effect, much less than acids or water. The water
absorbed by unplasticized PVC is about 1% at 25°C,
increasing with the temperature, and causes a milky
turbidity; it is influenced predominantly by the content
of residual emulsifiers. Aliphatic hydrocarbons and
alcohols do not attack PVC, whereas ethyl chloride,
toluene, styrene, benzene, chlorobenzene, and ketones
cause swelling or dissolution. Tetrahydrofurane or mixtures of acetone with tetrachloromethane or with carbon
disulphide are good solvents for PVC. Its resistance
to solvents is lowered by plasticizers, which are extracted up to an equilibrium content. Microorganisms do not
affect pure PVC.

The stabilization of PVC against heat and light is
an important aspect due to the wide application of this
polymer. The subject is dealt with in Sections 3.4.3
to 3.4.5.

The physical properties of PVC depend partly on the
method of production, and Table 30 gives some characteristic data on individual types produced industrially.

Table 30. Physico-chemical properties of industrially produced
poly(vinyl chloride) (PVC) /28/

Properties	Emulsion type PVC non-modified	precipitated	Suspension PVC	Block PVC
density (g cm^{-3})	1.39 ± 0.01	1.40	1.40	1.40
ash content (%)	0.5 to 2.0	0.05 to 0.5	0.01 to 0.1	0.03
chlorine content (%)	55 ± 1.0	56 ± 0.5	56.2 ± 0.3	56.2 ± 0.3
content of extractable portions (%)	2.0 to 5.0	0.2 to 2.0	0.1 to 1.0	0.1 to 1.0
water absorption (%)	<100	<10	<1	<1
thermal stability according to Meixner (min)	35 to 70	35 to 70	8 to 20	8 to 20

Poly(vinyl chloride) satisfies a number of demands
made of packaging materials. Unplasticized PVC is
highly impermeable to oxygen, nitrogen, carbon dioxide,
but its permeability to water vapour is of an order of
magnitude greater than that of high density polyethylene,
about twice that of low density polyethylene and slightly
higher than that of poly(ethylene terephthalate) (see
Table 17 on p.186). However, the gas barrier properties
of PVC are strongly impaired by plasticization which
is essential for many applications in packaging.

Health and safety acceptance of poly(vinyl chloride)
depends on the residual monomer content and on the
plasticizers, stabilizers, and fillers used.

Poly(vinyl chloride) has good mechanical properties,
especially tensile strength, and low flammability.

Its weldability by direct heat and especially by high
frequency heating makes many packaging processes
economically attractive.

Also significant for packaging is the good resistance
of PVC to many chemicals, its good printability, and
resistance to microorganisms. However, this microbiological resistance is impaired by additives and impurities,
like plasticizers, stabilizers, emulsifiers, and surface
contamination; therefore, suspension PVC is more resistant
than emulsion PVC, while plasticizers and stabilizers
act according to their chemical structure (cf. Sections
3.4.3 and 7.3).

Various bottles are made of poly(vinyl chloride) by
blow-moulding, while injection moulding is used for
cups and caps. Such bottles are unbreakable, attractive,
and much lighter than those of glass.

Plasticized PVC film in various formats is used to
manufacture covers, envelopes and wrappers very efficiently with high frequency welding.

Shrinkable caps of PVC are used for protecting bottle
stoppers and shrink films and stretch films are used for

a variety of packaging applications.

Poly(vinyl chloride) is produced in Europe under
the following commercial names: *Solvic* (Solvay et Cie,
France), *Vestolit* (Chemische Werke Hüls, FRG), *Geon*
(I.C.I., Great Britain), *Lacqvyl* (Aquitaine-Total-Organico, France), *Slovinyl* (ČSSR), *Neralit* (Chemopetrol, ČSSR).

4.2.3.1.1 Copolymers of Vinyl Chloride

The development of vinyl polymers was undoubtedly
accelerated by the discovery of copolymerization.
Vinyl chloride can be copolymerized with a large
number of other monomers; some of them, e.g. vinyl
acetate, vinylidene chloride, and acrylonitrile, may
be copolymerized in any ratios, whereas others, e.g.
esters of maleic, fumaric, acrylic, and methacrylic
acids, styrene, and ethylene are limited in this
respect. No copolymerization occurs with butadiene /28/.

4.2.3.1.2 Vinyl Chloride-Vinyl Acetate Copolymers

These copolymers are the oldest and best-known copolymers
of vinyl chloride. Their properties depend on the ratio
of the two components and on their molecular weight.
The copolymers with 80-95% vinyl chloride are of most
technical importance, but copolymers with less vinyl
chloride are used for coating compositions.

Compared with the homopolymer the copolymers of
vinyl chloride are characterized by a lower softening
point, higher transparency, and better solubility in
common polar solvents, the solubility improving with
more vinyl acetate. This is applied in the solution
polymerization of copolymers suitable for varnishes,
when the solvents used constitute part of the varnish
composition, e.g. acetone, ethyl acetate, butyl acetate,

ethyl methyl ketone, etc.; polymerization is initiated
by organic peroxides, of which dibenzoyl peroxide,
acetyl benzoyl peroxide, distearoyl peroxide, are
suitable /28/.

In suspension polymerization macromolecular compounds
like poly(vinyl alcohol), or the sodium salt of styrene-
maleic anhydride copolymer, are used as dispersants
and initiation is accomplished with organic peroxides.

The copolymers of vinyl chloride with vinyl acetate
have broad application in industry, especially for:

1. film and plate materials and gramophone records,
2. synthetic fibres (e.g. *Vinylon*),
3. varnishes and coatings,
4. as latices for impregnating textiles and paper.

4.2.3.1.3 *Vinyl Chloride-Vinylidene Chloride Copolymers*

Vinyl chloride and vinylidene chloride can be copolymerized in any ratio, under conditions similar to those
for vinyl chloride, usually by the emulsion or the
suspension methods. The properties of the products depend
on the proportions of the two monomeric components in
the macromolecule.

Both copolymers with low vinylidene chloride content
(max. 10%), e.g. *Geon 202*, and those with high content
(min. 80%), e.g. *Saran B*, *Diurit*, are useful: in the
former, vinylidene chloride improves flow properties,
whereas in the latter it increases the softening point
(up to 140°C).

In copolymers containing 40-60% vinylidene chloride
the plasticizing effect of vinylidene chloride reaches
its maximum, as in the 50% concentrated latices *Diofan
210 D*, *Vinitex* /28/. These are useful as coating and
impregnation materials on paper, leather, and textiles,
improving resistance to water, oils, and common chemicals
and impermeability to fragrant substances. Optimum

performance requires a short heating to 100°C.

The copolymers with vinylidene chloride content are characterized by crystallinity at room temperature. Such copolymers of crystalline character exhibit high impermeability to oxygen, water vapour, and carbon dioxide, and are therefore used in packaging for shrinkable films, hoses, and bags /29/, and for coating other polymers. The crystallinity decreases with increasing content of vinyl chloride until at about 50% the copolymer is practically non-crystalline.

4.2.3.1.4 *Vinyl Chloride-Styrene Copolymers*

The production of vinyl chloride-styrene copolymers was started only recently, because it was difficult to develop a homogeneous polymer. Copolymers containing up to 10% styrene have improved processability over poly(vinyl chloride), and are used for films, and dimensionally stable floor covering. They appeared on the market under the name *Gepolit*, and in contrast to the homo-polymers, they are soluble in acetone, acetic esters, and benzene /28/. Emulsion copolymerization, and initiation usually by oxidation-reduction systems, is used to obtain the desired styrene content in the copolymers.

4.2.3.1.5 *Copolymers of Vinyl Chloride and Acrylic*
Esters

The copolymers of vinyl chloride with esters of acrylic acid can be prepared by emulsion copolymerization with initiation by hydrogen peroxide or other suitable peroxides /28/. The preparation of homogeneous products with vinyl chloride predominant is rather difficult.

These copolymers are more flexible and more elastic than the homopolymer of vinyl chloride and with lower softening point.

The properties depend on the ratio of the two components in the polymer and on the particular ester
used.

The copolymers, due to their high prices and
difficult preparation, are used only for special
articles.

Those containing 80% vinyl chloride and 20% methyl
methacrylate (*Igelit MP-D*) are mainly used for production of films (*Astralon*) with excellent optical
properties, resistant to moisture, and with negligible
linear expansion. Transparent tubes were also used
as pipelines for milk, wine, etc. in the food industry
/25/.

4.2.3.1.6 Copolymers of Vinyl Chloride (VC) with Esters of Maleic Acid

4.2.3.1.6 Copolymers of Vinyl Chloride (VC) with
Esters of Maleic Acid

These copolymers were introduced under various names,
e.g. *Pliovic A* and *Pliovic AO* (vinyl chloride-dimethyl
maleate), *Luvimal M 20* (vinyl chloride-dimethyl maleate
80:20), *Vinoflex 452* and *Vinoflex 453* (vinyl chloride-
dimethylmaleate-diethyl maleate 80:10:10). Compared
with poly(vinyl chloride), they have slightly lower
softening points but are characterized by excellent
flow properties and dimensional stability /28/. They
are particularly used for deep-drawing films.

The copolymers of vinyl chloride with esters of
maleic and fumaric acids are prepared by polymerization
in emulsion or suspension, the former usually with
oxidation-reduction initiation.

4.2.3.2 Other Vinylic Polymers

Besides homopolymers and copolymers of vinyl chloride,
the group of vinylic polymers includes further
representatives which play quite a significant role in
packaging, viz. poly(vinyl acetate), poly(vinyl alcohol),

and poly(vinyl acetals).

4.2.3.2.1 Poly(Vinyl Acetate) (PVAC)

Poly(vinyl ,acetate) is· produced mainly by emulsion or
suspension polymerizations. Polymerization in solvents
(ethanol, benzene, toluene) gives polymers of lower
molecular weight depending on the solvent /25/.
Poly(vinyl acetate) is marketed solid (transparent
material), in solution in organic solvents, or as
aqueous dispersions – latices.

Poly(vinyl acetate) is comparatively stable even
at enhanced temperatures, it becomes soft above 80°C,
and resists oils, diethyl ether, and kerosene. It is
soluble in alcohols, benzene, toluene, ketones, acetic
esters and chlorinated hydrocarbons. It is sensitive
to water, especially at lower molecular weights,
and can absorb up to 20%. Poly(vinyl acetate) is
compatible with cellulose nitrate and chlorinated
rubber but not with cellulose acetate /6/.

Most poly(vinyl acetate) is consumed by coating
compositions and adhesives. Another application is
fabric finishing. In packaging it is used both as
wet (emulsion) and fusible adhesives /21/.

Commercial names are: *Mowilith* (Farbwerke Hoechst,
FRG), *Vinnapas* (Wacker–Chemie, FRG), *Elvacet* (Du Pont,
USA), *Vipal* (Lonza, FRG), *Vinamul* (Vinyl Products,
GB).

4.2.3.2.2. Poly(Vinyl Alcohol) (PVA)

Monomeric vinyl alcohol does not actually exist,
therefore, poly(vinyl alcohol) is produced by the hydrolysis of poly(vinyl acetate), and usually contains
a certain amount of the unhydrolyzed poly(vinyl acetate)
which acts as a modifier and imparts variable properties

to the product.

Pure poly(vinyl alcohol) is a white powder, insoluble in organic solvents, but soluble in water and more readily in warm water. Sensitivity to organic solvents depends on the vinyl acetate content.

In packaging, poly(vinyl alcohol) is used for water-soluble packages (e.g. for dyestuffs or agrochemicals), sometimes for impregnation of fabrics and paper, and exclusively as an adhesive in its own right or as component of poly(vinyl acetate) latices,

Commercial names include: *Polyviol* (Wacker-Chemie, FRG), *Moviol* (Farbwerke Hoechst, FRG), *Rhodoviol* (Rhône-Poulenc, France), *Vinol* (U.S.S.R.).

4.2.3.2.3 *Poly(Vinyl Acetals)*

Poly(vinyl alcohol) can be modified by combination with aldehydes, especially formaldehyde and butyraldehyde, to give poly(vinyl acetals) whose macromolecular chains are probably partially cross-linked which can explain the reduced thermoplasticity of these polymers.

Poly(vinyl acetals) can provide transparent packaging films which are less permeable than cellophane or cellulose acetate to both water vapour and water, their mechanical properties being very good and suitable for packing fruit and meat /6/.

The best-known commercial names are: poly(vinyl formal): *Mowital F* (Farbwerke Hoechst, FRG), *Plioform* (Wacker-Chemie, FRG), *Rhovinal* (Rhône-Poulenc, France); poly(vinyl butyral): *Mowital B* (Farbwerke Hoechst, FRG) *Rhovinal B* (Rhône-Poulenc, France).

4.2.4 ACRYLIC POLYMERS

The term acrylic polymers involves both the acrylic acid derivatives - acrylates and methacrylic acid

derivatives - methacrylates, both homopolymers and
copolymers /30/. Also polyacrylonitrile can be
classified as an acrylic polymer /25/.

The acrylic polymers are important beyond their
application as the organic glass known under the name
Perspex or *Plexiglas*. They are used extensively for
injection moulding or extrusion processes, and in the
form of aqueous dispersions for high-quality coating
compositions.

4.2.4.1 Polymerization Methods and Technologies

The esters of acrylic and methacrylic acid can be
polymerized by all the radical processes used in the
industry, according to the form required for application.
The polymers of these two series differ considerably
in their properties and applications.

The derivatives of acrylic acid are preferably
submitted to the emulsion and solution polymerizations,
whereas in the suspension and (more rarely) block
polymerizations they serve, first of all, as copolymerization monomers /30/.

4.2.4.2 Polyacrylates and Poly-methacrylates

Of the polymethacrylic esters the most widely used is
poly(methyl methacrylate) (PMMA). It resists dilute
acids, bases and salt solutions but not organic acids.
Concentrated acids hydrolyze the polymer. PMMA is
soluble in polar solvents, such as ketones, esters,
and chlorinated hydrocarbons, but resists aliphatic
hydrocarbons and mineral oils. It is subject to stress
cracking in many media including organic liquids and
aqueous solutions as well as in vapours.

Polyacrylates and polymethacrylates are thermally decomposed to a variety of products or to the monomers respectively at temperatures of 250°C.

Poly(methyl methacrylate) and other polymethacrylates and polyacrylates are stable to ultraviolet radiation and natural ageing.

The physical properties of polyacrylates depend on the chain length of the alcoholic residue in the ester substituent /3/ (Table 31).

Table 31. Comparison of some physical properties of poly(methyl α-chloroacrylate) and poly(methyl methacrylate) /3/

Properties	poly(methyl α-chloroacrylate)	poly(methyl methacrylates) Akrylon	Perspex
density (g cm^{-3})	1.49	1.10	1.19
flexural strength (VDE) (MPa)	170	82 – 120	140
tensile strength (MPa)	90	40 – 70	84
impact strength (kJ m-2)	20 – 25	10 – 14	20
hardness (VDE 5/50/10) (MPa)	280	1000 –1500	–
heat resistance (Vicat) (°C)	130 –135	105 – 120	110

The polyacrylates are mainly used in the form of dispersions as adhesives and in coating compositions.

There are broad fields of application for poly(methyl methacrylate) fabricated by injection moulding, casting, and extrusion to give tubes, extruded shapes, corrugated plates, etc. but in packaging, this polymer, being very expensive, is used only for packaging luxurious articles.

Commercial names include: polyacrylates: *Acronal* (BASF, FRG), *Plexigum* (Röhm and Haas, FRG); poly(methyl methacrylates): *Vedril* (Montedison, Italy), *Perspex* (I.C.I., Great Britain), and *Plexiglas* (Röhm and Haas, FRG), *Akrylon* and *Umacryl* (ČSSR).

4.2.5 POLYAMIDES

Polyamides (PA) are technically the most important polymers containing nitrogen in the backbone. They are characterized by the functional groups $-\overset{\displaystyle O}{\underset{\displaystyle \|}{C}}-\overset{}{\underset{\displaystyle |}{N}}-$.

Other nitrogen-containing polymers are characterized by the following functional groups:

$-O-\overset{\displaystyle }{\underset{\displaystyle O}{C}}-N-$ polyurethanes /31/

$-\overset{\displaystyle }{\underset{\displaystyle O}{C}}-\overset{}{\underset{\displaystyle |}{N}}-\overset{\displaystyle }{\underset{\displaystyle O}{C}}-$ polyimides /31/

Polyurethanes are next in importance after polyamides.

From the chemical point of view, polyamides are linear polycondensation products characterized by repeating groups, e.g. $-\left[-(CH_2)_5 CONH-\right]_n-$.

They are produced mainly by two types of reaction /1/:

1. The polycondensation of amino acids or of their cyclic amides (lactams). The technically important representatives of this group include: polycaprolactam known as nylon 6, poly(ω-aminoundecanoic acid) or nylon 11, and the polymer of lactam of 12-aminododecanoic acid called nylon 12.

2. The polycondensations of aliphatic diamines with dicarboxylic acids, e.g. hexamethylenediamine with adipic acid gives nylon 66, or hexamethylenediamine with sebacic acid gives nylon 610. The figures given characterize the starting compounds by the number of the carbon atoms in their molecules.

4.2.5.1 Production of Polyamides

Polyamides are produced industrially in the form of the granulated polymers for further processing directly from the melt for synthetic fibres and films.

They are produced either by pressure polymerization in autoclaves or by pressure-less polymerization in tubular reactors. Each of the processes can be continuous or discontinuous /32/.

4.2.5.2 Properties of Polyamides

Most polyamides crystallize easily (unless they have bulky side groups) due to the mobility of the $-CH_2-$ chain and easy formation of hydrogen bonds between neighbouring amide groups. Consequently, polyamides have a high crystalline content, and high and sharp melting points. They are practically insoluble in organic solvents except strongly polar ones (e.g. phenols). Polyamides are degraded by acids and strong bases especially at higher temperatures. Polycaprolactam is depolymerized, whereas polyhexamethyleneadipamide is degraded and cross-linked by heating.

Polyamides are white to yellowish materials with very good mechanical properties (Table 32). They are fibre-forming polymers which give easily-oriented fibres and films. They can be welded by all usual methods and easily printed.

Table 32. Some physical properties of polyamides (PA) /3/

Properties	PA 6	PA 66	PA 610
density (g cm^{-3})	1.12	1.13	1.07
melting temperature ($^{\circ}$C)	215	250 to 255	210 to 215
relative permitivity at 10^3 Hz frequency	5.0	4.2	3.5
factor of insulating material at 10^3 Hz frequency	0.04	0.03	0.025
modulus of elasticity in tension (MPa)	1300	1700	1250
water absorption (% by wt.)	11	10	4
heat resistance (Vicat) ($^{\circ}$C)	160 to 180	220 to 230	215 to 225
the maximum temperature:			
of permanent use ($^{\circ}$C)	80 to 100	80 to 100	80 to 100
of short-term use ($^{\circ}$C)	140 to 160	150 to 170	140 to 160

In packaging, the polyamides have found extensive applications especially in the form of films and injection moulded or blow-moulded products.

Polyamides are accepted for food contact, but some, especially those based on lactams, contain rather large amounts of free lactam (e.g. about 10% in nylon 6) /21/. Therefore, packaging employs mainly nylons 11 and 12 (*Rilsan, Vestamid, Grilamid, Zytel*) which are satisfactory in this respect and are impermeable to oxygen, aromatics, and particularly to water vapour. On the other hand, nylon 6 is hygroscopic absorbing about 3.5% water at normal relative humidity, and in aqueous media even up to 12%.

Polyamide dispersions are recommended for the modification of paper and can form part of adhesive compositions.

4.2.6 POLYURETHANES

Polyurethanes belong to a large group of materials which are combinations of esters and amides. They can be produced from various diisocyanates, glycols, and higher alcohols, which makes it possible to obtain very varied properties. The best-known are those made from 1,4-butanediol and 1,6-hexanediisocyanate.

It is possible to obtain linear as well as cross-linked polyurethanes which are both hard and soft. The basic properties include resistance to water, dilute acids and bases, good adhesion to metals, plastics, and other materials, and excellent abrasion resistance.

Polyurethanes very soon found extensive applications as cellular (expanded) materials, films, adhesives, and varnishes, all relevant to packaging.

Some commercial names are: elastomeric foamed material: *Moltopren* (Bayer, FRG), *Molitan* (Technoplast, ČSSR); varnishes: *Desmophen 1100, Desmodur TH* (Bayer, FRG);

adhesives: *Desmophen 800, Desmodur R* (Bayer, FRG);
casting resins and plastics: *Durethan U* (Bayer, FRG),
PU-1 (U.S.S.R.).

4.2.7 POLYESTERS

Polyesters form a large group of macromolecular
materials, many of considerable economic importance,
and include both natural products and very important
synthetic materials /33/.

The synthetic polyesters were used for a long time
exclusively for coating compositions, and they still
occupy a leading position in this area. Nowadays,
however, they play a significant role in many other
fields, including packaging. Technically important
varieties are the alkyds modified with fatty acids,
unsaturated polyester resins, poly(ethylene terephthalate), and maleic resins. The linear polyesters
which are important raw materials for packaging are
made by direct esterification of di-carboxylic acids
with di-hydric alcohols. Besides the acids themselves,
it is also possible to use their anhydrides, esters,
acyl chlorides, etc.

The polyesterification is a relatively slow reversible reaction /1, 33/:

$$n\,HO-R-OH + n\,HOOC-R'-COOH \rightleftarrows$$
$$\rightleftarrows H(O-R-OOC-R'-CO)_nOH + (2n-1)H_2O$$

The equilibrium and the molecular weight are affected
by the functionality of the starting monomers and by
equivalency of the ratio of the functional groups in
the reaction. The equilibrium can be shifted to the
right by removing water from the reaction medium.

Polyesters are characterized by molecular weight,
molecular weight distribution, melting point, glass
transition temperature, crystalline content, and the
kind and number of the functional end groups.

The ability of polyesters to crystallize is governed by their structure. Thus, polyesters lacking steric hindrance, e.g., the polyesters from terephthalic acid, crystallize very easily.

The linear polyesters can be oriented relatively easily, which considerably improves the strength, elasticity, and other mechanical properties, and is made use of in the production of packaging materials (cf. Section 5.4.1).

Tables 33 and 34 give a survey of the softening points of the polyesters based on aliphatic and aromatic dicarboxylic acids.

Table 33. Softening points, (^{o}C) of polyesters from aliphatic dicarboxylic acids and diols

Diol	Acid								
	Oxalic	Malonic	Succinic	Glutaric	Adipic	Pimelic	Suberic	Azelaic	Sebacic
1,2-ethanediol	159	-22	102	-19	47	25	63	44	72
1,2-propanediol	-	-	-2	-25	-25	-37	-41	-46	-34
1,3-propanediol	66	-25	43	35	36	41	47	46	49
1,3-butanediol	-4	-20	-15	-32	-36	-43	-	-52	-44
1,4-butanediol	103	-24	113	36	58	38	-	49	64
1,5-pentanediol	49	-26	32	22	37	39	43	46	53
1,6-hexanediol	70	-48	52	28	55	52	61	52	65
1,10-decanediol	76	29	71	55	70	63	70	67	71
1,20-eicosanediol	88	67	86	77	85	82	86	84	87
diethyleneglycol	5	-18	-11	-30	-29	-32	28	-36	44
triethyleneglycol	-14	-34	-24	-36	-39	-42	-41	-43	28
2,2-dimethyl-1,3-propanediol	111	67	86	-	37	-	17	0	26

Table 34. Melting points (^{o}C) of polyesters from aromatic dicarboxylic acids and diols /33/

Diol	Acid					
	Benzene-1,4-dicarboxylic	Benzene-1,3-dicarboxylic	Benzene-1,2-dicarboxylic	Biphenyl-4,4'-dicarboxylic	Biphenyl-3,3'-dicarboxylic	Biphenyl-2,2'-dicarboxylic
1,2-ethanediol	256	108	63	330	119	96
1,3-propanediol	217	92	–	246	76	70
1,4-butanediol	222	88	17	255	62	34
1,5-pentanediol	134	76	6	160	57	8
1,6-hexanediol	148	75	0	195	52	4
1,10-decanediol	127	34	-27	126	86	-7
1,20-eicosanediol	108	47	47	112	89	-18
diethyleneglycol	65	55	10	117	69	54
triethyleneglycol	60	60	-8	86	43	38
1,2-propanediol	106	80	45	130	93	39
1,3-butanediol	82	50	-8	125	85	36

Of the linear polyesters, the most significant is poly(ethylene terephthalate) (PET) /34, 35/. It is one of the most important plastics for packaging film (*Mylar, Melinex, Hostaphan*).

Poly(ethylene terephtalate) film is fairly low in permeability to moisture and gases, and possesses excellent strength (about 200 MPa); it retains excellent elasticity even at very low temperatures (down to -70°C), and resists temperatures up to 130°C /3/. It also possesses good resistance to fats, but is attacked by alcohols, aromatic hydrocarbons, strong acids and bases, and severely by aggressive chemicals like nitric

acid, ammonia, and amines. It exhibits good processability but is relatively high in price.

Some unsaturated polyesters, e.g. glycol maleates,
are used for glass reinforced plastics (GRP) of which
large capacity vessels and containers are made.

4.2.8 PHENOLIC RESINS AND AMINO RESINS

Phenolic resins and amino resins are the well-established
names for the synthetic resins or plastics based on
the condensation of aldehydes (most often formaldehyde)
with phenol or its homologues or an amino compound
(e.g. urea, thiourea, dicyandiamide, melamine) respectively. The plastics based on these condensation products
are hardenable resins which can be transformed by heat
or catalysts into an insoluble and infusible state.
They are used for the production of mouldings, laminates,
expanded materials, adhesives, binding agents, coatings
for modification of paper, and in various other applications.

4.2.8.1 P h e n o l i c   R e s i n s

Phenolic resins (phenol-formaldehyde resins) can be
divided into two types /36/:

Novolaks, or two-stage acid-catalysed thermoplastic
resins which are not hardened by heating alone, but
require additional formaldehyde, and

Resols, or single-stage base-catalysed resins which
are hardened by heat alone.

Table 35 gives a survey of properties of the three
novolaks most frequently produced.

The resols may be solid resins (for moulding
materials), aqueous emulsions or ethanolic solutions.

The phenolic resins are not suitable for food packing, because of their phenolic odour. In packaging
they are used only in the production of various boxes,
cups, caps and closures.

Table 35. Properties of novolaks currently produced /36/

Properties	1	2	3
softening point according to Nagel (^{o}C)	70 ± 5	95 ± 5	105 ± 5
free phenol content (%)	6 to 9	4 to 6	< 2
viscosity of 50% solution in ethanol (mPa s)	50 to 100	80 to 130	150 to 200
gelation rate at 150°C (s) (with 10% hexamethylenediamine)	110 to 160	100 to 150	60 to 110
water content (%)	1.5 to 3	1 to 2	0.5 to 1.5

4.2.8.2 Amino Resins

The amino resins can react further with heat or catalysts to give hardened materials.

These resins can be characterized by the number of components reacted together, the amino compounds used, the degree of polymerization, or the nature of the chemicals used for modification.

The chemistry and methods of production are complex, and from a wide variety of reactants a great number of different resins have been produced.

With respect to use, the amino plastics are divided as follows:

1. technical resins,

2. moulding materials,

3. laminates,

4. expanded (cellular) materials.

The technical amino resins include adhesives and binding agents, impregnation resins, resins for use in varnishes and for other purposes (e.g. casting resins). They are not cured or hardened until after application to other materials, e.g. paper for waterproof bags.

These moulding materials are more important than the phenolics for packaging because of their dyeability, absence of odour, and acceptability for food contact.

They can be used for food containers, bottle caps,
cups, and dishes.

Products made of amino resins are resistant to many
organic solvents (e.g. aliphatic or aromatic hydrocarbons, alcohols, or ketones), adequate with weak
acids and bases, but do not resist strong acids and
bases. The melamine derivatives are substantially
more resistant to chemicals than are the urea materials.

The assortment of the moulding amino resins is very
broad. Translucent and opaque materials of any colour
or colourless are produced. Some typical properties
of the moulding amino resins are given in Table 36.

Table 36. Typical properties of moulding materials - aminoplastics /37/

Properties	Moulding material based on	
	urea resin	melamine resin
tensile strength (MPa)	35 - 50	20 - 70
compressive strength (MPa)	100 - 300	150 - 250
flexural strength (MPa)	60 - 90	50 - 110
impact flexural strength (MPa)	~ 0.8	0.3 - 0.4
modulus of elasticity (GPa)	5 - 10	5 - 10
heat resistance (Martens) (^{o}C)	> 100	120 - 160
water absorption of aminoplastic mouldings (%) in cold water after 24 h	0.4 - 1.2	0.1 - 0.6

The laminated and expanded materials based on the
amino resins are used predominantly as decorative and
building materials, and their applications to packaging are insignificant.

4.2.9 EPOXIDE RESINS

Epoxide resins have been produced industrially for

about 40 years, during which time their use spread
very rapidly /38, 39/, due to their outstanding properties, especially excellent adhesion, chemical resistance, and low shrinkage during hardening. These resins
are widely used in industry and in packaging as
container and can internal lacquers and external decorative protective coatings.

Epoxide resins have macromolecular chains terminated
by at least one epoxide (ethylene oxide) group. They
are prepared usually by alkaline condensation of
2,2-bis(4-hydroxyphenyl) propane (Bisphenol A) with
epichlorhydrin or dichlorhydrin:

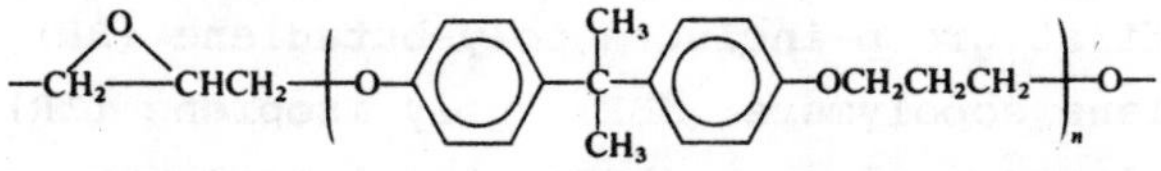

The ratio of reactants determines molecular weights,
which for the resins currently produced range from
380 to about 4000.

The epoxide resins are technically useful only when
cured as they contain hydroxyl groups, which make them
hygroscopic, they are not resistant to organic solvents
and possess neither film-forming properties nor sufficient mechanical properties. Cross-linking involves
reaction of the epoxide and hydroxyl groups and may be
affected at low temperature, below 100°C, or at high
temperature.

The usual hardeners are polyamines, amides of fatty
acid dimers and diamines and polyamides which, after
mixing, have long life at 20°C. Another group of
hardeners which initiate polymerization of the epoxide
groups, rather than reacting chemically with them,
include the Friedel-Crafts catalysts, strong mineral
acids, tertiary amines, aryl- and alkylmetallic compounds.

The epoxide resins generally have good chemical
resistance, which increases with increasing molecular
chain length and with increasing density of cross-
linking. Products hardened with anhydrides of dicar-

boxylic acids contain fewer hydroxyl groups and are not very hygroscopic. The thermal resistance of epoxide resins depends on both the resins and hardeners used and on the degree of cross-linking; tri- and tetra-functional resins resist temperatures 20-40° higher than for bifunctional resins; hardeners containing aromatic nuclei also impart higher thermal resistance.

4.2.10 SYNTHETIC RUBBERS (ELASTOMERS)

Synthetic rubbers fall /5, 40/ into two classes, those for general use replacing natural rubber, and special rubbers. The first group includes poly-butadiene (BR), styrene-butadiene copolymers (SBR), poly-isoprene (IR), ethylene-propylene copolymers (EPR), and terpolymers of ethylene, propylene and an unconjugated diene monomer (EPDM). None of these finds significant use in packaging other than in blends with other polymers to modify properties, e.g. blends of EPR with PE or PP to improve impact resistance.

The second group includes synthetic rubbers which surpass natural rubber in some property such as resistance to oils, oxygen, and ozone, low permeability to gases, higher thermal resistance, etc., and are usually more expensive than natural rubber. New types are continuously added, but the following are examples: butadiene-acrylonitrile copolymers, chloroprene homopolymers and copolymers, isobutylene-isoprene copolymers, acrylate rubbers, polysulphide rubbers, polyurethane rubbers, chlorosulphonated polyethylene, silicone rubbers, and fluorocarbon rubbers.

In packaging, these materials find small volume applications in seal and washers for container closures.

4.2.10.1 B u t a d i e n e   R u b b e r s

Syndiotactic 1,2-polybutadiene - a thermoplastic
polymer - was produced by the Japanese firm Synthetic
Rubber Co. Ltd. /41, 42/ employing solution polymerization catalyzed with a catalyst of the Ziegler type.
The polymer has more than 90% 1,2-addition configuration, adjustable crystallinity from 15 to 25%, melting point below 90°C, and brittle temperature of
-23°C. The polymer is photodegradable, but can be
modified with stabilizers. It is non-toxic and, hence,
suitable for applications in medicine and in the food
industry.

The polymer in the form of film is an interesting
packaging material. Due to its permeability to gases,
it is especially suitable for packing fresh fruit and
vegetables; it has transparency, good tensile strength
and ductility, high tear strength, elasticity and
flexibility, good weldability at relatively low temperatures, and is accepted for food contact. A thin
film is degraded in sunlight within several weeks,
especially in summer, offering a solution to the problem
of packaging litter.

4.2.10.2 B u t a d i e n e - A c r y l o n i t r i l e
          R u b b e r s

The outstanding properties of these nitrile rubbers
are resistance to swelling in oils and non-polar
solvents and to heat. They are produced by emulsion
polymerization, and contain 18% to 45% acrylonitrile.

Increasing acrylonitrile in the copolymer results
in increasing strength and solvent and heat resistance but reduced flexibility and resilience and higher
brittle temperature. During the ageing at enhanced
temperatures, the cross-linking reactions prevail over
the degradative ones, so that the vulcanized rubbers
become hard.

4.2.10.3 B u t y l   R u b b e r

The polymerization of isobutylene provides a rubber-
like polymer of high molecular weight. Its valuable
properties include chemical resistance, low permeability
to gases, excellent dielectric properties, and resistance to oxygen and ozone. Poor permanent set is a
drawback of polyisobutylene, but copolymerization
with 2-3% isoprene introduces double bonds capable of
vulcanization to improve permanent set without losing
other advantages. Such copolymers are known as butyl
rubbers, and have 0.6 to 3 % unsaturation. They are
produced by cationic polymerization at $-90^{\circ}C$ in methyl
chloride solution catalyzed with aluminium chloride.

Almost 80% of the butyl rubber produced is used for
inner tubes for tyres. In packaging butyl rubber is
a suitable material for lining large-volume vessels
and containers used in the transport of aggressive
chemicals.

4.2.10.4 P o l y u r e t h a n e   R u b b e r s

These are produced by polyaddition of higher-molecular
diols with diisocyanates. Depending on the starting
materials, their ratio, and the reaction conditions,
various products can be obtained (Section 4.2.6). The
solid types are processed on normal rubber equipment
and the liquid types are used for casting.

The reaction of isocyanate groups with water, which
liberates carbon dioxide, is utilized for the production
of expanded rubber or polyurethane foam (PUF). The
reaction is accelerated by basic catalysts, e.g.
tertiary amines, which control foaming and curing
rates.

The bulk of polyurethane rubbers is in the form of
foam or cellular rubber, where the high price of the

polymer is compensated by the low density of product. The shock-absorbing characteristics of polyurethane foams give them a place in packaging.

4.2.11 OTHER POLYMERS

This section covers polymers which have not been included in any of the previous groups of plastics.

4.2.11.1 P o l y c a r b o n a t e s

Polycarbonates are polyesters of carbonic acid and dihydroxy compounds. In particular, the name refers to polyesters of carbonic acid and 2,2-bis(4-hydroxy-phenyl)propane (*Bisphenol A*), which have the general formula /33/:

$$\left[-O-\!\!\left\langle \bigcirc \right\rangle\!\! \overset{\overset{\displaystyle CH_3}{|}}{\underset{\underset{\displaystyle CH_3}{|}}{C}} \!\!\left\langle \bigcirc \right\rangle\!\! -O-\overset{\overset{\displaystyle O}{\|}}{C}- \right]_n$$

These polycarbonates offer a unique combination of properties not encountered with any other thermoplastics. The main features are excellent mechanical and dielectric properties which remain constant over an unusually broad temperature range, low absorption of liquids, resistance to weather, and transparency. At the same time they possess a good dimensional stability, thermal stability, chemical resistance, and processability. There are two methods of manufacture, transesterification and direct phosgenation.

Transesterification of *Bisphenol A* and a diester of carbonic acid, usually diphenyl carbonate, is carried out at $150^{\circ}C$ to $300^{\circ}C$ in an inert atmosphere and under vacuum. The high viscosity of the melt limits both the scale of manufacture and the attainment of high mol-ecular weight: it is hardly possible to reach 50 000.

However, the advantage of the process is the formation
of polymer directly in the solid form suitable for
further processing.

Direct phosgenation is the reaction of *Bisphenol A*
with phosgene in the presence of compounds able to
react with the hydrogen chloride formed, usually
aqueous sodium hydroxide or pyridine. The solid resin
must be isolated and carefully purified; the advantage
of this method is that it can give polycarbonates of
high but controllable molecular weight under very
mild conditions in simple equipment.

Polycarbonates are soluble in some organic solvents
(methylene chloride, chlorobenzene, dioxane, pyridine,
tetrahydrofurane, *m*-cresol), fusible and stable for
a long period even above 300°C. They can be processed
as solution or melt to produce fibres, varnishes, film,
and blow mouldings as with other thermoplastic materials.
In packaging, polycarbonate film is used in special
circumstances and various cups, ampoules, and bottles
are made (especially for medical purposes) instead
of from glass. Suitability for repeated sterilization
also led to multi-trip containers for drinking water.

4.2.11.2 Polyoxymethylene

The polymerization of very pure formaldehyde using
amine or quaternary ammonium salts as catalysts gives
polyformaldehyde, or polyoxymethylene:

$$\ldots \;-O-CH_2-O-CH_2-O-CH_2-O-CH_2-O-CH_2-O-CH_2-\ldots$$

The molecular weight depends on the reaction conditions /6/, and this plastic was produced in the
U.S.A. as *Delrin*.

Other very efficient procedures start from trioxane
and use Lewis acids as catalysts /43/.

Polyoxymethylene possesses an excellent thermal and chemical resistance and extreme impermeability to gases /27/. As it retains its crystalline structure unchanged up to the melting point, it exhibits relatively negligible changes in mechanical and physical properties up to $120^{\circ}C$, which is unique among thermoplastics. Decomposition to formaldehyde does not take place until $220^{\circ}C$. It also shows negligible water absorption.

Polyformaldehyde is used in packaging for cosmetics, e.g. aerosol containers.

It is marketed under the names *Delrin* (Du Pont, U.S.A.), *Hostaform* (Farbwerke Hoechst, FRG), *Celcon* (Celanese Polymer Corp., U.S.A.).

4.2.11.3 Poly(Ethylene Imine)

The highly toxic monomeric ethylene imine is prepared by action of sodium hydroxide on 2-chloroethylammonium chloride in aqueous solution at $90-95^{\circ}C$. Gaseous ethylene imine is liberated and dried over sodium hydroxide. The polymerization of the monomer is carried out in the presence of carbon dioxide, being catalyzed with ethyl chloride. Poly(ethylene imine) is a structural analogue of polyoxyethylene /6/:

$$\ldots -NH-CH_2-CH_2-NH-CH_2-CH_2-NH-CH_2-CH_2-\ldots$$

It is particularly suitable for paper refining, which makes paper washable and abrasion resistant without increasing its stiffness and creasing.

REFERENCES

/1/ Stevens M.P., Polymer Chemistry - An Introduction, Addison-Wesley, Massachusetts, 1975.

/2/ Rodrigues F., Principles of Polymer Systems, McGraw-Hill, New York, 1970.

/3/ Kovačič L ., Bína J., Plasty - vlastnosti, spracovanie, využitie, Alfa, Bratislava, 1974.

/4/ Janda M., Bláha K., Rosmus J., Skála V., Bioorganická chemie, Prague Institute of Chemical Technology, Prague, 1972.

/5/ Champetier G., Monnerie L., Introduction à la chimie macromoléculaire. Masson, Paris, 1969.

/6/ Champetier G., Les macropolymeres et leurs applications, Centre de Documentation Universitaire, Paris, 1963.

/7/ Franta I. et al., Gumárenské suroviny, SNTL, Prague 1979.

/8/ Tomis F. et al., Polyethylen, SNTL, Prague, 1961.

/9/ Ziegler K., Holzkamp E., Breil H., Martin H., Angew. Chem., 1955, 67, 426.

/10/ Natta G., Pino P., Gorradini P., Danusso F., Mantica E., Mazzati G., Moranglio E., J. Amer, Chem. Soc., 1955, 77, 1708.

/11/ Peters R.F., Evering B.L. (Standard Oil Co.), USA pat. 2692 251, 1954.

/12/ Feller M., Field E. (Standard Oil Co.), USA pat. 2717 889, 1955.

/13/ Clark A., Hogan J.P., Blanks R.L., Lanning W.C., Ind. Eng. Chem., 1956, 48, 1152.

/14/ Peters E.F., Zletz A., Ind. Eng. Chem., 1957, 49, 1879.

/15/ Feller M., Field E., Ind. Eng. Chem., 1959, 51, 155.

/16/ Anonym, Plastics, 1966, 31 (340), 117.

/17/ Anonym, Int. Bottler Packer, 1967, 41 (12), 120.

/18/ Anonym, Emballages, 1969, 39 (260), 201.

/19/ Anonym, Emballages, 1969, 39 (264), 222.

/20/ Robinson D., Paper Film Foil Convert, December 18, 1966.

/21/ Billmeyer Jr. F.W., Textbook of Polymer Science, Wiley, New York, 1971.

/22/ Doležal B., Odolnost plastických hmot a pryže, SNTL, Prague, 1980.

/23/ Simon E., Ann., 31, 265, 1839.

/24/ French pat. 849 926 (taken from Ref. /37/).

/25/ Švastal S., Lím D., Kolínský M., Úvod do chemie a technologie plastických hmot, Práce, Prague, 1954.

/26/ Plato G., Schröter G., Kunststoffe, 1960, 50, 163.

/27/ Štěpek J., Zpracování plastických hmot, SNTL, Prague, 1966.

/28/ Kubík J., Gřunděl F. et al., PVC, výroba, zpracování a použití, SNTL, Prague, 1965.

/29/ Wallenberk E., Verpack. Rundsch., 1964, 15, 1264.

/30/ Marek O., Tomka M., Akrylové polymery, SNTL, Prague, 1964.

/31/ Doyle E.N., The Development and Use of Polyurethane Product, McGraw-Hill, New York, 1969.

/32/ Veselý R., Sochor M. et al., Polyamidy, jejich chemie, výroba a použití, SNTL, Prague, 1963.

/33/ Mleziva J. et al., Polyestery, jejich výroba a zpracování, SNTL, Prague, 1978.

/34/ Korshak V.V., Vinogradova S.V., Polyesters, Pergamon Press, New York, 1965.

/35/ Goodman I., Rhys J.A., Polyesters, Iliffe, London, 1965.

/36/ Hudeček S., Fenoplasty, SNTL, Prague, 1963.

/37/ Šňupárek J., Černý J., Aminoplasty, SNTL,
Prague, 1963.

/38/ Lidařík M., Kincl J., Roth V., Bring A.,
Epoxydové pryskyřice, SNTL, Prague, 1961.

/39/ Potter W.G., Epoxide Resins, Springer, New
York, 1971.

/40/ Pech J. et al., Syntetický kaučuk, SNTL, Prague,
1971.

/41/ Takeuchi Y., Sekimoto A., Abe M., A New Thermo-
plastic Syndiotactic 1,2-Polybutadiene.I.
Production, Fundamental Properties and
Processing, ACS Symposium Series No. 4, New
Industrial Polymers, pp. 15-25. American Chemi-
cal Society, 1974.

/42/ Takeuchi Y., Sekimoto A., Abe M.. A New Thermoplastic Syndiotactic 1,2-Polybutadiene. II.
Applications. ACS Symposium Series No.4, New
Industrial Polymers, pp. 26-36. American Chemical Society, 1974.

/43/ Furukawa J., Saegusa T., Polymerisation of
Aldehydes and Oxides. Wiley-Interscience, New
York, 1963.

5

Processing of Plastics with Special Reference to Mechanized Packaging Procedures

J. ŠTĚPEK, V. DUCHÁČEK

5.1 PREPARATION OF MIXTURES

Plastics processing involves several stages and
modifications of materials prior to completion of
the required products. Usually, plastics are processed with various auxiliary materials, e.g. plasticizers, heat and light stabilizers, curing agents, and
other macromolecular materials, which impart properties
which are not exhibited by the plastic alone. Some
additives, like pigments, fillers, extenders, improve
the appearance of the products or make them cheaper.
Other substances may be added to facilitate processing
(e.g. solvents, lubricants, plasticizers) /1/.

5.1.1 MIXING AND KNEADING

Mixing and kneading play very important parts in
preparing and processing plastics, and the methods used
affect very much both processing through-put and product
quality. Dispersion is best if large shearing forces
are operating, i.e. if the material is consistent,
the volume kneaded is small, and no "dead areas" are

present. It is well-known that mixing on a laboratory scale (when the amounts are smaller) needs shorter times than the corresponding industrial process giving the same final results.

The mixing operation can be considered according to the physical state of the materials: dry, viscous, and liquid. Energy consumption is least if the polymeric substance forms a powder, liquid, or solution (or suspension), as in the case of lacquers and dipping suspensions. In such cases simple mixing apparatus is used, and the energy consumption is 3 to 80 kJ kg^{-1} /1/.

The simplest apparatus for mixing dry materials consists of rotating drums with internal arms of various shapes (e.g. like a plough share), or the drums themselves can have various shapes (cylinder, barrel, cube, V-shape, double cone). Ball mills are used, if simultaneously the material has to be ground; they achieve very efficient mixing, but their energy consumption is high; a conical shape improves their efficiency. They are used for thorough grinding and fine dispersing of pigment particles especially in preparing pigment master batches for poly(vinyl chloride) pastes and particularly for surface coatings which only have the thickness of several microns (e.g. 20 μm). The grinding fineness must at least correspond to this magnitude.

Viscous materials are mixed with two-arm mixing machines having Z-shaped arms or planetary mixing machines which are useful for mixing wetted solids.

If the plastic material is highly viscous, the equipment must be more efficient in mixing and grinding, and heat may be required. The energy consumption for mixing more viscous materials is about 100 kJ kg^{-1} /1/. For pigments, special machines like a grinding three-roll calender, a two-roll machine with grinding blade, or a turbine mixer are required.

5.1.2 COMPOUNDING AND GELATION

The homogenization of mixtures of polymers and
additives, without solvents and liquid plasticizers,
by mechanical shearing is called compounding. It
involves melting and kneading to eliminate density
differences, internal stress, or phase boundaries
between individual particles, and is especially
important in the case of spray dried poly(vinyl chloride).
Poly(vinyl chloride) "dry blend" grades can be processed rapidly and economically.

Compounding is carried out with calenders, Z-blade
mixers or extruders with attention to temperature and
time so that undesirable degradation of the material
is avoided.

Kneading machines with friction screws are especially
suitable for continuous operation: they have both
threads of normal width and ones with wider friction
surface, the threads having also a wedge-shaped scarf
in the rotation direction.

The energy consumption for compounding plastics is
substantially lower than for rubbers, e.g. for poly(vinyl chloride) about 700 kJ kg^{-1} and for rubber about
1700 kJ kg^{-1}, the compounding process itself consuming
about 75% of the energy supplied /1/. The efficiency
of the process is partly dependent on mixing time,
so with an extruder, better mixing is achieved with
screws of higher length diameter ratio.

Gelation is an important process in processing
a polymer with plasticizers. A gel structure occurs
with a macromolecular compound in a solvent, when the
concentration of the disperse phase is high enough.
In a dilute solution, or sol, the macromolecules have
sufficient mobility to change their shapes, but with
increasing concentration the number of macromolecules
which are in mutual contact increases, their inter-

molecular forces lead to chain entanglement and, if
mutual affinity of the chains is greater than their
affinity for solvent, the intermolecular forces will
produce cohesion between chain segments. This cohesion
depends both on concentration and on temperature. In
dilute solutions or at higher temperatures, the contacts
are broken by thermal motion with fresh contact made
elsewhere. When the number and stability of cohesive
contact are sufficient, due to high concentration of
the disperse phase or to decreased temperature, thermal
motion does not overcome cohesion and a gel structure
is formed. The sol-gel conversion is a reversible
process.

The gel and the solution are colloidal dispersions
system, but in the gel, both the dispersing medium and
the disperse phase are continuous, the latter forming
a macromolecular network.

A practical example is the gelation of a poly(vinyl
chloride) paste or mixture, i.e. heterogeneous dispersion (suspension) of poly(vinyl chloride) powder
in plasticizer. The dissolution of the polymer would
proceed most readily above the melting temperature of
poly(vinyl chloride), but this lies above the decomposition temperature, so gelation must proceed at a
lower temperature aided by heat stabilizers when
decomposition does not take place, i.e. at about
160-180°C. At this gelation temperature, all the
poly(vinyl chloride) is progressively dissolved in the
plasticizer, with simultaneous gel formation, but
formation of a permanent solid elastic gel occurs only
after cooling to room temperature. At the gelation
temperature there is sufficient gel content to prevent
spontaneous deformation, but by the application of
pressure (e.g. during embossing, extrusion, etc.)
shapes may formed which are made permanent by cooling.
The physical network imparts elasticity and rubber-like
character to the plasticized poly(vinyl chloride).

This is attributable to cohesive cross-links which
are thought to involve crystallites arising from
isotactic and syndiotactic segments in the macromolecules; 1% crystalline content in technical poly(vinyl
chloride) can produce a sufficient network /1/.

The gelation procedure must be followed with care
to obtain high-quality products. It affects the
strength, elasticity, and tackiness and, to some
extent, the migration and "sweat-out" of plasticizers
and additives. The selection of machines depends on
energy consumption, space requirement, output rate
investment, manning, and whether a continuous process
or a discontinuous one is preferred for a particular
purpose. In general, discontinuous two-roll mixing
mills are advantageous for smaller amounts and
frequent changes in recipes. Long production runs with
small changes suit a continuous process.

5.1.3 PELLETING, GRANULATION, AGGLOMERATION (DRY
       BLEND MAKING)

For many forming procedures it is advantageous to
pre-form the starting material into pellets, or to
granulate it into free-flowing, uniform sized crushed
material.

The tabletting process has many advantages, because
the tablet-making machines produce tablets of a
definite weight, so that dosing of the moulding
materials is easier, avoiding losses due to large
spues or to short moulding; also most of the air is
removed so that production of smaller articles does
not necessitate venting of the mould. Pelleting is
applied to powders (especially thermosetting moulding
resins), usually to give small flat cylinders. For
fibrous materials uniform pressure is applied in
three directions in special moulds, to give prisms.

Granulation is performed in granulators, pelletizers, or crushers. A uniform shape of granule is advantageous for filling the hoppers of extrusion and injection machines, because it enables a regular supply of the material to be fed to the screws without undesirable hold-up of the material as occurs with powders. The granulated material should have sufficiently large surface area but should entrain as little air as possible. Thus in processing unplasticized poly(vinyl chloride) on extruders, which is an exacting procedure, granules of lens shape with 1-2 mm diameter enable good heat transfer and efficient introduction into the screw.

Powdered polymers cannot be processed easily on the single-screw extruders. Small particles of powder and large amounts of entrained air make the friction between individual grains insufficient.

Twin-screw extruders, with intermeshing screws fitted into overlapping barrels, are more efficient in creating friction between particles. Efficient gelation and/or plasticizing of powdered poly(vinyl chloride) in single-screw extruders is possible with dry blend or agglomerates made in high-speed mixers. Most frequently adopted is the fluidized-bed mixer, in which poly(vinyl chloride) powder is mixed with plasticizers, stabilizers, pigments, or lubricants at elevated temperatures (above $100^{\circ}C$ for dry blend and about $160^{\circ}C$ for agglomerates) /1/. Temperature load is dictated by the poly(vinyl chloride) type and the amount of plasticizers. These factors also determine the shape of the final agglomerate. Suspension poly(vinyl chloride) with up to 30% plasticizer content gives a powdery dry blend. If the plasticizer content is higher and, in particular, if emulsion poly(vinyl chloride) is used, then agglomerates are formed of up to 4 mm diameter. In the both cases, the poly(vinyl chloride) is pre-gelled, which

increases the throughput of extruders, injection
moulding machines, etc.

An important procedure is the dispersion of coloured
pigments or carbon black in plastics, for which master-
batches are usually employed. These are pre-made
concentrates of the additives in polymer or plasticizer. The dispersion is performed in mills of various
types, more recently kneading machines, and is both
energy- and time-consuming. The dispersion of carbon
black in crystalline polymers, e.g. polyethylene,
must be carried out at a controlled temperature,
because the softening temperature range of these polymers is rather narrow; above a certain temperature the
melt viscosity is too low and at a lower temperature
they are not sufficiently plastic.

5.2 PROCESSING OF LIQUID DISPERSION SYSTEMS

Plastics processing methods depend on whether the
material is in the liquid or in the plastic state. The
processing of liquid dispersions relates to latices,
pastes, low-viscosity melts and solutions. Processing
in the plastic state applies if the flow of the material
is achieved by increased temperature or pressure. Solid
plastics are processed mainly by machining.

Liquid systems are used for dipping, coating, casting,
lacquering, varnishing, and lining. These procedures
provide plastic coatings for permanent modification or
protection of surfaces of they provide a means of
obtaining a desired shape as in moulding, dipping, and
casting /1/.

5.2.1 DIPPING

The dipping former of the required shape is immersed
in the liquid and when a sufficient amount is adhering,
the former is slowly removed from the vessel. The
layer formed is subjected to a suitable thermal treatment
such as evaporation of liquid components, curing, gelation, or cooling. The aim of the dipping process is
to produce a uniform, strong coating on the former,
whereupon the coating is removed from the former.

In some organic liquids and plasticizers, poly(vinyl
chloride) powder forms stable dispersions called pastes.
The dispersing medium must have at room temperature only
a limited solvation effect on the polymeric particles.
Paste-forming poly(vinyl chloride) is used for dipping,
casting, coating, and for obtaining expanded plasticized poly(vinyl chloride). Its properties are determined
by the shape, size, and size distribution of the particles
and the character of their surface.

Oval particles are most suitable and obtained by
spray-drying poly(vinyl chloride) latex. The maximum
diameter is 50 μm, and the optimum about 2 μm. The size
distribution should give minimum air inclusion and the
dispersing medium should have the lowest possible viscosity. The surface structure of the polymer particles
is determined primarily by the drying conditions, and
should not be porous to prevent excessive penetration
of the plasticizer into the polymer.

Such properties of the poly(vinyl chloride) pastes
can be adjusted with diluents, by suitable combinations
of grades according to solvation and size distribution.

A correct choice of plasticizer is essential for
poly(vinyl chloride) pastes: the plasticizer, being the
dispersing medium, influences its flow properties, as
well as the flexibility, behaviour at low temperatures,
and durability of the final product.

Suitable plasticizers have negligible solvation
effect on the polymer at room temperature, especially
if the paste has to be kept for a longer period, but
solvation should rapidly increase with temperature for
good gelation. In practice paste viscosity increases
with time. The viscosity of a good paste should not
increase by more than 100% during a month's storage.

Further components of pastes include stabilizers,
fillers, and pigments which are dealt with in Sections
3.3 and 3.4. The primary function of fillers is to make
the products cheaper. Sometimes they are used to change
the flow properties of the paste and/or to obtain a
smooth surface of the product. The presence of a high
content of fillers or pigments in a paste is not desirable, because they tend to absorb the plasticizer. This
absorption can be prevented with high-viscosity plasticizers or with fillers exhibiting low oil absorption.
A good paste exhibits neither thixotropy or dilatancy.
Thixotropy is manifested by the paste turning more
viscous or even gelling on standing because of sol-gel
transition, but is reversed on stirring. Dilatancy is
viscosity increase with increased shearing due to the
displacement of plasticizer surrounding the particles
which thus come into close mutual contact increasing
the resistance of the system to flow. Additives have
been introduced which prevent these phenomena. Thus
methylcyclohexene decreases thixotropy by diluting the
polar effect of the plasticizer. Dilatancy can be prevented with plasticizers which increase solvation.

5.2.2 CASTING PROCESS

The casting process is used for both thermosetting
resins and thermoplastics. The main thermosetting resins
used are phenol-formaldehyde, urea-formaldehyde, epoxide
and polyester resins; the main thermoplastics are

poly(vinyl chloride) pastes, cellulose derivatives,
poly(methyl methacrylate), polyamides, polyolefins,
and polycarbonate.

Casting of poly(vinyl chloride) pastes is used in
packaging predominantly for the sealing ring of tin-plate
closures /2/.

The casting process is also used for the production
of films from cellulose materials. Cellophane films
(*Cellophan, Transparit, Heliocel,* and in ČSSR *Prosvit*)
are made from viscose (aqueous solution of cellulose
xanthogenate) which is extruded through a die into a
precipitation bath containing sodium sulphite. The
coagulated film of regenerated cellulose passes through
further washing baths containing 15-20% sodium sulphide
(where sulphur is completely removed), 15-20% ammonium
sulphate, 5% hydrochloric acid, water, 2-3% sodium
hypochlorite bleaching solution, and finally water. The
final phase consists in drying on steam-heated cylinders.

The films are produced two ways:

1. The "Cellophane" method (Fig. 42a) in which the
viscose passes through a die into the precipitation bath
on a cylinder, and is then transferred to further baths
supported by a conveyer belt because the film is not
strong. Before one of the sodium sulphite baths there
is a press cylinder which modifies the film surface.

2. The "Transparit" method (Fig. 42b) in which a
horizontal cylinder of about 3 m diameter rotates in
the precipitation bath, and a thin film of coagulated
viscose is formed at its surface; the film is unwound
from the cylinder and processed further as in the
previous case.

Both methods necessitate apparatus made of high-
quality corrosion-resistant materials. The cylinder
surface must be perfectly smooth as with glass or
ceramic cylinders with finely polished surface. The
viscose films usually have a thickness of 0.02 to 0.05 mm,

but it is possible to produce films of up to 0.95 mm
thickness. Thicker films are made by pressing together
several thinner films. As the film shrinks during
production, synchronization of the draw-off velocities
is essential for uniform tension in the film. The
velocity may be as high as 40 m min^{-1}.

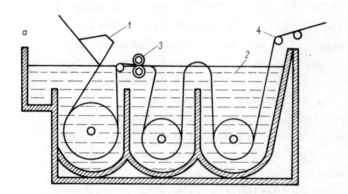

Fig. 42. Schematic representation of production of
cellophane film /1/
a - the "Cellophan" method: 1 - viscose container with
shifting bottom, 2 - precipitation bath, 3 - the
cylinder for wide stretching of the film, 4 - the film
on the guide rollers

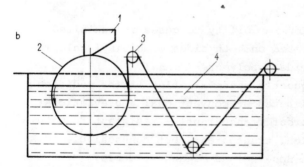

b - the "Transparit" method: 1 - viscose container,
2 - casting cylinder, 3 - guide roller, 4 - precipitation
bath

The mechanical properties of cellophane vary considerably with the content of absorbed water. Very dry cellophane is brittle but wet cellophane loses its strength. The optimum water content is 7% to 10%, so cellophane for use at 30-70% relative humidity is modified with 10-15% glycerol.

Cellophane is not thermoplastic, so it is usually joined by glueing with gelatine or zinc chloride /2/; it may be coated with another polymer enabling heat sealing to be applied and, at the same time, decreasing the high permeability of cellophane to water vapour. Without coating, cellophane would be inappropriate for packaging hygroscopic products. It is coated mostly with nitrocellulose varnishes, with vinyl chloride-vinylidene chloride copolymers (*Saran*), or aqueous or alcoholic solutions of carbamide resins. Other combinations consist in coating the film with extruded polyethylene after applying an anchoring interlayer, and in laminating cellophane with a polyethylene film using microcrystalline way or another adhesive.

Dry cellophane transmits very little oxygen and carbon dioxide, retains organic vapours, and resists fats and oils.

The highest impermeability to gases is exhibited by cellophane coated on both sides with vinyl chloride-vinylidene chloride copolymers (*Saran*).

The most frequently used varieties of the modified cellophane are denoted by conventional abbreviations:

MT (Moistureproof Transparent) - resistant to water vapour penetration, transparent, but non-sealable;

MST (S - Heat Sealing) as MT and weldable;

MSAT (A - Anchored Water Resistant Coating) - resistant to water and water vapour, weldable, transparent.

Sometimes, the abbreviations are combined with numbers denoting film thickness.

5.2.3 ADHESIVES AND BONDING IN PACKAGING

5.2.3.1 The Demands Made on Adhesives

Joining packaging materials with adhesives is practically as old as packaging itself. At the beginning, paper and board were glued to give a variety of bags and boxes, to attach labels, and to seal filled packages. Later this technology was extended to wooden packages (boxes and crates), and more recently to polymeric materials which have become indispensable to modern packaging, both industrial and retail.

However, adhesive bonding is used not only for the production of packages but also for the production of an increasing assortment of packaging materials possessing many useful properties which cannot be obtained directly. Some of these laminated packaging materials (laminates) can also be produced in other ways (e.g. coextrusion and coating), but in many cases adhesive bonding is unavoidable. This is particularly true of the production of aluminium foil laminates.

This brief introduction already indicates that the demands made on adhesives depend on the way they are used. Nevertheless, there are several basic requirements which must be met by adhesives used for packaging materials. An ideal adhesive should fulfil the following requirements:

1. sufficiently strong bond,
2. sufficient adhesion in the liquid state,
3. adequate temperature-resistance of the bond,
4. a bond resistant to water, water vapour, fats, oils, and microorganisms,
5. odourless,
6. acceptable for food contact, and
7. cheap.

No one adhesive fulfils all these requirements;
selection and compromise are necessary with respect
to particular circumstances.

Individual types of adhesives (which will be dealt
with in more detail below) must meet further special
requirements connected with technological peculiarities
of their application, e.g. whether the bonding process
is manual or mechanized, whether the adhesive should be
water soluble (as in the case of removable labels), etc.

The strength of bond is determined by the physical
and chemical properties of both the adhesive and the
materials bonded. Both the adhesion (i.e. the affinity
of two substances) and the cohesion (which concerns
the one substance, such as adhesive) are important.

The adhesion due to intermolecular forces is called
specific. The adhesion depends also on the physical
character (e.g. roughness) of the surfaces of the substrate and adherend, and this is differentiated as
mechanical adhesion.

Important parameters of adhesives used as aqueous
dispersions (solutions, latices) are the drying time
and "stringiness" /3/. The time of drying varies from
5 to 20 s for dispersion adhesives, 30-100 s for dextrin
adhesives, up to 240 s for starch adhesives. The
"stringiness" (elongation at break) of the wet adhesive
depends on its rheological character and is assessed
by the "string" length which may be short (15 mm), medium
(up to 50 mm), long (up to 90 mm), or extremely long
(above 90 mm).

5.2.3.2 Types of Adhesives

Adhesives can be classified according to various criteria,
such as their physical state (e.g. dispersions), or their
mode of application (e.g. pressure-sensitive or hot-melt

adhesives).

Another criterion is their origin, that is, based on natural substances or based on synthetic materials. The natural adhesives can be further divided into: those of plant origin and those of animal origin. However, as all these substances are of polymeric nature, origin can be considered a less significant criterion.

The first adhesives for bonding paper and similar light materials were the glue obtained by boiling raw hide and hooves. Further development continued through natural resins and starch to casein obtained from milk /2/.

Gelatine or glue containing only about 25% water /4/, melts at 80°C to a viscous, rapidly solidifying material which has sufficient adhesion and cohesive strength, the latter being further increased by subsequent loss of water. Gelatine is also applied as a good transparent adhesive for cellophane.

Modern adhesives are based on synthetic resins and latices of natural or synthetic rubbers. Packaging also makes use of the "rediscovered" bituminous adhesives which were used in ancient Egypt.

Adhesives based on starch and dextrin (cf. Section 4.1.2), being of natural origin, are easily attacked by microorganisms, and must be preserved by means of formaldehyde or pentachlorophenol.

Starch adhesives can be made water resistant by reaction with urea-formaldehyde resins /5/. The reaction takes place in acid medium, being catalyzed usually with ammonium chloride.

Dextrin glues (in contrast to the starch adhesives) exhibit considerable adhesion immediately after application and, if not much diluted, dry out quite rapidly; they are suitable for manual glueing as well as for machine glueing of bags and collapsible boxes; they

are also used for labelling and glass bottles and for
producing less strong adhesive paper tapes.

The main advantage of the adhesives produced from
starch and dextrin lies in their low price. They are
now used in less exacting applications and are often
a source of disagreeable odour which can be easily
transferred to the package content. Therefore, they
are being replaced by synthetic adhesives, especially
those based on vinyl polymers /6/.

Low prices also characterize adhesives based on
sodium silicate, which have been used for many years.
The glue joint obtained is relatively strong, and they
are fast adhesive because bond strength increases
rapidly with viscosity on losing a small amount of
water. The dry joints resist microorganisms and the
adhesives are odourless and non-toxic /5/.

However, they have some drawbacks. The dry adhesive
is brittle, especially in thin layers. At high humidity the alkali metals present tend to migrate into
the paper and to destroy the bonds between fibers.
This migration can cause decolourization of the fibres
or deterioration of dyestuffs which are sensitive to
alkali substances.

Industrial production of adhesives is moving more
and more from starch and dextrin to synthetic vinyl
dispersions, and from natural rubber to synthetic
rubbers.

Although synthetic adhesives exhibit good and rapid
adhesion, resistance to microorganisms, and often
resistance to water, their use is limited by a relatively
narrow range and higher prices than for natural adhesives.

A transition between natural and fully synthetic
adhesives is represented by semi-synthetic adhesives
based on cellulose; these include both water-soluble
types (methylcellulose, carboxymethylcellulose) and
types soluble in organic solvents (cellulose nitrate
and acetate, alkylcellulose).

Synthetic adhesives can be classified as thermoplastic and thermosetting.

Thermoplastic adhesives are most often applied in
packaging as dispersions (solutions, latices), or they
form the basis of fusible adhesives. When the adhesives
are applied as solutions, the choice of solvent is
important, because it determines wettability of the
surfaces being glued, the rate of evaporation, and the
viscosity of the solution which affects the thickness
of the layer.

The thermoplastic adhesives most frequently used in
packaging are poly(vinyl acetate) and polyacrylate
latices which exhibit excellent adhesion to various
substrates, and are suitable for machine glueing of
bags, collapsible boxes, for cellophane, paraffin-
and wax-coated paper, as well as for labelling packages
made of poly(vinyl chloride) and polystyrene. After
drying out these adhesives are resistant to water.

Poly(methyl acrylate) solutions are used for
glueing inserts into crown caps. Also applied as adhesives are solutions of poly(vinyl formal), poly(vinyl
butyral), poly(vinyl acetate), chlorinated poly(vinyl
chloride), copolymers of vinyl chloride with vinyl
acetate or vinylidene chloride, and modified polyamides.

Excellent adhesive properties, which are especially
used in pressure-sensitive adhesives, are found with
poly(vinyl ethers), of which poly(isobutyl vinyl ether)
is used most frequently.

Thermosetting adhesives /7/ (phenol-formaldehyde,
resorcinol-formaldehyde, melamine-formaldehyde, epoxide,
polyurethane adhesives, etc.) find very limited application in the packaging technology.

The adhesives based on elastomers or rubbers are
solutions or latices of natural or synthetic rubbers.
The adhesives from natural rubber have sufficient adhesive
power, whereas the adhesive power of synthetic rubbers

must be usually increased by addition of *Koresin* or
Rubresin (a tert. butylphenol resin), as with *Alkapren*,
based on chloroprene rubber. Rubber adhesives are used
in pressure-sensitive adhesives.

The adhesives based on the combination of rubber
latices with casein have been used for laminated packaging materials, especially aluminium foil to paper.
The types based on polychloroprene, butadiene-styrene
copolymers, and natural rubber are suitable for this
purpose. The adhesives based on polychloroprene latices
are characterized by slight but characteristic odour,
low toxicity, and good resistance to water, oils and
heat.

Adhesives based on wax have been used for many years.
Their main advantage lies in that they do not need to
be dried. Wax forms a bond immediately on cooling. Wax
adhesives are used exclusively for laminated packaging
materials which are rendered impermeable by the wax.
Their drawbacks are brittleness and low bond strength
but these can be remedied partially by addition of
polyisobutylene. Generally, microcrystalline waxes are
more suitable adhesives than paraffins, due especially
to their greater adhesive power, flexibility, and
resistance to fats, water, and water vapour. Elastomers
like polyisobutylene, butyl rubber, or rubber hydrochloride are sometimes blended with the microcrystalline
waxes, to a maximum of 10%. The wax adhesives form
a transition to hot-melt (fusible) adhesives which are
dealt with in a later section of this chapter.

Pressure-sensitive adhesives are used in packaging
especially for tapes and labels. The mode of action of
pressure-sensitive adhesives is given in Fig. 43. The
resistance against detachment of the surfaces joined
by a thin layer of liquid adhesive is due to the forced
flow of the adhesive to the centre of the joint when
the surfaces are parted. This resistance is larger

the smaller the distance between the surfaces and the
more viscous the liquid. At the same time, however,
the viscosity of the adhesive is limited by the requirement of sufficient adhesion which is decreased with
increasing molecular weight. As specific chemical
effects are not so important as these, pressure-sensitive
adhesives are applicable to surfaces which cannot
otherwise be easily joined with adhesives, e.g. fluorinated plastics.

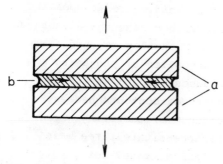

Fig. 43. Schematic representation of action of
pressure-sensitive adhesives

a - the materials to be joined, b - the pressure-
sensitive adhesive

Pressure-sensitive adhesives are substances which retain
their character of highly viscous liquid throughout
their service life. Chemically they are degraded or
low-molecular weight rubbers, polyisobutylene, poly(vinyl
ethers), or acrylic copolymers. The adhesive power or
tack of these polymers is usually increased by the
addition of colophony, or tert. butylphenol resin.

Pressure-sensitive tapes are produced on the basis
of paper and textiles, and various polymeric materials,
e.g. cellophane, poly(vinyl chloride), polypropylene.
They have an adhesive layer usually on one side; if
on both sides a release interlayer is inserted during
winding the rolls. The single-sided adhesive tapes have

release properties on the reverse side. The double-sided
tapes are useful as anti-slip devices for bags and
boxes on pallets /7/.

In packaging, pressure-sensitive tapes made of
non-woven textiles and unplasticized poly(vinyl chloride)
are used for sealing heavy (up to about 100 kg) cardboard boxes, while paper tapes are used with light
collapsible boxes. Polypropylene tapes for sealing
light and medium-weight cardboard boxes have replaced
cellophane tapes. Pressure-sensitive tapes are also
used for binding purposes, e.g. holding rod shapes in
bundles, closing gathered bags. They have the advantages
of rapid handling and reducing manual work, and facilitate mechanization and automation of packaging
processes; they replace wet adhesives and stitching
methods, and guarantee security of contents because
removal of tape provides evidence of tampering.

Pressure-sensitive labels necessitate three main
layers: the printed surface (paper, metal foil, polymeric film, etc.), pressure-sensitive adhesive, and
a supporting release material which is removed
immediately before use.

In packaging, these pressure-sensitive labels are
mostly on a continuous backing strip and used as price
labels in self-service stores; they are also used for
product identification and specification labels on
packages.

Hot-melt (fusible) adhesives are thermoplastic
materials which by definition are transformed by heat
into a molten or plastic state before use. Those based
on synthetic polymers were introduced about twenty-five
years ago, and their use has spread widely. They are
delivered to users in various forms, e.g. pellets,
granules, powder, tapes and wires wound on spools,
and are used to bond both similar and different materials
in packaging applications /6/.

These adhesives do not absorb moisture, and are
therefore often coated onto porous packaging materials
to make them impermeable to water and water vapour.

Hot-melt adhesives have two main applications in
packaging. First they are used to bond packaging materials based on paper and board for box and carton
construction. The second application field is continuous coating on reel-stock materials for laminating
(bonding over the entire area), or as stripes for subsequent use in the production of thermally sealed
bags, sacks, and containers.

The polymers suitable for formulating hot-melt
adhesives include ethylene-vinyl acetate copolymers,
polyethylene, and polyurethanes /5/, also polyamides
and polyesters; waxes are also used /2/.

Bonding is achieved by solidification of the molten
adhesive layer. This determines both the advantages
and drawbacks of these adhesives. The advantages
include rapid bonding and the possibility of joining
porous materials. The main drawback lies in the necessity to maintain relatively precisely the temperature
of the hot-melt adhesive during the process. That is
why applications of hot-melt adhesives demand attention
to variations of the processing temperature and to factors
affecting it /8/.

The "processing temperature" means the temperature
at which the hot-melt adhesive is coated onto the
surface of the packaging material. The "processing
temperature" affects the viscosity of the adhesive
during application and the overall time of the bonding
operation. With increasing temperature, the solid adhesive
softens at first and becomes rubber-like. Further heating
converts the polymer to the state of plastic flow. Only
above the "flow temperature" can the hot-melt adhesive
wet perfectly the substrate and exhibit its adhesive
properties. Therefore, this flow temperature must not

be confused with softening temperature. Depending on
the polymer used, the hot-melt adhesive exhibits a
narrow or broad softening temperature range. A typical
case of dependence of viscosity on temperature of
a hot-melt adhesive is presented in Fig. 44.

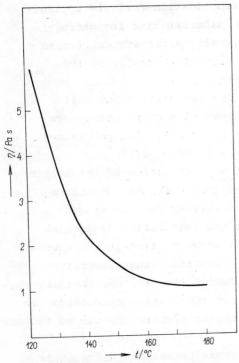

Fig. 44. Dependence of viscosity of a hot-melt adhesive
on the processing temperature /8/

The "flow temperature" is also determined by the
polymer used. For perfect wetting of the substrate
the "processing temperature" is chosen slightly above
the "flow temperature". The difference between the
"processing temperature" and the "flow temperature"
affects the time of the bonding operation, which is a
sum of the time needed for spreading the adhesive and
the time necessary for resolidification of the adhesive

layer. Therefore the "processing temperature" should not be very much higher than the "flow temperature". Moreover, an excessive increase of the temperature above the "flow temperature" does not bring any further viscosity decrease (cf. Fig. 50), but increases the danger of thermal degradation.

The time needed for solidification of the applied hot-melt adhesive layer (also the time for which the bond must remain under pressure) depends on several factors:

- the distance of the head of the spreading machine from the packaging material
- the ambient temperature
- the rate of spreading
- the thermal conductivity of the packaging material
- the thermal conductivity of the hot-melt adhesive
- the viscosity/temperature behaviour of the adhesive.

Lower processing temperature is sufficient for packaging materials with low heat conductivity, whereas higher temperatures must be chosen for materials with high heat conductivity. With respect to the heat conductivity of the adhesive it is essential to maintain constant layer thickness, which usually presents no problems with disc or roll spreaders where the high-pressure dosing of adhesive is relatively simple. Excess adhesive is removed from the discs or rolls by means of a knife, so that a constant amount of adhesive is transferred to the substrate. A more complicated situation is that of nozzle spreaders, where layer thickness depends on the number and size of the nozzles and on the design of the spreader.

Soft types of hot-melt adhesive generally have a broader solidification range than that of the hard or brittle adhesives which can solidify just below the flow temperature.

5.2.4 COATING COMPOSITIONS IN PACKAGING

Coating compositions play two roles, functional and
decorative, in the packaging. Functionally they may
protect packaging materials against the effects of the
contents or they may protect the contents (e.g. foodstuffs) against deterioration by substances liberated
from the package or permeating from the environment.

5.2.4.1 T y p e s   o f   C o a t i n g   C o m p o -
s i t i o n s

Coating compositions can be classified according to
the main components from which they are made:

Bituminous, hot-melt compositions of bitumen,
ceresines, natural and synthetic waxes and resins are
easily coated but rather soft and of low heat resistance. Also their price is higher than that of synthetic
materials.

Cellulosic, most frequently used in packaging as
the nitrate, acetate, and ethers, are applied to modify various materials, especially cellophane and
aluminium foil, also paper.

Ethanol-soluble compositions based on shellac
(a complex natural mixture of polyesters, derivatives
of hydroxy acids, especially 9, 10, 16-trihydroxyhexa-
decanoic acid /9/, a hardened resinous secretion of the
lac insect (*Kerria lacca*)), plant resins (e.g. *Manila
copal*), linear phenol-formaldehyde resins, are still
used as protective coatings for containers. Shellac
was also used to improve the lustre and hardness of
some synthetic coating compositions. Natural shellac
is replaced by acetaldehyde resins, the so-called
synthetic shellac (*Synthelac, Wacker shellac*), which
can be modified /10/ to increase the softening point
or combined with drying oils.

Oil coating composition contain drying oils as the main component sometimes combined with natural or synthetic resins (mainly phenolics or alkyds).

Alkyd resins are polyesters of three-dimensional structure formed by condensation of polyfunctional acids or anhydrides with polyfunctional alcohols /11/, the most important starting materials being phthalic anhydride and glycerol giving the name of the resins glyphthal resins:

phthalic   glycerol
anhydride

cross-linked resin

The alkyd resins are usually modified by drying oils, non-drying oils, or rosin. The drying oils are glycerol esters of higher unsaturated fatty acids, e.g.

oleic acid $CH_3(CH_2)_7CH=CH(CH_2)_7COOH$, $C_{17}H_{33}COOH$

linoleic acid $CH_3(CH_2)_5(CH=CH)_2(CH_2)_7COOH$ $C_{17}H_{31}COOH$

linolenic acid $CH_3(CH_2)_3(CH=CH)_3(CH_2)_7COOH$ $C_{17}H_{29}COOH$

Such modifications are usually carried out by direct reaction of the oil with phthalic acid and the corresponding amount of glycerol at $200^{\circ}C$, giving resins containing double bonds, and free hydroxyl and carboxyl groups; they can be air dried or used as baking varnishes. Air drying represents oxidation of the double bonds of the acidic component by oxygen and consequent cross-linking polymerization. The coatings thus produced are hard. If non-drying oils are used (e.g. castor oil or palm oil, which contain predominantly the saturated fatty acids, palmitic or ricinoleic), the resins are not air drying, but are suitable as plasticizers for nitrocellulose lacquers and other coating compositions.

Rosin, which is used for the modification of glyphthal resins, must be "hardened" by heating at 200°C with calcium hydroxide or zinc oxide. The acid[1] is transformed into its higher melting salt, and the acid value is decreased, which improves compatibility with alkaline pigments.

The modification, based on esterification of the free hydroxyl groups of the primary condensates with higher monocarboxylic acids, makes the resin insensitive to water and soluble in various organic solvents (ketones, esters, alcohols, and aromatic or aliphatic hydrocarbons); depending on the acid used, the derived coating films may have varied mechanical properties.

Modified glyphthal resins are also classified according to the oil content /12/:

- greasy, which contain 65% to 80% drying oils, are mainly used for air drying coatings
- medium, contain 50% to 65% oil
- dry, contain 30% to 50% oil.

The oil varnishes are used for metal packages. The baking gold varnish is still important for the internal lacquering of tin-plate cans.

Synthetic coating compositions are based predominantly on a large group of thermoplastic and thermosetting resins, of the latter the most widespread are polyesters (alkyds) modified with oils, phenolic and alkylphenolic resins, melamineformaldehyde resins, epoxides, and polyurethanes; the most frequently used thermoplastic resins are poly(vinyl acetate), polyacrylates, polyethylene, copolymers of vinyl chloride with vinyl acetate or vinylidene chloride.

---

[1] *Abietic acid is the basis of American rosin, whereas French rosin is predominantly α-pimaric acid.*

The phenol-formaldehyde coating compositions are
modified with oils, rosin, rosin esterified with
polyfunctional alcohols, or fatty acids.

Alkylphenolic resins provide films with better
light stability. They are condensates of formaldehyde
and p-substituted phenols, e.g. propyl-, butyl-,
pentyl-, benzyl-, and most frequently p-tert. butylphenol. According to their modification they yield
air drying or baking enamels for internal coatings
of tin-plate cans.

The unsaturated polyester resins can also form a
basis for solvent-free lacquers and are used for
coating metal containers especially for foodstuffs.

Epoxide resins give excellent baking varnishes
for tin-plate cans and aluminium tubes.

Melamine-formaldehyde resins are used most often
in combination with glyphthals as baking varnishes
for metals.

Thermoplastics are predominantly applied in the
form of dispersions (solutions and latices).

The polymers important for packaging include
poly(vinyl acetate), polyacrylates, and poly(vinylidene
chloride) latices or solutions for paper, foil and
film or sheet materials used in heat sealable packages.

5.2.4.2 M e t h o d s   o f   A p p l y i n g
         C o a t i n g s

If a coating dries by evaporation of solvent, the
choice of the solvent is as important as that of the
polymer. Usually solvent mixtures are used, the
individual solvents having different vapour pressures
so that they evaporate successively, which facilitates
their diffusion through the coating layer and prevents
the formation of bubbles and unevenness.

The coating compositions are applied by means of
a brush, paint-spraying gun, by dipping, glazing,

wiping off, etc.

The polymers which give low-viscosity melts at elevated temperatures (mainly the crystalline polymers like polyethylene and polyamides) can be applied as the melt. For low-molecular weight polyethylenes it is possible to use machinery of the same design as for dispersions; for higher-molecular weight polymers doctor knife application, as for poly(vinyl chloride) pastes, may be used.

5.2.5 PRINTING OF PACKAGING MATERIALS

Every package must satisfy the main requirement of protection of the goods packed and, besides that, it has further functions of no less importance: to provide customer appeal, product information, instructions on handling and use, and any requisite warnings. In this respect the printing of packages is indispensable.

The ways of printing plastics were adapted from the printing of paper, textiles, and leather /13/. Both rigid containers and flexible substrates, especially thermoplastic films, are printed. However, as polymers have different surface properties from those of paper and textiles, both machines and inks had to be suitably adapted.

Nowadays, five basic methods are used for application of ink onto plastics /13 - 16/:
1. typographic printing (relief printing),
2. intaglio printing (gravure printing, photogravure),
3. screen printing (serigraphy),
4. planographic printing (lithography),
5. graphical printing (letterpress, typography, flexography, offset).

The choice of method is governed by the type of packaging material, area of printing, the pattern, number of colours, surface pretreatments, thickness and width of film, rate of printing, number of meters per pattern, and final treatment of the film. Technical considerations are:

- character of the material printed, i.e. soft (paper, cellophane, plastics, metal foil) or hard (metal sheet, glass); in the latter case screen printing and offset printing only can be applied

- behaviour of the material printed, i.e. whether the printing ink is absorbed (paper) or not (metals, glass) or even repellant (polyethylene); in the last instance a surface treatment must be applied before printing (e.g. surface oxidation of polyolefins). Sometimes special demands are made on the printing ink, e.g. resistance to boiling water, high sterilization temperatures, etc. The capital and maintenance costs of printing rollers, essential to the choice of the printing method, affect the final price of the package.

The printing inks must exhibit primarily good adhesion to the substrate. This property can be checked in various ways /15/, but one of the simplest practical methods uses adhesive tape which is applied to and then pulled off the surface. If the removed tape bears no traces of the printing ink, the adhesion can be considered satisfactory /16/. The pigments used can be both organic and inorganic. They must, however, be compatible with the binding agent and must be insoluble in the plasticizers used in the production of the corresponding film, otherwise the colour would migrate into the substrate and the intensity of the printing would decrease. Solvents and diluents in printing inks should be acceptable for health and hygiene. Most often used are alcohols, esters and ketones, less

frequently chlorinated and aromatic hydrocarbons.
The following further requirements must be met by a
satisfactory printing ink /13/:

1. light fastness,
2. resistance to bases and fats,
3. rapid drying, to prevent migration and
blurring of colours during subsequent colour printings,
4. the drying rate must relate to printing rate, to
avoid drying out on printing rollers or in screens,
5. the printing ink must not cause permanent
distortion of the substrate,
6. the viscosity must be suitable for the printing
method chosen.

5.2.5.1 Typographic Printing

The print is transferred to the substrate from inked
raised areas. Depending on the quality of the film, up
to 12 colours can be printed simultaneously at a speed
of 60 m min^{-1}. The film is stretched very little, thus
avoiding print distortion.

The printing machine can   contain up to 12 printing units placed at the periphery of a large cylinder.
Each unit has an independent printing roller made of
light alloy (aluminium), which takes the printing
ink from a textile support supplied from a dipping
roll located in the ink reservoir. The doctor blade
is immovable and wipes off excess printing ink from
the textile support and not from the printing roller
(as it is the case with gravure printing). Therefore
the rollers can print up to 5,000,000 meters of film
without getting worn.

5.2.5.2 Gravure Printing

This uses a recessed printing roller and is the

opposite to typographic printing. The machine
consists of a rubber-coated metal cylinder around which
the printing units are placed, viz. ink reservoir,
transfer roller, printing roller, and steel doctor
blade. The transfer roller in the reservoir transfers
ink uniformly to the printing roller; excess ink is
removed from the printing roller with a ground steel
blade. For multicolour printing, a stretchproof support
made of rubber-coated textile is used. Gravure printing
is the most widespread method of printing plasticized
poly(vinyl chloride) film.

5.2.5.3 S c r e e n   P r i n t i n g

Screen printing is applied increasingly to printing
plastic packages used in the foodstuff industry,
pharmacy, and similar fields. It allows printing of
flat as well as cylindrical and conical surfaces. The
original simple manual apparatus was gradually improved to the present state of highly efficient
automatic machines with a capacity as high as 6000
packages per hour (printed in several colours simultaneously).

The most important part of the equipment used for
the process is the screen made of metal or polyamide
and having 50 to 100 mesh mm^{-2}: the ink is pressed
through the screen by means of a squeegee (made
usually of polyurethane) onto the substrate. The
printing screens for industrial purposes are prepared
by photomechanical methods to provide the pattern of
blocked mesh apertures.

5.2.5.4 P l a n o g r a p h i c   P r i n t i n g

This method is used especially for printing packages
made of plasticized poly(vinyl chloride) film. The

pattern is printed first on paper with a thin layer
of alkyd resin or other suitable plastic. Then, poly-
(vinyl chloride) paste is poured on this paper to a
thickness of 0.05 to 0.1 mm. The paper with the paste
is placed in a drier where the poly(vinyl chloride)
layer is gelled. The printing is transferred from
the paper to the film which is then separated from
the paper (therefore, this technique is also called
transfer printing). The paper remains perfectly clean
and can be reused.

5.2.5.5 Polygraphic Methods of Printing

Letterpress printing (relief printing) is the oldest
printing technique /14/ and makes use of metal plates
(blocks). Although the method allows printing of
various packaging materials (paper, metal, cellophane,
plastics, etc.), it is only applicable to flat surfaces (sheets).

Flexography uses rubber plates with patterns which
are transferred directly to the film by pressure. The
machines are similar to those used in relief printing.
The term aniline printing is used, when the method
applies aniline printing inks: these inks are low
viscosity liquids containing either suspended or
completely dissolved pigments, the diluent present
being rapidly evaporated (i.e. in several seconds).
As the printing plates (or cylinders) are made of
rubber, the solvents chosen (predominantly alcohols)
must not attack rubber. The method allows up to six-
colour printing at a speed of 50 m min^{-1}. The ink
does not distort the film (which is often the case
with typographic and gravure printing). For some
purposes, e.g. printing of polyethylene, thermal inks
are also used /2/, which comprise pigmented low-melting

resins or waxes. Flexography is very frequently used
in packaging because the rubber plates can be easily
and quickly changed /16/; it is particularly economical for printing polyethylene film /14/.

Offset printing employs a smooth surface and makes
use of the incompatibility of fats and water. Those
parts of the printing surface which have to accept
and transfer the ink are covered with fat, the other
areas being covered with water. The printing ink is
transferred to the film either directly or by means
of an intermediate rubber roller. Offset printing
can produce up to six-colour prints, the rate being
the highest of all the methods described /14/.

5.2.6 EXPANDED PLASTIC PACKAGING MATERIALS

5.2.6.1 Basic Methods for Production of Expanded Materials

The materials used in packaging have densities from
0.4 to 7.8 g cm^{-3}, e.g. paper or wood, polymers, glass,
light metals, steel. The density of many polymeric
materials can be reduced substantially by techniques
giving expanded (cellular or foamed) materials. These
materials contain many small cavities (cells or pores)
of various sizes and shapes.

The expanded polymeric materials are mainly produced on the basis of homopolymers and copolymers of
styrene, polyolefins, polyurethanes, natural rubber,
synthetic rubbers, and phenol-formaldehyde and urea-
formaldehyde resins /17/. More recently, attention
has been focused on expanded poly(vinyl chloride)
which has the significant advantage of low inflamma-
bility.

The foaming process has three phases: initiation,

growth, and stabilization of cavities /18/. A porous
structure can be produced by thermal, catalytic, or
radiation decomposition of blowing agents, by chemical
interaction of individual components of the mixture, by
evaporation of low-boiling liquids, by means of inert
gases, or mechanically. Extraction of water-soluble
salts previously dispersed in the plastic has many
drawbacks, especially with thick-walled products,
and so has been largely abandoned /1/.

The porous structure produced in the polymer can
be stabilized either by cooling the polymer below its
glass transition temperature or chemically by cross-
linking. Either of the two processes must be programmed
so that stabilization is neither premature (before
sufficient foaming of the material) or delayed (the
foam formed loses its regular structure or shrinks).

5.2.6.1.1 Mechanical Foaming

A pre-polymerized material or partially condensed
resin of low viscosity is submitted to rapid stirring
to produce a foam which is then stabilized by completing the polymerization or polycondensation with
heat. This procedure is applied to urea-formaldehyde
resins; an aqueous solution of the resin is mixed
with a hardener (phosphoric or naphthalenesulphonic
acid), emulsifier, and foam stabilizer, and the foam
produced by stirring (whipping) is poured into a mould
where it is hardened at room temperature and finally
dried at 40-60°C /1/.

Poly(vinyl chloride) plastisols are also mechanically foamed. The fine air bubbles formed are stabilized with foam-forming agents, i.e. surfactants /19/.

5.2.6.1.2 Blowing with Inert Gases and Low-Boiling Liquids

This method is used especially for expanded polyolefins, polystyrene, and poly(vinyl chloride) /1/.

Expanded polystyrene (EPS) is made in the following way: a blowing agent (e.g. ethyl chloride) is dissolved in the molten bulk polymer at elevated pressure; on cooling, the rubber-like mixture ("dough") obtained is ground into granules which are pre-foamed in hot water at about $96^{\circ}C$ (with up to 30 times volume increase). The dried pre-foamed granules are then re-expanded in perforated moulds of the required shape by applying heat and steam to form the final product.

Cellular poly(vinyl chloride) from plastisols uses besides mechanical foaming saturation of the polymer with an inert gas (air, nitrogen, carbon dioxide, or combinations of gases like methane, propane, ethylene, propylene, helium, argon, air, nitrogen), or with low-boiling liquids, e.g. pentane, hexane, heptane, octane, hexene, pentene, ethanol, acetone, ethyl methyl ketone (2-butanone).

The blowing of poly(vinyl chloride) with inert gases can proceed either under pressure or without pressure. As the pressure methods are more expensive and demanding with respect to machinery, the pressureless process is preferred. The most important factor is to maintain the paste viscosity which affects operation of the equipment. The plastisol is saturated at about $-5^{\circ}C$. The temperature must be maintained carefully, because it affects the amount of gas dissolved and, hence, the average density of the expanded polymer; the saturated paste is poured onto a conveyer belt or into moulds where the expansion (foaming) takes place. Gelation is carried out continuously with high-frequency dielectric heating, or in the case of moulded products in a hot-air tunnel.

The pressure method is used for both soft and hard expanded poly(vinyl chloride). The technology is practically the same in the two cases, the only difference being in the consistency of the starting mixture. In the pressure method, saturation with gas is carried out in autoclaves or air-tight moulds at about 165°C and at 10 to 50 MPa followed by cooling. The polymer used for the saturation can be e.g. in the form of film. The material is then expanded at normal pressure by reheating at $70-100^{\circ}$C. The saturation can be similarly performed with dry blend or with poly(vinyl chloride) pastes.

5.2.6.1.3 Blowing with Chemical Components

This method is applied mainly to expanded polyurethanes. It makes use of the facile reaction of isocyanate groups with water or organic acids to produce carbon dioxide:

$$R-N=C=O + H_2O \longrightarrow R-NH-CO-OH \longrightarrow RNH_2 + CO_2$$
$$R-CO_2H + O=C=N-R' \longrightarrow R-CO-NHR' + CO_2$$

1,6-Diisocyanatohexane is used for soft linear polyurethanes and the isomeric mixture of 2,4- and 2,6-diisocyanato-1-methyl-benzenes for hard expanded polyurethanes. Both are physiologically hazardous, but they are advantageous in being easily produced and, hence, cheap. 1,5-Diisocyanatonaphtalene exhibits the highest reactivity, whereas 1,1´-methylenebis(4-isocyanatobenzene) is most favourable with regard to physiological aspects. A number of other diisocyanates are also used. The trade name Desmodur is used for diisocyanates.

The second reactive component is a low-molecular

weight polyester or polyether obtained by reaction
of dicarboxylic acids with polyfunctional alcohols,
giving excess hydroxyl groups. The number of free
hydroxyl groups in the polyester (*Desmophen*) may be
varied and affects the properties of the final polyurethane (from elastic types to rigid materials). The
polyesters used are based on phthalic, adipic, and
sebacic acids, and diols (ethyleneglycol, butylene
glycol) or triols (glycerol, hexanetriols, trimethyl-
olpropanes, trimethylolethanes). Besides the hydroxyl
groups, free carboxylic groups are also present and
they react with isocyanates to give carbon dioxide.

The consistency of the expanded product is determined
by the ratio of polyester and diisocyanate used. If
the amount of polyesters is greater than stoichiometric,
the product is soft, the cross-linking is limited,
and unbound polyester acts as a plasticizer. If
excess diisocyanate is used, hard products are formed
which contain residual isocyanate groups able to
undergo further reactions.

Production of expanded polyurethanes requires the
addition of water for foam formation. The carbon
dioxide formed by the reaction of isocyanate groups
with water blows the reaction mixture before the
hardening, which must follow immediately. Excess water
causes the foam to break and collapse. The water can
be added in the form of vapour or crystalline salts,
but usually it is added as liquid.

Catalysts also play an important part: they are
essential for correct foaming and hardening. Whereas
acidic substances hinder the reaction, bases, e.g.
triethylamine, diethanolamine, hexahydrodimethylaniline,
pyridine, quinoline, have an accelerating effect.
Tertiary amines containing at least one ester group
exhibit an equivalent catalytic effect; their advantages lie in their solubility or emulsifiability in

water, and low odour. Examples are 2-(diethylamino)-
ethyl acetate and 2-(dimethylamino) ethyl acetate. The
amount and type of catalyst control the foaming rate.
The usual ratio is 0.5 to 2.0 wt. parts of catalyst
per 100 wt. parts of polyester.

Homogenization of the components and foam stability
depend on the presence of emulsifiers which affect
shape and size of the cavities. Typical emulsifiers
are sulphonated castor oil, silicone oil, and triethyl-
amonium oleate.

Powdered or fibrous inorganic fillers are added to
increase the average density and strength. The powder
fillers also aid nucleation to promote formation of
uniform cellular structure. Expanded polyurethanes
can be coloured by suitable pigments added before
foaming.

Successful production of expanded polyurethane
necessitates rapid preparation of a homogeneous reaction
mixture and pouring on a conveyer belt where foaming
and hardening take place.

The mixture can be prepared from three streams:
polyester resin (*Desmophen*), diisocyanate (*Desmodur*)
and a mixture of catalyst, emulsifier, and water.

In the two-component system, one component is a
mixture of the polyester resin with catalyst, emulsifier, and water, while the other component is the
diisocyanate. A more successful two-component system
contains a pre-polymer of all the diisocyanate and
part of the polyester as one component, the other
component being the rest of the polyester with the
other additives. Thus, due to small viscosity difference between the two components, their homogenization
is achieved more quickly, the increased cross-linking
resulting from use of pre-polymer increases the final
strength of the expanded material; also this method
can use polyesters which have very high viscosities.

The foaming of polyurethanes is carried out usually at atmosphere pressure. The heat of the exothermic reaction is sufficient for hardening the foam. Both foaming and hardening are finished within a few minutes, but the required mechanical properties are reached only after 6-10 hours. Heating to about $70^{\circ}C$ accelerates the final hardening of the foam.

5.2.6.1.4 Use of Chemical Blowing Agents

The applicability of a chemical blowing agent for a given polymer and conditions of aftertreatment of the plastic are determined by the decomposition temperature at which the blowing agent liberates gas. The required properties of blowing agents as well as the main types are given in Section 3.4.7.

5.2.6.2 The Factors Affecting Behaviour and Applications of Expanded Materials

Expanded polymeric materials are increasingly used in packaging because of their low density at relatively good strength, and very good insulating properties. Frequent use is made of expanded polystyrene, polyurethane, and polyethylene for protection againts impact during transport and handling, and against temperature variations. The products include trays, supports for collated packages, bottle crates, etc. An interesting application consists in the direct spraying of polyurethane foam from a pressure aerosol container on the article placed in a box or crate (the Mini-Foam method) /20, 21/. Both dynamic and static properties of expanded materials are determined by the polymer as well as by the structure of the foam, the latter factor being no less important.

According to the character of the cavities (cells
or pores) we can differentiate soft moss-like materials
which have small uniform closed cells filled with air
or some other inert gas and soft sponge-like materials
whose cavities are interconnected pores; these two soft
types have their hard counterparts, viz. foamed materials
with interconnected cavities.

The structure of expanded polymeric materials depends
on a number of factors:

1. the blowing method chosen,
2. the polymer and the additives (i.e. polymeric
mixture),
3. the blowing agent,
4. the viscosity of the polymeric mixture within the
temperature range appropriate to the blowing agent,
5. the temperature and duration of blowing.

The decomposition of a blowing agent can be controlled
by addition of kickers which are described in Section
3.4.7. The available information on their use /22, 23/
indicates that fast kickers give structures with more
closed cells, whereas slow kickers lead to more open
cavities.

The way in which the composition of the polymeric
mixture affects the structure of the expanded material
will be illustrated with poly(vinyl chloride). For
rapid blowing and stabilization of the foam, the best
results were obtained with polymers having K values
below 70 /22, 23/. The plasticizer is also important.
Low-molecular weight poly(vinyl chloride) and a plasticizer exhibiting solvation form a low-melting mixture,
and if the blowing agent is finely divided, the amount
of gas liberated will be small and closed cells will
result. The soft expanded material obtained also has
lower density than those obtained with plasticizers

showing weaker solvation. These plasticizers give
higher-melting mixtures, and if the blowing agent is
coarser, the amount of gas liberated can break the
cell walls and penetrate through the melt, the resulting material being soft with open pores.

Table 37. Properties of some soft expanded materials /1/

Properties	Expanded material based on		
	rubber latex	plasticized poly(vinyl chloride)	poly-urethane
mean density (g cm^{-3})	0.109	0.112	0.086
resistance to fire	combustion	melting	melting
tensile strength (MPa)	0.23	0.22	0.17
elongation at break (%)	455	190	250
deformation: statical at			
70°C (%)	6	10	8
dynamical (%)	15	35	22

Table 37 gives physico-mechanical data on some
expanded materials. The highest strength is attained
with cellular plastics having closed cells. The
dependence of compression strength on the density and
cell structure is given in Fig. 45, and Table 38, which
relate some properties of hard expanded polyurethane
to its density. As varying cell sizes impair the mechanical properties of expanded materials, it is always
desirable to secure uniform cell size. With respect
to chemical composition, expanded materials prepared
by polycondensation (e.g. polyester resins) have lower
strength than those obtained by cure (expanded rubber),
polymerization (polystyrene), or polyaddition (polyurethanes).

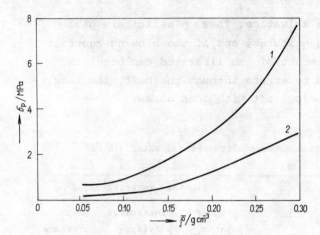

Fig. 45. Dependence of the compressive strength on the mean density and structure of cells of a cellular material /1/

1 - spherical closed cavities, 2 - fibriform open cavities

Table 38. Properties of hard expanded polyurethanes /1/

mean density (g cm^{-3})	0.050	0.100	0.150	0.200
modulus of compression (MPa)	6	18	40	70
flexural strength (MPa)	0.4	1.0	2.1	3.2
impact strength (kJ m^{-2})	0.2	0.4	0.6	0.8
tensile strength (MPa)	0.2	0.7	1.2	2.0
shear strength (MPa)	0.1	0.5	0.9	1.5
compressive strength (MPa)	0.3	0.9	1.8	2.8
water absorption (%)	2.0	1.75	1.5	1

Expanded plastics are used in packaging mainly as space-filling materials protecting goods sensitive to impact or pressure. They have two main functions:

- to compensate for the unevenness of bearing sur-faces, which is particularly important in the case of fragile articles and those with a sensitive surface (porcelain, glass, fine mechanics, fruit, etc.);

- to dampen, i.e. to weaken shocks and impacts by extending the deceleration period; their application is important in packing articles with sensitive inner construction, e.g. measuring and reproduction apparatus, clocks, etc.

An example is the expanded high density polyethylene produced by Dow Chemical Co. under the name *Ethafoam*. It can be vacuum shaped and sealed by heat /24/.

Other examples are the short strips of expanded polystyrene, which can be joined by high-frequency heating into one block around the article packed /25/, or an expanded polystyrene block with cavities of the shape of the articles packed. Expanded polystyrene is also used for trays, transport crates, cups, etc.

Aerosol polyurethane sprays (*Mini-Foam*, USA) can be used to produce foam directly around the packed articles, e.g. TV picture tubes, electric motors.

5.3 PLASTICS CONVERSION PROCESSES FOR PACKAGING MATERIALS

The plastic state is utilized in such methods of processing as compression moulding, injection moulding, extrusion calendering, extrusion coating, and blow moulding.

As the temperature behaviours of thermoplastics and thermosetting resins are different, the conditions of processing differ also. Thermoplasticity and thermo-reactivity have their advantages and drawbacks. It is advantageous that there is no chemical reaction during the heating of a thermoplastic which would influence melt consistency, so the material can be kept longer in its plastic state as compared with thermosetting resins, unless significant thermo-degradation takes place. On the other hand, thermo-plastics require cooling as part of the forming

process, to prevent deformation occuring, e.g. during
removal from a mould.

Thermosetting resins are usually heavily filled (40
to 60%) to modify or improve the properties of the
final product, whereas thermoplastics are processed
without fillers or with only small amounts which modify
the appearance or reduce the price of the product.
Polyolefins represent an exception: they are filled for
some applications either with mineral powders, particularly those based on magnesium, calcium, or aluminium
silicates, or fibrous materials, especially asbestos
and chopped glass fibers (up to 40%); this not only
reduces cost but also increases stiffness, hardness,
distortion temperature, and creep resistance. However,
ductility is reduced, and powdered fillers also reduce
toughness and surface gloss.

Polymers for processing should contain and absorb
water as little as possible, because moisture makes
processing difficult and impairs the quality of the
product. Some types of plastics and additives have to
be dried completely prior to use.

The pressures needed, particularly for the injection
moulding of thermoplastics, depend on the shape of
product, the grade of polymer, the length and cross-
section of injection channels, etc. Pressures of 24
to 150 MPa are used for thermoplastics. The shape of
filler particles also affects the pressure needed when
processing thermosetting resins.

Thermoplastics generally shrink less than thermosetting resins during the transition from liquid to
solid (by less than about 0.5%, see Table 39). Greatest
shrinkage occurs with high-molecular weight polymers
which, on transition from liquid to solid, undergo
crystallization to a high degree; this increases their
density as with polyamides and polyethylenes. On

Table 39. Shrinkage of some thermoplastic materials on transition from melt to solid state /1/

Polymer	Shrinkage (%)
cellulose acetate	0.2 - 0.3
ethylcellulose	0.4 - 0.7
polystyrenes	0.2 - 0.5
polyacrylates	0.2 - 0.3
poly(vinyl chlorides)	1.0 - 1.7
polyamides	< 3.5

average, shrinkage is higher (about 1%) with the thermosetting resins due to considerable increase in molecular weight and consequent increase in desity (e.g. phenol-formaldehyde resin without filler 0.9-1.1%).

In contrast, the linear thermal expansion of amorphous macromolecular compounds is lower with cured thermosetting resins than with thermoplastics whose intermolecular bonds between chains are released by heat, increasing both mobility and volume (Table 40).

Thermoplastics are processed above their softening point which varies according to polymer type, for polyethylene varying from $120^{\circ}C$ to $200^{\circ}C$, for polyamides from $200^{\circ}C$ to $260^{\circ}C$. The processing temperatures for thermosetting resins are usually within the range from $140^{\circ}C$ to $180^{\circ}C$, and the temperature determines the pressure needed.

Table 40. Linear expansion (α) of some plastics /1/

Polymer	Linear expansion, $\alpha \times 10^6$, K^{-1}
Thermoplastic materials:	
ethylcellulose	100 – 140
polyethylene	240
polystyrene	70 – 100
poly(vinyl chloride)	80
poly(methyl methacrylate)	82 – 130
Thermosetting resins:	
casting phenol-formaldehyde resins	29 – 95
carbamide resins	40 – 50
aniline resins	45

5.3.1 PRODUCTION OF FLAT PACKAGING MATERIALS BASED ON PLASTICS

The term plastic flat packaging materials includes
especially films, lay-flat tubing, sheets, and laminates.

Plastic films are made by a number of processes,
particularly using screw extruders with a die, the blown
tube process, and calendering.

The extrusion blowing process can produce very wide,
thin-walled tubes, which are suitable for conversion
to sacks and bags in a simple and cheap way.

These bubble extruded films are melt-stretched in
both longitudinal and transverse directions during the
process, and have relatively homogeneous properties.
In contrast, slot die-extruded films are stretched
longitudinally only, and consequently they have higher
strength in this direction than in the transverse
direction. Air-blown film is more slowly cooled than
film extruded into water, which causes a higher degree
of crystallinity in the former. This is undesirable in

some cases, because crystallinity is accompanied by
properties unsuitable for some packaging applications,
for example, blown polyethylene films are opalescent
rather than transparent.

From the processing point of view, however, it is
the bubble process which is often more attractive
than slot die extrusion, particularly for very wide
film. Production of the blown film is, however, associated with greater risk of film defects due to impurities
in the polymer (foreign matter, insufficiently plasticized polymer - the so called "fish eyes", etc.) which
can cause holes, tearing, and rapid bubble deflation
during the process. The rate of production of blown
film for a given size of extruder is, however,
substantially lower than that using slot die extrusion,
because of the lower temperature employed (usually
bellow 180°C), and the low efficiency of air cooling.

In cases where the rate of production has to be
increased or where the tubular blown film would be
disadvantageous in the packaging application, slot die
extrusion is preferred. This process is also suitable
for sheet materials. The difference between films and
sheets is not very distinct; the term film is used
for materials up to 1 mm thickness, whereas sheets are
1-18 mm thickness. Besides extrusion, the main process
for sheet production is moulding.

The terms multi-ply and laminated are very wide even
in the field of packaging, since increasing demands
for special properties have initiated their rapid
development in both quantitative and qualitative terms.
The available processes are very varied: The main
methods of producing multi-ply packaging materials are
extrusion coating, coextrusion, and laminating.

5.3.1.1 Slot Die Extrusion Process for Films and Sheets

Many thermoplastic films and sheets are produced by slot die extrusion process. In this way, pellets or polymer granulates are transformed into flat packaging materials directly. Although this technology requires considerable investment, it ensures high production rates and both uniform and very high quality products.

Formerly, films were extruded vertically into a water cooling bath, and after drying they were wound up to rolls. This method is simple in principle but is connected with some engineering problems. One of these is the difficulty of maintaining uniform water temperature within the full width of the cooling bath. Uniform cooling is necessary to obtain uniform properties of the extruded film. Another problem concerns complete removal of water from the film surface. For high take-off speeds, a wringing and doctor blade arrangement must be installed before the final wind-up drum. Figure 46 shows this arrangement schematically.

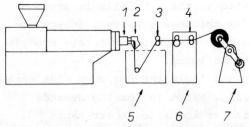

Fig. 46. Schematic representation of a production line for vertical extrusion of film through a slit die /26/

1 - extrusion die head, 2 - slit die, 3 - wringing rolls, 4 - doctor rolls, 5 - cooling water tank, 6 - doctor blade equipment, 7 - wind-up unit

In this way, however, only relatively thin films can be produced. With thicker films "undulation" occurs caused by uneven cooling.

The Chill-Roll Method was developed to obviate partially these difficulties: it is based on taking-off of the extruded film by two rolls arranged similarly as in the production of films by the calendering process (Fig. 47) /27/. The rolls ensure both perfect

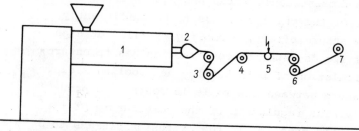

Fig. 47. Schematic representation of a production line for flat polypropylene film /27/

1 - screw extruder, 2 - extrusion die head, 3 - chill rolls, 4 - trimming equipment, 5 - surface modification for printing, 6 - take-off rolls, 7 - wind-up unit

film surface and adjustable and controlled cooling which ensure uniform quality. The speed of rotation of the rolls, which is related to the film cooling rate, can be adjusted according to production demands. This technology also extends to extrusion coating, cf. Sect.   5.3.1.4   /26/. Figure 48 shows the principle

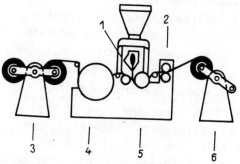

Fig. 48. Schematic representation of a production line for heat lamination on the extruded film /26/

1 - extrusion die, 2 - doctor unit, 3 - input equipment, 4 - preheating unit, 5 - lamination equipment, 6 - wind-up unit

of this process. Such a production line can also be
linked with the surface modification of films which
is sometimes required for the subsequent packaging
process, e.g., the surface of polyolefins has to be
oxidized before printing, see Section 5.3.1.7.1.

Slot die extrusion is exacting with respect to the
design and maintenance of the production line. The
diameter of the chill roll has to be sufficiently
large to ensure effective cooling of films at high
take-off speeds, and its construction has to ensure
fast circulation of both heating and cooling media.
The pressure between the rolls is controlled pneumatically. The regulation of the temperature of the
thermoplastic material in the extrusion die has to be
very precise; temperature indication is mainly by
thermo-couples installed in both the middle and sides
of the feed channel to the die. Pressure indicators
in the same positions are very useful also. Micro-
metric screws are used for controlling the die gap.
Most modern equipment includes apparatus for automatic
thickness checking and a β-ray thickness gauge recorder.

The quality of extruded film is evaluated according
to its appearance, strength, ductility, and dimensional stability /15/. The chill-cast process is
suitable for producing films from most thermoplastic
materials. The main types include polyethylene, polypropylene, poly(vinyl chloride), and cellulose acetate.
The film thickness lies in the range of 0.01 to 0.1 mm.
The maximum width of films is normally 2 m, but polyethylene films of up to 6 m width are produced /26/.

Sheets whose thickness does not allow winding up
into rolls are extruded as in the production line shown
schematically in Fig. 49 /15/. For their production
various types of polystyrene, polyethylene, plasticized
and unplasticized poly(vinyl chloride), polyamides,
acrylonitrile-butadiene-styrene terpolymers, and

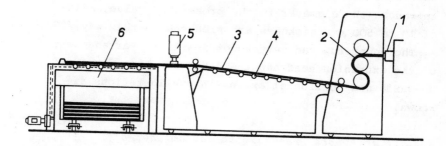

Fig. 49. Sheet production line /15/
1 - extrusion die head, 2 - take-up rolls, 3 - sheet,
4 - roller conveyer, 5 - cutter, 6 - stacker

cellulose derivatives can be used. They are produced
from 0.1 to 13 mm thickness and up to 2 m width./26/.

5.3.1.2 Blown Extrusion of Film and Tubing

The blow extrusion process is another continuous method
of production of both thermoplastic films and tubing.
This method /15/ is based on screw extrusion of thin-walled
tubes through a circular die. The tube is hauled off
through a pair of rubber cylinders (pinch rolls) which
close the tubing so that the air expanding it cannot
escape. The pressure of the air introduced into the
extruded tube through the centre of the mandrel is very
small: 0.15 to 1 Pa is sufficient. The air is introduced
at the beginning of operation, and rarely has to be
augmented for several hours; it is automatically controlled using a hydraulic valve.

The blowing increases the diameter of the tube, while
the haul-off rate controls elongation in the machine
direction. For automatic film width adjustment, detectors
are used which operate by a photocell. When the tube

diameter increases above the given limit and interrupts
the light beam, an electromagnetic valve lets out air
to reduce the diameter to the prescribed value. Films
of 10 to 500 µm thickness are produced in this way /27/.

The apparatus can vary according to the arrangement
of the extrusion head and the haul-off: thus down-ward,
up-ward, and horizontal extrusion configurations are
known.

5.3.1.2.1 Bubble Process with Bottom Haul-off

Bottom haul-off is especially useful for small diameter
tubing hauled off at a speed of up to 30 m min^{-1}. This
version avoids overfilling of the circular die with
melt, which is very important, particularly with low-
viscosity melts (e.g. polyethylene with a high melt
index, polyamide, poly(ethylene terephthalate)). The
process has the drawback that the tube is stretched by
its own weight. The blown tube is cooled through a
4-5 m distance between the die and the pinch rolls by
an external stream of cold air introduced through an
annular cooling ring around the die. The extruder can
be placed on an upper floor, and the tube taken off
through a hole to the lower floor.

5.3.1.2.2 Vertical Bubble Process Upward

When using this method in contrast to the previous one,
the whole line can be placed on one floor of 6 m ceiling
height. A typical setup for this version is shown in
Fig. 50. In this arrangement, however, the circular die
is very often overfilled, especially so when low-viscosity melts are used. There are problems with efficient
cooling of the film bubble, because it is heated by warm
air rising from the extruder head. Feeding the film
between the pinch rolls also is more difficult.

In spite of these disadvantages, this technique is usually preferred, particularly for polyethylene, because it is possible to view and control the whole operation from one position, and extruder feeding is easier.

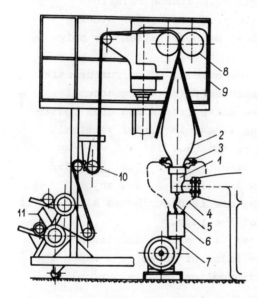

Fig. 50. Schematic representation of the bubble process /15/

1 - bubble head (cross head die), 2 - film, 3 - air cooling ring, 4 - cooling air inlet, 5 - air inlet, 6 - pressure rectifier, 7 - ventilator, 8 - take-off rolls (nip rolls), 9 - plateau, 10 - cutting equipment, 11 - wind-up unit

5.3.1.2.3 Horizontal Bubble Process

In this arrangement, the horizontally extruded tube has to be supported with a conveyer. An advantage of this method is in its applicability under low ceilings. Also the construction of the direct extruder head used here is simpler than that of the cross head used in both vertical methods.

The drawback lies in the tendency to non-uniform
film thickness due to unsymmetrical supporting and
cooling of the tube, which is why the horizontal bubble
process is not very widespread. It has been successfully
used e.g. for vinyl chloride-vinylidene chloride
copolymers (Saran) /28/ film production, for blowing
of hard poly(vinyl chloride) or polyethylene or polyamides /27/.

In order to eliminate the effects of circumferential
gauge variation in the extruded film (which accumulate
during winding up and make uneven reels and difficulties
on automatic thermoforming lines) in either vertical
arrangement the extruder or the haul-off and wind-up
assembly may be rotated, or oscillated (Fig. 51) /29/.
The frequency of rotation is 0.2 min^{-1}, and any
unevenness produced at the die is distributed all
round the circumference of the tube.

Recently, bubble-process film production lines have
been developed to produce films of up to 12 m width for
special purposes (e.g. for applications in agriculture).

5.3.1.3 F i l m   P r o d u c t i o n   b y   M e a n s
        o f   C a l e n d e r i n g

The calendering process is used in the production of
plastic films and sheets. The calenders are generally
classified according to the number of rolls (two to
five). The machines having more than five rolls are
used only exceptionally. The two-roll calenders are
used for both mixing and gelling blends of plastics.
Films are produced only on multi-roll calenders of
various roll arrangements.

5.3.1.3.1 *Calenders*

The most widespread machines are four-roll calenders,

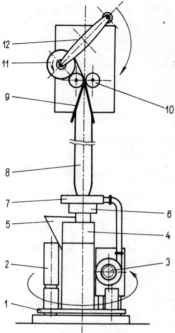

Fig. 51. Vertical arrangement
of bubble process unit with
turning extruder /27/

1 - turn plate, 2 - screw extruder drive, 3 - ventilator,
4 - screw extruder, 5 - hopper,
6 - direct (horizontal) head
for tubular film, 7 - cooling
ring, 8 - trapped air bubble,
9 - collapsing guide, 10 - take-
off rolls (upper nip rolls),
11 - roll of the final film,
12 - arm of the revolver
wind-up unit

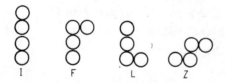

Fig. 52. Schematic representation of various
arrangements of calender rolls: I - four-roll superimposed calender, F - four-roll inverted L-type
calender, L - four-roll L-type calender, Z - four-roll
Z-type calender /13/

whose basic types are shown in Fig. 52. The oldest were
mainly I-types (superimposed), but they were not fully
satisfactory either from the design aspect (difficult
adjustment of the precise distance between the rolls)
or with respect to the total height of the machine,
or its accessibility for the operator. These were
gradually replaced in turn by the L, F, Z, and inclined
Z-types. The Z-types have lower height and less cylinder
deflection   with improved thickness control of the
film produced. However, roll adjustment is less easy
than with the L and F-types, and assembly more difficult.
In these respects, the arrangement of the rolls on inclined Z-type represents an improvement.

The Z-type also has the advantage that it obviates
"floating" of the roll which in the other types allows
rheological changes of the plastic melt to produce
thickness variations.

However, the Z-type four-roll also has its drawbacks,
particularly with regard to assembly. In this respect
both the F and the L-types are better.

The five-roll calenders are arranged similarly, and
another type - C-type - has top and bottom rolls set
out of centre. Three-roll calenders are usually built
as the I-type, and sometimes the rolls are slightly
set out of centre, so that the position of the middle
roll may be stabilized.

5.3.1.3.2 Haul-off and Winding Equipment

The film produced is taken from the calender and to
a cooling unit of several water-cooled rolls. After
the first cooling roll, the edges of the film are
trimmed off by rotating disc knives. If a number of
such knives are placed side by side, the film can be
cut into several bands. The edge-trim is recycled
through heated rolls to the calender.

The film cut to the prescribed width and partially
cooled passes to a conveyer where the residual stress
is released. After this shrinkage, the film is reeled
up on cylinders (drums) or tubes. Drums are employed if
the film is to be used in the form of cut sheets:
these are made by cutting the film wound on the drum.

If the film is needed in the form of continuous web,
it is wound up on tubes. It is essential for this
operation, that the film be transversly stressed all
the time; the commonest method uses rollers with
thread grooves which are left-hand on one half of the
roller and right-hand on the other.

5.3.1.3.3 Production Lines

As in most industrial production, modern machine
design is directed to continuous production with full
mechanization and automation.

Screw extruders are usually employed for continuous
mixing and Figure 53 gives an example of a production
line arranged on this basis /13/. The overall design
of the line depends on the type and variety of plastic
mixtures to be processed. Film thickness is measured
either mechanically (with a thickness gauge pressing
the film to the roll), or electrically (by condenser
capacity changing with film thickness), pneumatically,
or by means of radioisotopes /25/.

The polymers used in calendering include elastomers
and plastics exhibiting rubber-like character over
a sufficiently broad temperature range, especially
poly(vinyl chloride), copolymers of vinyl chloride
with vinyl acetate, mixtures of low density polyethylene with at least 25% polyisobutylene, high density
polyethylene (usually with small amount of transformer
oil), etc. /15/.

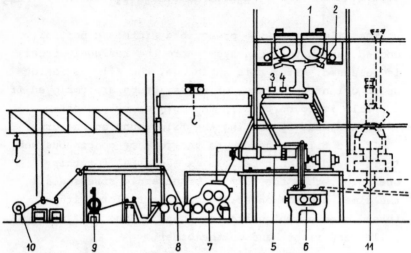

Fig. 53. Schematic representation of a production line
for calendering of film materials with continuous
preparation of mixtures in a screw extruder /13/

1 - mixer, 2 - vibrator, 3, 4 - metal detector (for
removal of iron particles), 5 - gelation screw machine,
6 - mill rolls, 7 - four-roll calender, 8 - cooling
drums, 9 - winding up of unfinished film, 10 - winding
up and finishing, 11 - an alternative arrangement with
pressing kneader

5.3.1.4 Methods of Production of Laminated Materials

Sheet materials composed of two or more layers are
widely produced for packaging: they are called laminates. The individual layers are formed either from
various kinds of plastics or produced by combination
of thermoplastics with other materials, e.g. paper or
aluminium foil, to meet specific requirements.
Properties and characteristics of sheet packaging
materials are discussed in Section 5.3.2.5.2.

Laminates are produced in different ways depending
on the properties of the materials used and on process
cost.

For coating polymer webs with dispersions (solutions,
latices) or melts (low-molecular weight easily melted
plastics) it is usual to adapt the equipment used in
the paper or textile industries, e.g. a three-roll
coater (Fig. 54) /15/. Common coating materials are
solutions of cellulose nitrate, latices based on
poly(vinyl chloride) or copolymers of vinyl chloride
and vinylidene chloride, and melts of low-molecular
weight polyethylene. In the case of dispersions, one

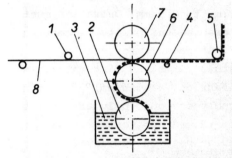

Fig. 54. Schematic representation of a three-roll
coater /15/
1 - guide rolls, 2 - wading roll, 3 - melt or dispersion bath, 4 - supporting stick, 5 - chill roll,
6 - coating roll, 7 - pinch roll (press roll),
8 - substrate

disadvantage lies in the necessity of using an efficient
drying tunnel (which takes much space), and in the
case of solutions it is necessary to include expensive
exhaust or solvent regeneration equipment, usually
explosionproof, whose installation is only economic
for efficient production lines.

More viscous systems (pastes, melts, or high-molecular weight polymers) can be applied by doctor blade
coaters.

For coatings of wide thickness range it is usual
to use the versatile extrusion coating process which
was mentioned in the description of slot die extrusion
(Sect. 5.3.1.1). The production line (Fig. 46) consists

of unwind and rewind stations, a slot die extruder, and a pair of nip rolls. The polymer melt is extruded through the die as a film which is pressed on the substrate by a smaller cooled metal roll with a polished surface, the larger cold backing roll being rubber coated. The line speed is considerable; with polyethylene coating, it is possible to reach 200 m min^{-1} without difficulty. The advantages of this process lie in high productivity and the ability to produce coatings of thickness 0.1 mm to 0.9 mm. Operation must be continuous to achieve precise uniformity of coating.

This method is used advantageously for coating thermoplastics onto infusible films and flat materials (e.g. cellophane, paper, fabrics, aluminium) or coating relatively lower-melting polymers onto polymeric films with higher softening temperature (polyesters, poly-amides). To increase adhesion between the layers it is sometimes recommended to use primers (tie coats, or anchoring layers) which are applied (e.g. as lacquers) immediately after unwinding the substrate and dried before extrusion coating.

Roll coaters may take the place of extruders. Thermoplastic material is melted on the rolls and the coating film is taken off through the gap between the rolls (Fig. 55) /15/.

If several plies of the same or of different materials are extruded together (using several extruders) in two or more streams and fed into a multi-channel die for slit-die or blown film extrusion, then the final product is a multi-layer material (film, sheet, tube), and the process is called coextrusion. This single-step pro-duction of multi-layer materials was developed in USA about 20 years ago and offers (in contrast to the other routes) some substantial advantages. It can achieve thinner individual plies and better inter-layer adhesion, because the materials are joined prior to coming into

contact with air, which would decrease adhesion by
surface oxidation /30/. On the other hand, there are
some difficulties, such as the impossibility to recycle
scrap material, and the requirement for the melting
temperatures of the individual materials to be within

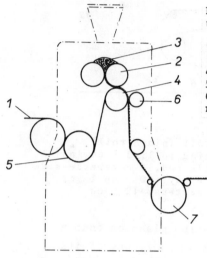

Fig. 55. Schematic representation of a roll coater /15/

1 - paper, 2 - mill rolls
(grinding), 3 - polyethylene,
4 - supporting roll,
5 - preheating roll,
6 - polishing roll, 7 - chill
roll

a specified range.

Therefore, blown film coextrusion is more economical
as it does not produce the waste of edge-trim. When
homogeneous materials are coextruded, e.g. two polyethylene films (film tubes for milk packages), the
individually extruded plies are inseparable, and perforations (and porosity) minimized. If the polymers used
are chemically dissimilar, adhesion of the layers may
be insufficient, requiring an interlayer of sufficient
adhesion to both layers. Thus polyethylene was successfully combined with polyamide using ionomer resin, which
is more expensive but can be applied in very thin but
effective layers (5 μm to 10 μm).

   Doubling of films made of thermoplastics can also
be carried out in multistage presses (which are most
often used for solid materials) or in drum presses
(Fig. 56) the choice depending on whether subsequent
use requires continuous film or sheets.

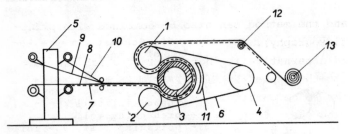

Fig. 56. Drum press /1/

1 - drawing roll, 2 - press roll (pinch roll),
3 - heated drum, 4 - stretching roll, 5 - unwind
unit, 6 - conveyer, 7 - textile pad, 8 - reverse side
layer (bottom ply), 9 - middle ply, 10 - up layer
(front ply), 11 - additional heaters, 12 - chill
roll, 13 - wind-up unit

   Another possibility of coating plastics onto a
support is the so-called backing /30/. One of its
alternatives is the technique of "dry backing": an
adhesive in dispersion form (solution or latex) is
coated on the support film by means of a spreading
machine, and, after drying in a tunnel, the second
film is pressed on it in a two-roll machine. The usual
backing adhesives are solutions or latices based on
rubbers, polyurethanes, poly(vinylidene chloride) or
copolymers of vinylidene chloride and vinyl chloride,
poly(vinyl acetate), poly(vinyl alkyl ethers), etc.
   Another alternative is backing by means of waxes
which is predominantly used for doubled cellophane
film (the "non-rustling" *Duplofolie*). As this method
uses melts of various waxes, microcrystalline, paraffins,
or low-molecular weight polymeric adhesives (e.g. the
copolymer of ethylene and vinyl acetate - *Elvax*), no
drying tunnel is necessary, and the corresponding

equipment takes less space than the previous one. The
adhesive is melted in a heated bath and coated on the
film by means of rubber-coated rolls. The other film is
pressed thereon by means of the backing rolls, and
finally the doubled film is cooled on cooling rolls and
wound up. Correct maintenance of the temperature is
important and the method can also be combined with printing (e.g. flexography).

The third alternative is "hot backing" which can only
be applied to thermoplastic films. In this method (Fig.
57) the films to be combined are led to a wedge-shaped
heating device where they are heated with hot air or by
radiant heat (on the sides which are to be joined),
pressed together by backing rolls, cooled, and wound up.
This method, however, loses its importance, as coextrusion
expands.

The quality of laminates not only depends on the starting film, but also on the type and amount of the adhesive
and method of application /26/.

Lacquers, fusible adhesives, emulsions, and latices
can be used as the adhesive components. In producing packaging materials, the fusible (hot-melt) adhesives are
sometimes avoided, because they do not ensure a non
porous layer. Lacquers have the advantage of rapid
drying, which shortens the production time, but drawbacks
are the flammability and volatility of the solvents used
(which necessitates strict fire precautions), higher
prices, and atmospheric pollution. These drawbacks are
circumvented by emulsions and latices, though these are
unsuitable for water-sensitive substrates. Their main
advantages are high solid content and usually lower prices
compared with lacquers. In most applications the thickness
of adhesive layer is in the 10^{-2} mm range, which corresponds to 4-8 g m^{-2}.

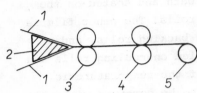

Fig. 57. Schematic representation of heat backing process
1 - film, 2 - heater,
3 - backing rolls,
4 - chill rolls, 5 - the
wound-up film

Drying the adhesive coating is an important part of
the lamination process. Lacquer coatings are most often
dried with intensive air circulation or by means of
infrared radiators, typically at 110-120°C. Latex
coatings are dried at slightly lower temperatures.

5.3.1.5 Properties and Character-istics of Homogeneous Films and Laminates

5.3.1.5.1 *Simple (Homogeneous) Films*

Films, sheets, or plates made of plastics are flat
products of thickness ranging from 2 μm to 5 mm (Fig.
58) /31/. A feature of these packaging materials are
their barrier properties, which differ among the
individual types, and these differences must be taken
into acount when choosing the appropriate packaging
material for any product to be packed. The choice of
a packaging material must also take account of the
following properties:
- resistance to both high and low temperatures
- transparency, translucency, or - on the contrary -
opacity
- vacuum packaging or inert gas filling
- shaping and shrinking
- mechanical strength, weldability, printability
The flexibility of unplasticized materials depends
on layer thickness. Packaging materials are classified

as rigid or flexible, and the former include the films for shaping, cups, support trays, blown or injected vessels. The generally used and suitable materials are polymers whose glass transition point lies above the temperature of use (homopolymers and copolymers of vinyl chloride, polystyrene), polymers with sufficiently high crystallinity (polyolefins, polyamides), or cured resins (phenolic plastics, amino plastics). Another important factor is their cost in competition with classical materials based on paper, metals, and glass.

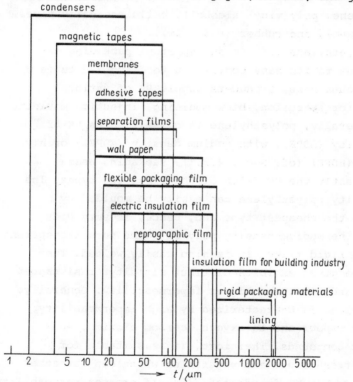

Fig. 58. Fields of application of plastic films and sheets of the thickness t /31/

Unmodified flexible packaging materials are mainly used for films, bags, and tubes. They have been used in industrially advanced countries for a long time; and new

developments consist in various physical modifications
(shrinkability, ductility, orientation, etc.). The
flexible plastics packaging materials are fundamental
to modern packaging and marketing. They made possible
convenient and hygienic packing of foodstuffs with
short shelf life, with consequent application in
self-service shops. Mostly they are produced from
polyolefins, poly(vinyl chloride) and copolymers of
vinyl chloride and vinylidene chloride (*Saran*),
poly(ethylene terephthalate) (*Mylar*), polyamides,
polystyrene, poly(vinyl alcohol), cellulose derivatives
(cellophane), and rubber (*Pliofilm*).

Polyethylene (PE) is an important packaging material due to its easy conversion to films and tubes
and its cheapness. Producers supply PE of various
grades, for extrusion, blow-moulding, injection moulding,
etc. Generally, polyethylene is classified as PE of
low density (LDPE), with medium density (MDPE), or high
density (HDPE) (cf. Sect. 4.2.1). Packaging uses
predominantly the PE films of the first two types. The
low-density polyethylene can easily be moulded by
heat, is the cheapest type and, hence, is very widespread. The medium-density polymer forms more transparent
and more rigid films. It is also easily welded. The
last type given exhibits distinct rigidity, resistance
to heat and oils, and higher impermeability. Generally,
polyethylene is characterized by high impermeability
to water vapour and to oxygen, carbon dioxide, and
aromatic compounds. Therefore, it is suitable for
packing fresh fruit and vegetables which can "breathe"
without becoming dehydrated /15/. If greater permeability
is required, the film is perforated.

The permeability of polyethylene films decreases
with decreasing temperature and increasing relative
humidity. This behaviour is utilized in packing frozen
foodstuffs (meat, fruit, vegetables, ready-made meals).

First the polyethylene bags are filled, then evacuated, sealed, and put on freezing supports. The frozen packages are packed into collapsible paperboard boxes which are suitably printed.

As the polyethylene film does not protect the goods packed from the influence of oxygen and carbon dioxide, it cannot be used for packing fat and oil-containing products which could easily become rancid by oxidation; it can even swell in oils. Bags made of LDPE cannot be used for packing at elevated temperatures, because at about 80 $^{\circ}$C, e.g. hot jam, they are not sufficiently strong.

Polyethylene bags are especially suitable for short-term consumer packages. In this field, their transparency and elasticity are advantageous. For some foodstuffs, e.g. pickled gherkins or sauerkraut, bags proved suitable just for short-term storage (10 days at most).

Polyethylene bags and sacks are also used as inserts or liners in casks and barrels, boxes and crates, sacks of other materials, containers, etc. The insert protects the content against water and contamination, and is used with chemicals, hygroscopic powders, sugar, hops, malt, etc.

Polyethylene bags and sacks are closed e.g. with special clips made of polyethylene, poly(vinyl chloride), or aluminium, or by welding (heat-sealing), by adhesive tape, or by binding with a tie. In ČSSR polyethylene bags are used for sleeve packing of milk. Such direct packing into the sleeve was introduced by the Swiss firm Officine Bertoglio SA for automatized packing into polyethylene bags *Polipack*. Other variants use a simple film which is welded at first longitudinally, filled, and welded transversally: the BTH system (Blanické strojírny Vlašim, ČSSR), *Thimopack* (Thimonnier, France), *Hassia* (FRG), and *Prepac* (Leab, France) with the sealing backed with paperboard /2/. For larger milk consumers

5 1 to 25 1 bags are used which are placed in a paperboard package of prismatic shape "Bag in Box".
Well-known are, e.g., the packages *Polygal* and *Pergall*
used in USA   and in Great Britain, respectively /32/.

Specially transparent polyethylene film *Dixopack* is
used for packing bread, meat, frozen foodstuffs, fresh
vegetables and textiles /33/.

Bags of HDPE are used for boiling foodstuffs ("Boil
in the Bag") and for vacuum packing (*Rigidex*) /34/.
HDPE in the form of a very thin film has replaced most
grease-proof paper for packing fatty foodstuffs /35/.

Polypropylene (PP) is used increasingly due to its
valuable properties and its low density (0.89 g cm^{-3})
giving high area yields: an important application is
for oriented heat-shrinkable films. Unoriented types of
polypropylene films are only produced by the Italian
firm Montedison under the designations *Moplefan SX* (for
vacuum forming) and *Moplefan B* (for lamination). Bags
made of unoriented polypropylene serve for packing
bread, bakery products, pastas, dry fruit and vegetables,
tobacco, and hygroscopic products (sugar , salt), as
"Boil-in-the-Bag" tapes, and for packing textiles (e.g.
shirts, sweaters).

Poly(vinyl chloride) (PVC) is used in packaging in
both the unplasticized and plasticized forms, the starting material for both the types being emulsion, suspension, or bulk polymer grades. Unplasticized poly(vinyl chloride) is fairly impermeable to oxygen, nitrogen,
and carbon dioxide, whereas its permeability to water
vapour is higher than that of high density polyethylene
and low density polyethylene and of the same order as
that of poly(ethylene terephthalate) (Table 41) /36/.

In view of these barrier properties, unplasticized
poly(vinyl chloride) cannot be used for packing fresh
meat, fruit, and vegetables. It is used to a lesser
extent for packing bread, cakes, and other foodstuffs
which need a low permeability package. Plasticization

Table 41. Gas permeability of thermoplastic films of the same thickness /35/ (relative values)

Material	oxygen	nitrogen	carbon dioxide	water vapour
vinyl chloride-vinylidene chloride copolymer (Saran)	0.0094	0.053	0.29	14
polychlorotrifluoroethylene	0.03	0.10	0.72	2.9
poly(ethylene terephthalate) (Mylar A)	0.05	0.22	1.53	1 300
polyamide (Nylon 6)	0.10	0.38	1.6	7000
unplasticized poly(vinyl chloride)	0.40	1.20	10	1 560
cellulose acetate	2.8	7.8	68	75 000
linear polyethylene (density 0.954 – 0.960 g cm^{-3})	2.7	10.6	36	130
polystyrene	2.9	11	88	12 000
polypropylene (density 0.910 g cm^{-3})	–	23	92	680
branched polyethylene (density 0.922 g cm^{-3})	19	55	352	800

and orientation change significantly the properties
of poly(vinyl chloride) films in both mechanical and
barrier properties. Generally, poly(vinyl chloride)
exhibits increased permeability to carbon dioxide and
water vapour after plasticizing (cf. Table 7 on p. 81)
making it suitable for packing fresh fruit, vegetables,
cheese, etc.

For packing machinery parts, foodstuffs, pharmaceuticals, and cosmetics both unplasticized and plasticized forms are used as welded bags or sachets /13/. The
filled bags are water-tight when sealed by welding. The
welding may be by high-frequency current, heated bars,
or thermal impulse. Heated bars apply a temperature of
140-160 $^{\circ}$C attained by resistance heating. Packing in
bags is very similar to sachet or cushion packing in
which the package is formed by transverse welding of
tubing of plasticized poly(vinyl chloride). Such packages are usually used for packing cosmetic products,
especially shampoos, and also oil, cleaning and chemical preparations.

Films plasticized with the usual plasticizers generally
possess a characteristic odour, which is undesirable
with some products, especially foodstuffs. Odourless
plasticized film can be obtained with special plasticizers,
e.g. monoisopropyl citrate, stearyl citrate, triethyl
citrate, higher-molecular weight polyester plasticizers,
or nitrile rubber which does not impair the permeability
to water vapour and even improves flexibility at low
temperatures. Tabel 42 compares the results obtained
by storage of goods in bags made of plasticized poly(vinyl
chloride), cellulose acetate, and cellophane. Plasticized
poly(vinyl chloride) film without nitrile rubber also has
proved useful for packing metal components exported to
subtropical areas.

Very important in packaging are the films of vinyl
chloride-vinylidene chloride copolymers produced by the

Table 42. Comparison of permeability of various packaging materials during storage of various products /13/

Product packed [a]	Packaging material [b]	Weight increase (or decrease) of the package (%) after:			
		24 hours	6 days	13 days	21 days
A	plasticized PVC	–	0.17	0.31	0.66
A	cellophane	–	3.15	5.18	soaked
A	cellulose acetate	–	2.67	5.77	14.61
B	plasticized PVC	–	-0.99	-3.19	-5.66
B	cellophane	–	-3.45	-9.08	-11.31
B	cellulose acetate	–	-6.08	-9.07	-11.93
C	plasticized PVC	0.00	1.00	3.20	–
C	cellophane	4.00	25.00	44.00	damaged by mildews

[a] A – candied fruit kept at 20°C at 90% relative humidity; B – candied fruit kept at 37°C in a drying atmosphere; C – dried mushrooms kept at 20°C at 90% relative humidity. [b] Plasticized poly(vinyl chloride) (PVC) cannot be applied in ČSSR in direct contact with foodstuffs even if it contains hygienically unobjectionable plasticizers

firm Dow Chemical Co. under the name *Saran* which has
almost become a general term. The Japanese firm Kureha
Chemical Industry Co. marketed this copolymer under
the name *Krehalon*. The films are characterized by high
impermeability to oxygen, carbon dioxide, and water
vapour (cf. Table 46) which surpasses that of polyester
films. Due to their barrier properties, these films
are quite unsuitable for packing fresh meat (except
for the ripening period of meat quarters in slaugh-
terhouses), fresh fruit and vegetables. Their main
applications include shrinkable films and laminates.

Polystyrene is one of the important plastics in
packaging, all three basic types being used, viz. the
standard (unmodified), toughened, and expanded polystyrenes. Flexible films involve only the unmodified
type, characterized by crystal clarity even in thick
layers, but with poor mechanical properties. Its barrier
properties (Table 46) exclude its application to
packing more sensitive foodstuffs for long-term storage,
but it can be used for packing fruit and vegetables.
Broader application is found in rigid packages for
the modified and expanded types or for physically
modified films (oriented, shrinkable). Polystyrene
packages can be easily printed.

The greatest problem connected with polystyrene
in packaging is its free monomer content (cf. Chapter 7).

Polyamide (nylon) packaging films are produced
predominantly from polyamide 11, *Rilsan* (produced by
Aquitaine-Total-Organico). They are clear, weldable,
impermeable to oxygen, water vapour (in which respect
they differ from polyamide 6 which is very permeable),
and aromatic substances, and can be sterilized with
steam or hot water.

Some types of polyamide films withstand prolonged
exposure up to 100 $^{\circ}$C and short exposure up to 160 $^{\circ}$C.
They are strong, can be vacuum formed (unoriented), and

welded by the usual methods. Also they are easily printed
without special surface treatment. Films based on polyamide 6 contain the free monomer (caprolactam), whose
bitter taste prevents application to the packing of some
foodstuffs.

Bags of polyamide film are also used for packing
bread, and the perforated bags are used for rice and pastas
("Boil-in-the-Bag" types). Tubular film can replace
other sausage casing materials. Polyamide films have found
applications in multi-layer and oriented packaging
materials.

Poly(vinyl alcohol) film found an interesting special-
ised application in the form of water-soluble packages
suitable for washing and cleaning preparations, agrochemicals, dyestuffs, chlorinated lime, photographic
developers, dirty linen (e.g. in hospitals), and packed
doses for precise dosing of various substances.

The Japanese firm Nippon Goshei produces these films
under the name *Hi-se-lon* (S-type - quickly soluble in
both cold and warm water, H-type - soluble above $65^{\circ}C$
and insoluble in cold water) /37/.

Polyester films based on linear unsaturated polyesters
of the poly(ethylene terephthalate) type are fairly
impermeable to oxygen and carbon dioxide, but not to
water vapour. They are produced only by some bigger
firms like Du Pont (*Mylar*), Imperial Chemical Industries
(*Melinex* - non-shrinkable type only), or Kalle AG (*Hostaphan*). They are used predominantly for "Boil-in-the-Bag"
and vacuum packing application with foodstuffs.

Films of rubber hydrochloride are produced by the
firm Goodyear Tyre and Rubber Co. under the name *Pliofilm*.
They are permeable to oxygen and carbon dioxide, but
less permeable to water vapour. Therefore, their uses
are similar to those for polyethylene film but they
cannot compete effectively because of cost. Also they
have insufficient resistance to light ageing, and are

most often incorporated in multi-layer laminates with
an aluminium foil as protection against light.

5.3.1.5.2 *Laminates*

Simple polymer films often do not have the properties
needed to meet all the requirements of flat packaging
materials, especially package performance. This problem
is solved by combining several materials in such a way
that their individual advantages may be utilized and
their drawbacks overcome. Such laminates are made by
coating a plastic onto another material, by adhesively
laminating two or more film or web materials, or by
coextrusion.

Suitable combination of two plastics can bring
useful improvements in the following properties /26/:

1. tensile strength,
2. impact strength (toughness),
3. ductility (elongation at break),
4. flexibility,
5. chemical resistance,
6. heat resistance,
7. permeability to water vapour,
8. gas permeability,
9. surface modifications,
10. ease of processing or converting (e.g. weldability).

Laminating can also serve for the production of high-
quality thicker films from one kind of material. The
reason is that directly produced thicker films have the
same faults as thin films - particularly they contain
microscopic cracks which lower the functional properties
of the material. In the case of a laminated structure
it is improbable that the cracks in one layer would
coincide with the faults in the other layer.

During the last decade, the applications of laminates
to packaging have been growing continuously. The demands

made of packages are steadily increasing and this leads
to a proliferation of packaging films. The situation
is also influenced by established production of traditional materials and commercial competition.
Therefore, this section will deal only with examples
exhibiting significant properties and frequently used
in packaging. The flexible materials include /31/ :
poly(vinylidene chloride) lacquer/cellophane, poly(vinylidene chloride) lacquer/polyethylene, polyamide/
poly(vinylidene chloride) lacquer/polyethylene, and
polyester/aluminium foil/polyethylene.

Large-scale production of these materials led to
extensive modernization of equipment and research
into the properties and possibilities of laminates.
Polyethylene became one of the most important components.

The combinations polyamide/polyethylene and poly(ethylene terephthalate)/polyethylene as well as polypropylene/
polyethylene proved useful for packing liquids. If
light-sensitive products (e.g. fruit juices containing
vitamin C) have to be protected, aluminium foil laminates are useful and can be processed by fully automated
packing machines. Such materials are marketed in the
form of self-supporting bags produced by the *Doypack*
system.

In the last few years, metallized films have found
growing use. The metallization with aluminium is carried
out in vacuum, and various films can serve as the
support, e.g. polypropylene, polyester, polyethylene,
poly(vinyl chloride), or cellophane. Generally, the
metallized films exhibit almost 100% opacity for ultraviolet and infrared radiation and are especially suitable
for packing goods exported to areas with exacting climate
or for packing frozen goods /38-40/.

Properties of the combined films, however, make
quite considerable demands on the packing process,
especially in fully automated deep-drawing packing
lines. The packages are often evacuated before sealing.

For these purposes it is recommended to use deep-rawing
laminates based on poly(ethylene terephthalate) with
polyethylene, and polyamide with polyethylene whose
carrier films are coated e.g. with a lacquer based on
vinyl chloride-vinylidene chloride copolymers (*Saran*),
if enhanced barrier requirements have to be met.

If the package content is very hygroscopic or sensitive to oxidation, polymer combinations are chosen
to reach the maximum possible impermeability to water
vapour (polyolefins) and oxygen (poly(vinylidene chloride) and copolymers of vinyl chloride and vinylidene
chloride, poly(ethylene terephthalate), cellophane,
and polyamides) (cf. Table 41).

The laminates for use in bag-type packages sterilizable at 120°C are based predominantly on combinations
of poly(ethylene terephthalate), polyamides, and aluminium foil with polyethylene.

The long-term protection of sensitive foodstuffs
must also take into account the risk of penetration by
microorganisms, and therefore, the film for packages
has a thickness above 12 μm /31/.

Polyethylene-coated paper is a widespread packaging
material /15/. The two materials combine their advantages
and eliminate their drawbacks. Whereas the polyethylene
coating makes the package greaseproof, waterproof, and
resistant to chemicals, paper makes it strong and rigid.
Paper with polyethylene coating is less permeable to
water vapour than a polyethylene film of the same
thickness. A coating thickness of 8 g m^{-2} is used for
common packaging purposes, but for frozen foodstuffs the
coating is about twice as thick. The inner coating of
multi-layer paper sacks has about 30 g m^{-2}. Polyethylene-coated paper can be welded by direct contact of
hot bars on the paper side. The paper can be printed
before overcoating with polyethylene.

The polyethylene-coated paper is particularly

suitable for packing hygroscopic goods or, on the
contrary, goods where dehydration has to be prevented.
In the chemical industry it is used for packing
chemicals, insecticides, glues, ion exchangers,
dyestuffs, latex, photographic chemicals, and especially fertilizers and cement. Double-layer or
four-layer polyethylene-coated paper sacks can be
equivalent to and replace sacks made of seven paper
layers alone. Although the polyethylene-coated sacks
are thinner, they resist both water and chemicals and
in this respect are better than bitumen-coated
sacks. In the pharmaceutical industry, polyethylene-
coated paper is used for packing medical and cosmetic
preparations. In the foodstuff industry, this type
of laminate represents a suitable material for packing frozen goods, biscuits, milk, butter, syrups, milk
powder, dried eggs. For packing sugar, salt, and other
hygroscopic substances, jute sacks are provided with
inner polyethylene coating; such sacks are perfectly
tight, and the jute fibers cannot contaminate the
product.

Polyethylene-coated paper has also proved useful
in packing beverages into tubes of this material (e.g.
for milk). A very widespread package based on this
material is called *Tetra-Pak* (cf. Sect. 1.1.2, Fig. 2)
which is produced from reel stock and filled continuously,
thus meeting hygiene requirements.

A relatively new method uses a double polyethylene
film (the outer layer is white and the inner black,
their thicknesses being 7 and 4 μm, respectively) for
packing milk. This material is produced by coextrusion
and sterilized by ultraviolet radiation. It is opaque
to daylight and preserves the riboflavine and vitamin
C content in milk /41-43/.

Some products are also packed in polyethylene-coated
cellophane. The cellophane, pre-coated with a layer of

cellulose nitrate or poly(vinylidene chloride), is
combined with about 50 μm low density polyethylene
by extrusion coating /26/, providing chemical resistance, impermeability to water vapour, and heat weldability.

The adhesive laminating of cellophane uses either
solvent adhesives or hot-melt adhesives, giving lower
permeability to water vapour. The following are the
most significant cellophane laminates:

cellulose nitrate/cellophane/polyethylene: *Cryovac*
NCL (W.R. Grace and Co., USA), *Metathene D* (Shorko-
Metal Box Ltd., Japan), *Viscothen* (Wolf Walsrode AG,
FRG), *Sidathene MD* (British Sidac Ltd., Great Britain);

cellophane lacquered one side with vinyl chloride-
vinylidene chloride copolymer/polyethylene: *Cryovac*
XCL (W.R. Grace and Co., USA), *Metathene X* (Shorko-Metal Box Ltd., Japan), *Viscothen DX* (Wolf Walsrode AG,
FRG), *Sidathene XD* (British Sidac Ltd., Great Britain);

cellophane lacquered both sides with vinyl chloride-
vinylidene chloride copolymer/polyethylene: *Cryovac*
XCXL (W.R. Grace and Co., USA), *Viscothen MX* (Wolf
Walsrode AG, FRG).

All three types are also produced by the firm
Kalle AG under the name *Cellophan-PE*.

These laminates are used in the form of bags for
packing cut sausages, bacon, smoked meat, smoked fish,
cut cheese, fine salads, olives, mayonnaise, pickled
gherkins, etc.

The cellophane laminates also include the films
bonded with microcrystalline wax: they are composed
of two layers of lacquered cellophane with inner
printing (*Cellophan-Weka* produced by Kalle AG, FRG)
and are used for packing dried fruit, potato chips,
fine bakery products, sweets (non-rustling "theatre"
bags), etc.

The firm British Cellophane Ltd. produces metallized cellophane films with a heat-seal coating used

for packing biscuits, cigars, cosmetic preparations, frozen foodstufs, bakery products, etc.

Polyester films alone or coated with poly(vinylidene chloride) are used for various laminates. A heat-sealable material is obtained by bonding or extrusion coating a polyethylene layer. The combination polyester/copolymer of vinyl chloride and vinylidene chloride (coating)/polyethylene has high impermeability to oxygen. The laminate composed of a polyester film and a poly(vinyl chloride) film has excellent strength and chemical resistance. It is used especially for packing pharmaceutical products, but cannot be welded by conventional methods and needs a special procedure; the material is rapidly heated to the welding temperature and only then pressure is applied in the weld position and maintained until the material is cold /26/.

A laminate of polyester and rubber hydrochloride is used for the same purposes; its chemical resistance is slightly less, but its heat sealability is good. An analogous laminate of cellulose acetate instead of polyester has similar properties. If cellulose acetate is combined with poly(vinylidene chloride), the resulting laminate exhibits excellent resistance to chemicals, and in other properties resembles the laminate of poly(vinylidene chloride) and polyester; the main application of this material is the packing of pharmaceutical products.

Polyester laminates are also suitable for vacuum or inert gas packing. If, in a polyester laminate, the layer of low density polyethylene is replaced by high density polyethylene, the material withstands sterilization at 130°C for 20-30 min.

The following are important polyester laminates:

poly(ethylene terephthalate)/polyethylene: *Hostaphan-PE* (Kalle AG), *Combitherm P* (Wolf Walsrode AG), *Extruester 12/50* (Heinrich Nicolaus GmBH, FRG), *Metathene M*

(Shorko-Metal Box Ltd., Japan), *Cryovac EL* (W.R. Grace
and Co., USA); they are particularly used for packing
bacon, smoked fish, frozen meat, and ready-made
"Boil-in-the-Bag" meals, and for packing pickled
vegetables, mustard, etc.;

poly(ethylene terephthalate)/vinyl chloride-vinylidene
chloride copolymers(lacquer)/polyethylene: *Hostaphan-PEX*
(Kalle AG), *Combitherm PX* (Wolf Walsrode AG),
Extruester X 15/50 (Heinrich Nicolaus, FRG), *Cryovac AXL*
(W.R. Grace and Co., USA); packages for cut sausages
and bacon, smoked and fresh fish, cut hard cheese,
pickled vegetables, and mustard;

poly(ethylene terephthalate)/vinyl chloride-vinylidene
chloride copolymers(lacquer)/polyethylene for vacuum
forming: *UK 6130* (Kalle AG), *Combitherm PX* (Wolf Walsrode AG), *Cryovac EXT 1550, Cryovac EXT 2250* (W.R.Grace
and Co., USA); packaging laminates of polyester and
polyethylene, also with a coating of vinyl chloride-
vinylidene chloride copolymer, are also produced by
the firm Du Pont de Nemours as *Mylar* and *Mylar H*,
respectively.

Laminates of polyester (often metallized) and poly(vinyl chloride) are used for decoration purposes; poly
(vinyl chloride) layer serves as a reinforcing support.

Coatings of poly(vinyl chloride) or copolymers of
vinyl chloride with other monomers are often applied
to other packaging materials to increase their impermeability to water vapour and resistance against chemicals /13/. The usual support materials are paper,
cardboard, cellophane, or metal foil. Production of
packages from these materials makes use of the fact that
the coating of vinyl chloride homo- or copolymers can
act as both thermoplastic coating and heat-seal adhesive.

Coatings obtained by spreading or dipping processes
are less permeable than those made by calendering or
heat-bonding. Films of unplasticized poly(vinyl chloride)

are heat-bonded to the substrate (especially paper)
with great difficulty, so a poly(vinyl acetate)
dispersion is used as an adhesive interlayer.

Laminates of unplasticized poly(vinyl chloride)/poly-
ethylene are little used. The include *Cryovac VL* (W.R.
Grace and Co., USA) and *Vinyskin* (Soplaril, France);
both can be vacuum formed. They are made by adhesive
bonding and not by polyethylene extrusion coating.

Polyamide laminates for packaging have found
increasing use in the last few years. They are produced
by bonding of polyamide and polyethylene films with
polyurethane adhesives, or by extrusion coating (most
often polyethylene onto polyamide film) or by blown
film coextrusion. Adhesion between layers is improved
by anchor coats (primer) or by surface treatment with
electrostatic discharge. These combinations make use
of the impermeability of the polyamide layer to gases
and that of the polyethylene layer to water vapour.
Gas impermeability can be further improved in both
adhesive bonding and extrusion coating with an inner
coating of vinyl chloride-vinylidene chloride copolymers.

Polyamide laminates based on unoriented films are
more suitable for vacuum forming than polyester laminates. Their main application is for tray-shaped packages, and their barrier properties make them useful
just for vacuum and inert gas packing. Their heat
resistance is very good (from -50 to +90°C); if high
density polyethylene is used they can be sterilized
at 120°C. They are used particularly for packing
pre-boiled peeled potatoes, carrots, bacon, pickled
meat, cut sausages, smoked and fresh fish, cut soft
and hard cheese, milk powder, ready-made frozen "Boil-
in-the-Bag" meals, pickled vegetables, mayonnaise, etc.

The polyamide/polyethylene laminates are marketed as:
Rolform (Soplaril, France), *Cryovac ML* (W.R. Grace and
Co., USA), *Alkorn PA-PE* (Alkor-Oerlikon Plastic

GmbH, FRG), *Combitherm PA*, *Combitherm PAX* with inner
coating (*Saran*); and *Combitherm SPA* with heat resistance
up to 120°C   (Wolf Walsrode AG, FRG).

Polypropylene laminates based on unoriented polypropylene films are less common. The Italian firm
Montedison recommends them for packing bread, bakery
products, fried potato chips, sweets, sugar, salt, and
for "Boil-in-the-Bag" packages /44/.

Although laminates based on aluminium foil are
more expensive packaging materials, they are very
important due to their air-tightness and opacity. They
increase the shelf life of some packed foodstuffs,
and if other components of the laminate are properly
chosen, the packed goods can be sterilized, e.g., oriented polypropylene/aluminium foil/unoriented polypropylene).

As all aluminium foil laminates have three layers
(the outer layers protect the aluminium), the lamination
process is carried out by adhesive bonding, which is
more economical than double extrusion coating.

These materials are used particularly for packing
foodstuffs sensitive to light and oxygen, e.g. ketchup,
sauerkraut, ground coffee (vacuum packed), soup and
sauce powders, smoked meat, nut kernels (vacuum packed),
instant-soluble powder foodstuffs.

In common use is the paper/aluminium foil/polyethy-
lene laminate. Useful properties are exhibited by the
combination of polyester/aluminium foil/modified polyolefin film which is used for packing ready-made
meals from vegetables and meat; sterilization is
possible at 120°C for 30 min. in an autoclave, which
ensures 1 year shelf-life.

These laminates are produced by the following firms:
Kalle AG (FRG):

Cello-Metall - unmodified cellophane/aluminium
foil/thermoplastic coating,

Cello-Metall/PE - unmodified cellophane/aluminium foil/polyethylene,

Hostaphan-Metall-PE - polyester/aluminium foil/ polyethylene;

and Wolf Walsrode (FRG):

Aluthen - cellophane with cellulose nitrate lacquer coating/aluminium foil/polyethylene,

Aluthen P - polyester/aluminium foil/polyethylene,

Aluthen PX - polyester with vinyl chloride-vinylidene chloride copolymer coating/aluminium foil/polyethylene,

Aluthen PP - polypropylene/aluminium foil/polyethylene,

Aluthen PPX - polypropylene with vinyl chloride-vinylidene chloride copolymer coating/aluminium foil/polyethylene,

Aluthen PA - polyamide/aluminium foil/polyethylene,

Aluthen S - special laminate resisting temperatures up to 120°C.

5.3.2 PRODUCTION OF HOLLOW CONTAINERS FROM PLASTICS

5.3.2.1 Production of Bottles by Blow Moulding

Blow moulding for the production of hollow containers (bottles, tubes, cans, canisters, drums, etc.) has two variants: extrusion blow moulding (i.e. from a parison or tube) and injection blow moulding (i.e. from a preform). Most vessels are produced by the first variant, and bottles represent the majority. The technology allows production of bottles in ranging capacity from several milliliters to 600 liters, the production rate being up to 15,000 pieces per hour in the case of smaller products.

The extrusion blow moulding process is based on extrusion of a plastic material through a circular die; the tube extruded is closed in a mould (usually consisting of two parts) and blown by air pressure to adopt the shape of the mould. A very important improvement of

this technology was the introduction of programmed
regulation of the extrusion so that the thickness of
the tube extruded is inversely proportional to the
blowing ratio, improving both the quality of products
and economy of material.

Extrusion at a rate below 100 kg h^{-1} is carried out
with single-screw extruders as in simple extrusion,
but with blow moulding somewhat different demands are
made on the flow properties of the polymer used. In
the case of poly(vinyl chloride), the types used have
lower K values (between those usual for injection
moulding and those for simple extrusion). The working
temperature must be chosen so that the parison is not
stretched by its own weight, but so that sufficient
fluidity of the material is attained to ensure filling
of the mould and a product of correct shape with a
smooth surface.

The working procedure of bottle blowing is represented
schematically in Fig. 59. The air for blowing is delivered
most often through the extrusion head, but to form a
reliable weld at the bottle base in the mould it is
better to introduce the air from the bottom so that
the weld is formed in the warmest part (top) of the
parison.

The parison diameter is determined by the requirement
that it should not be increased more than three times
by blowing and by the size of the bottle neck. The wall
thickness is calculated from the volume difference
before and after the blowing procedure.

The tube extruded from the head must be uniformly
heated, and the extrusion rate must be high enough for
the temperature to be sufficient for welding the base.
Variations in parison temperature cause irregular
blowing, non-uniform wall thickness and, hence,
different rigidity at different places.

Non-uniformity of the wall thickness of the tube

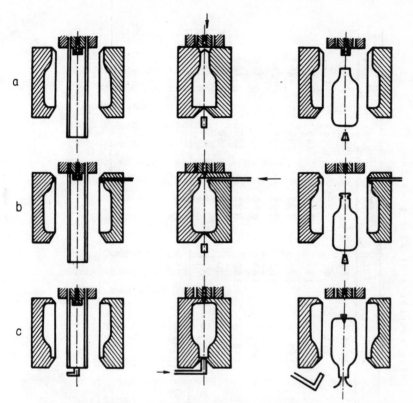

Fig. 59. Schematic representation of the procedures
used for bottles blow-out from tubes /15/
a - top air inlet through the neck, b - side air inlet
through a neck adapter, c - bottom air inlet through
the neck

and warping can be caused by incorrect adjustment of
the die gap. Wall thickness uniformity depends also
on the pressure drop, extrusion rate, and flow properties
of the polymer. Table 43 gives technological extrusion
blow moulding data for the plastics commonly processed
this way /45/.

In contrast to injection moulds, the movement of
individual parts of the mould must be precisely synchron-
ized during closing. The construction must ensure
perfect parallelity of mould contact surfaces even at

Table 43. Technological parameters of blowing processes of some plastics /43/

Parameter	1PE	PVC	PS	PA6	PC
screw length do diameter ratio	20:1	25:1	18:1	20:1	15:1
compression ratio	3:1	4:1	2:1	2:1	1.5:1
temperature of the inlet zone ($^{\circ}$C)	140	145	135	245	270
temperature of the compression zone ($^{\circ}$C)	160	150	155	240	265
temperature of the extrusion zone ($^{\circ}$C)	175	165	185	235	260
temperature of the head ($^{\circ}$C)	180	165	180	250	270
temperature of the slot die ($^{\circ}$C)	165	160	185	220	250
temperature of the mould ($^{\circ}$C)	40	20	20	50	60
blowing pressure (MPa)	0.5	0.4	0.2	0.2	0.6

1PE - linear polyethylene, PVC - poly(vinyl chloride), PS - polystyrene, PA6 - polyamide 6, PC - polycarbonates

short cycle times and considerable closing pressures
(about 100 MPa). Bottles of more than 5 l capacity
are usually produced with the aid of ram accumulators.

Economical and reliable operation of extrusion blow
moulding needs full uninterrupted utilization
of the capacity of the screw extruder. A number
of construction arrangements are possible:

- the moulds are arranged in a circle (Fig. 60)
- the apparatus has two heads and two moulds;
the plastic material is prepared continuously in the
screw extruder, and the melt is introduced into the
first or into the second head by means of an automatic
distributor /27/

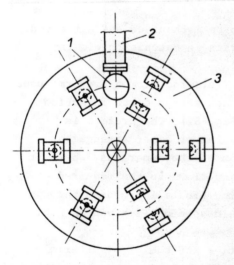

Fig. 60. Extrusion blowing equipment with circular
arrangement of the moulds

1 - extrusion head, 2 - extrusion chamber, 3 - rotating
table with the moulds

- larger vessels (30-50 l capacity) are produced
by means of a combination of a screw extruder and a
transfer moulding device (Fig. 61). During cooling of
the filled mould, plastic is extruded into the cylinder
of the transfer moulding device, which is later

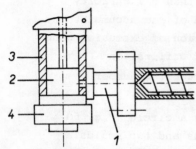

Fig. 61. Schematic representation of a transfer moulding
equipment for blowing of
large-capacity vessels

1 - extruder head,
2 - pneumatically operated
plunger of the transfer
moulding equipment,
3 - transfer moulding
cylinder, 4 - tube extrusion
head

transferred in tube form into the empty mould /27/.

Extrusion blow moulding is used particularly for HD
and LD polyethylenes, poly(vinyl chloride), polyamides,
and polycarbonate.

The whole procedure can be mechanized and automated.
An important problem consists in matching cycle rate
to cooling capacity.

Injection blow moulding (blowing from a preform) uses
a tube (preform) which is not extruded but injection
moulded on a metal core (Fig. 62). The preform is
transported by a turret system or other arrangement
towards the blowing section of the apparatus, where it
is placed into the mould and blown to the required
shape (Fig. 63). The blowing air is introduced through
the core. A modern automatic version is given in Fig.
64.

As far as the construction and choice of material
are concerned, the current injection machines and moulds
are fully satisfactory for injection of the preforms,
the only difference being in that aluminium cores are
used in the case of large-capacity bottles (easier
handling). Quality of the final products is determined
by a number of factors, e.g. injection conditions,
construction of the injection mould and core, blowing
ratio, flow properties and ductility (elongation at
break) of the polymer, homogeneity of the melt, tempe-

rature of the preform, and shrinkage of the material.

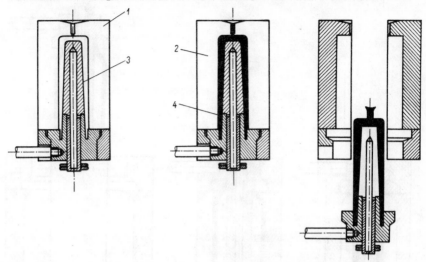

Fig. 62. Injection moulding of preform for blown-out bottles /15/
1,2 - the injection mould, 3 - mandrel, 4 - the preform

The technique has the advantages that the final products have a smooth surface, precise dimensions, homogeneous volume and weight and are seamless; the geometrical shape of the neck is independent of that of the vessel and can be precisely maintained, and no waste material is produced. As the preform can be biaxially oriented during blowing, the product then has better mechanical properties than a non-oriented product, which leads to weight decrease of the package. The wall thickness of the bottle can be determined by the shape of the preform.

The procedure compared with extrusion blow moulding, has also its drawbacks such as the limitations on shape and volume, higher initial costs and operating expenses, non-uniform wall thickness at high length-to-diameter ratios, and the impossibility to produce asymmetrical shapes, and multi-neck vessels. In spite of all these drawbacks and relatively small contribution to total

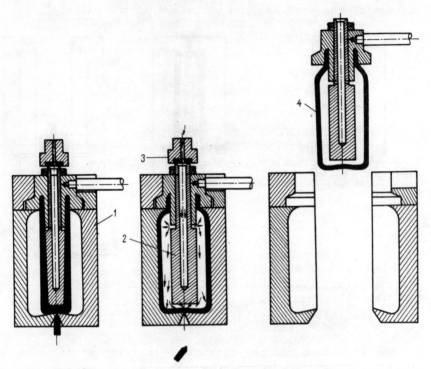

Fig. 63. Blowing-out of bottles from a preform /15/
1 - the blowing mould, 2 - mandrel, 3 - pressure air
inlet, 4 - bottle

bottle production, the technique holds an important
position among the rest.

Both versions of blow moulding utilize polyolefins,
polystyrene, acrylonitrile-butadiene-styrene terpolymers,
polycarbonate, polyacrylates, and polyesters.

5.3.2.2 P r o d u c t i o n o f H o l l o w
         C o n t a i n e r s a n d C r a t e s b y
         I n j e c t i o n M o u l d i n g

Injection moulding was introduced at the end of the last
century /26/. The present state of the technique results
from the parallel developments of equipment and thermo-

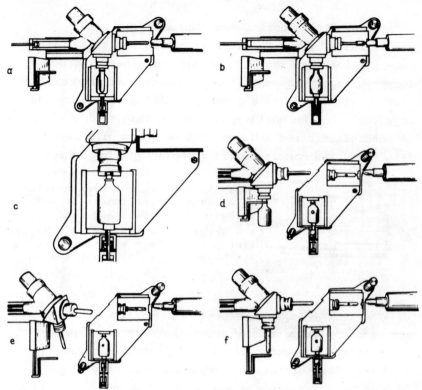

Fig. 64. Technological cycle in production of bottles by injection blowing

a - closing joint, b - simultaneous tube (preform) injection and blowing-out of the preform from the proceeding cycle, c - the blown-out bottle, d - opening of the mould and taking out of the bottle, e - turning of the mandrel, f - a new preform on the turned mandrel

plastic materials which started to accelerate in 1930´s.

Injection moulding is a method of processing plastics in which the material (supplied usually in granulated form) is melted in a plasticizing chamber and injected by pressure into cooled metal moulds. As plastics are poor heat conductors, the machine must ensure that the material is heated over as large a surface and in as thin a layer as possible.

Accordingly injection moulding machines are
classified as without pre-plasticizing and those with
pre-plasticizing. They are of various types. In the
former category the material is plasticized (homogenized
- cf. Sect. 5.1.2) in a chamber (melting cylinder) and
injected into the mould by a plunger (Fig. 65), or it
is both plasticized and injected by means of a screw
(Fig. 66). The machines with pre-plasticizing, which

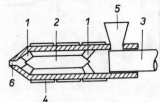

Fig. 65. Schematic representation
of a plunger-type injection unit
without preplasticizing /45/

1 - struts, 2 - torpedo, 3 - injection plunger, 4 - electric heating,
5 - hopper, 6 - injection nozzle

are used for large volume shots, have separate plasticizing and injection units. The plastic material
is plasticized either in a separate melting cylinder
(Fig. 67), or in a screw extruder (Fig. 68), then
transferred by pressure into the injection cylinder and
then into the mould by plunger.

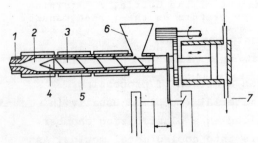

Fig. 66. Schematic representation of a screw injection
machine without preplasticizing /45/

1 - injection nozzle, 2 - the melt reservoir, 3 - dosing
section of the screw, 4 - ram part of the screw,
5 - electric heating, 6 - hopper, 7 - hydraulic screw
control

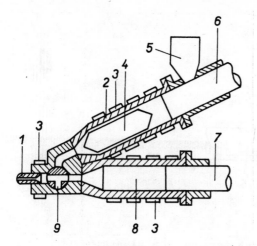

Fig. 67. Schematic representation of the plunger-type
injection machine with preplasticizing in a melting
chamber with a plunger /27/

1 - the injection nozzle, 2 - the melting chamber
for preplasticizing, 3 - electric heating, 4 - torpedo,
5 - hopper, 6 - feeding plunger of heating cylinder,
7 - injection plunger, 8 - injection cylinder,
9 - three-way valve

The screw injection moulding machines are well balanced with respect to heat transfer and flow of the
plastic material, so that they can be used for processing polymers which cannot be injected by plunger,
e.g. unplasticized poly(vinyl chloride), polytrifluorochloroethylene, polyvinylcarbazole /27/.

Plasticizing in plunger machines without pre-plasticizing is adequate up to a shot weight of 500 g.
Larger volumes would be non-homogeneous. The thermal
gradient in the melting chamber of plunger machines
is much higher than in screw machines, and, therefore,
the working temperature must be about $30^{\circ}C$ lower than
in screw machines (max. $230^{\circ}C$).

In the screw machines the material is plasticized
by heat transferred through the wall of the melting
chamber and by the heat of friction, the plasticizing
zone being divided into separately temperature con-

trolled sections.

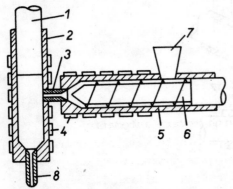

Fig. 68. Schematic representation of a plunger-type injection machine with screw preplasticizing /1/

1 - injection plunger, 2 - injection cylinder,
·3 - extrusion channel, 4 - electric heating,
5 - heating cylinder of screw extruder, 6 - screw,
7 - hopper, 8 - injection nozzle

The lower temperatures in plunger machines and the resultant higher melt viscosity necessitate higher pressure and energy consumption, and thus higher operating expense. The pressures necessary for injection of various types of plastics are higher by 40-70% than those in screw extruders.

Injection mouldings have weight of 10 g to 20 kg. So the machines range in applied pressure up to 250 t. All injection moulding machines have the following basic units /15/:

1. melting chamber with injection mechanism,

2. a device for dosing the material,

3. a mechanism for closing the mould,

4. hydraulic control,

5. electrical control.

Modern injection moulding machines are equipped with hydraulic control, and only some smaller ones are controlled mechanically.

The operating cycle is determined by the speed of
the injection and mould closing and opening mechanisms
and by the period of mould closure. These intervals
are adjusted according to the thermoplastic material
used, wall thickness, and shot weight using electronic
time relays.

Generally this cycle lasts from several seconds to
several tens of seconds. Thin-walled components are
produced very efficiently with modern automatic
machines at 5 to 15 pieces per minute with one operator
controlling several machines. Such automation increases
the advantages of modern injection moulding in the
field of packaging for thin-walled hollow containers
(cups, lids, boxes, pails etc.).

High quality of the articles produced is ensured
by maintaining optimum working parameters like constant
pressure, temperature of the chamber and moulds,
simultaneous and complete filling of all mould cavities,
and synchronization of individual operations in the
cycle. With regard to mould filling, fluidity of the
material, and hydrodynamic resistance of the flow
paths, it is necessary that the flow channels to all
cavities of the mould be closely balanced by design
of length and diameter.

For packaging, injection moulding is also applied
to the production of small components, e.g. closures.
As every producer tries to reach the highest possible
economy of production, such products are frequently
made by injection into multiple cavity moulds, the
number of cavities being limited by their individual
volumes and the shot capacity of the machine. Multiple
cavity moulds are more expensive, and adjustment of
operating parameters at the beginning of production
is exacting and lengthy, if the highest possible quality
of product is to be obtained.

Increasing demands are made on injection moulding
machines to reach maximum productivity and product

quality. The first of these demands requires efficient
plasticizing of the material in the melting chamber
with high melting capacity and small thermal gradient.
The injection force must be adjustable within broad
limits. The closing force in the mould clamp unit
must be relatively large to accomodate large products.
No less important are the area and weight of the
machine, and the installed cost, which all have to be
a minimum.

In principle the injection moulding process can be
applied to all thermoplastics /26/, but each type has
its favourable and less favourable properties which
dictate the most suitable machine construction and
operating conditions.

For example, vinyl chloride polymers and copolymers
tend to be degraded by prolonged high temperature and
this fact must be taken into account, especially in
moulding the unplasticized polymers which necessitate
processing temperatures close to their decomposition
point. Also the corrosive properties of decomposition
products (especially hydrogen chloride) demand chemical
resistance for the construction material of the melting
chamber. The poor flow properties of unplasticized
poly(vinyl chloride) are also inconvenient /25/, so
its processing by injection moulding is not so common
as with polystyrene, polyethylene, and polyamides,
although certain screw extruders can be employed.

Polyamides have a very narrow temperature range of
rubber-like character, and molten polyamides (like
some polyethylenes) have such a low viscosity that
they flow spontaneously out of the nozzle of the
melting chamber, and special closable nozzles are used.

Some polymeric materials easily absorb moisture
from the air, especially in the cases of cellulose
derivatives, vinylic polymers and copolymers, and
styrene-acrylonitrile copolymers, polyamides, poly-

esters and polycarbonate. These materials must be
dried prior to injection moulding and driers are sometimes used for this purpose, but most often it is
sufficient to remove moisture by preheating in a
ventilated feed hopper.

Polystyrene and polyacrylates absorb moisture
relatively little, and only in extraordinarily humid
conditions does water condense on the surface of the
granules, so that they must be dried, but the time
needed for this is shorter than with absorbed water.

Polyethylene often sticks to the walls of the
melting chamber and diminishes its effective diameter,
so that some granules may pass the plasticizing zone
without being melted, and spoil the product. Such
faults are usually remedied by increasing the nozzle
temperature of the injection unit.

However, the operating conditions are far from
being controlled only by the polymer used. Often more
important is the presence of lubricants, plasticizers,
fillers, stabilizers, and other additives which also
affect the quality of the final product and are chosen
according to the requirements of the package.

Packaging utilizes many products made by injection
moulding: besides hollow containers and transport
crates various closures are made predominantly from
polyethylene, polypropylene, poly(vinyl chloride),
and melamine-formaldehyde resins (transfer moulding,
cf. Sect. 5,3.3.1).

In recent years, injection moulding expanded in
the production of transport crates and boxes. The
materials used for this purpose are mainly polyethylene,
polypropylene, expanded and toughened polystyrene, and
acrylonitrile-butadiene-styrene terpolymer. These
packages are produced in various shapes to suit the
goods transported. As polymers are relatively cheap
in industrially advanced countries, they are also used

for non-returnable transport packages (e.g. from expanded polystyrene) for packing foodstuff products,
fruit, vegetables, etc. /46, 47/.

The materials used for transport crates must meet
the requirements of high impact strength, weather
resistance, resistance to cleaning agents, and some
others. If they are for foodstuffs, they also must
be accepted for food contact, washable, and with low
dirt pick-up; for the last, antistatic agents are
incorporated.

If the boxes and crates are designed for shock-
sensitive articles, their construction combines load-
bearing materials with cushioning, e.g. expanded
polystyrene or polyethylene, or air-filled cushions
which are either connected to form flat interlayers
or used individually. Sometimes, the articles are
also supported by polyurethane foam directly sprayed
from an aerosol pressure bottle /48/.

Injection moulding is used also for cups (predominantly from polystyrene) and pails, etc.

Pallets and containers can sometimes be shaped by
the relatively inexpensive vacuum forming technique
(cf. Sect. 5.4.2) using multi-purpose machines with
short cycle times.

5.3.2.3   Uses   and   Characteristics
          of   Hollow   Containers   and
          Vessels   Made   of   Thermo-
          plastics

The best-known systems for producing polyethylene
bottles by extrusion blow moulding include *Botiplast*,
Botiflio (Breil Mertel, France), *Bottle Pack* (Romme-
lag, Switzerland), or *Becum* (FRG). The Botiplast
system employs two parallel moulds from which the
bottles are transferred to the filling station. The

bottles are closed by flat bar sealing of the neck /2/.
The Botiflio system /32/ is promoted for aseptic
filling and the Becum system is similar. In the Bottle
Pack system /49/, milk is filled into hot mouldings
in the mould.

Capped and heat-sealed bottles of both LD and HD
polyethylenes have proved suitable for packing
detergents and cleaning agents, photographic chemicals,
and other chemical products. They are also used as
non-returnable bottles for milk, sauces, etc. HD
polyethylene bottles allow sterilization at 120°C,
useful in the pharmaceutical industry for packing oils
and medicines.

The bottles can be decorated by screen printing,
offset, or stamping. They can also be metallized.
The adhesion of dyes and inks to polyethylene is
improved by surface treatment by flame oxidation or
corona discharge.

Blow moulding with HD and LD polyethylenes is used
for containers of up to 2000 l capacity for the
transportation and storage of liquid chemicals and
oils /50/.

The use of poly(vinyl chloride) for bottles was
hindered by its poor frost resistance, and, in the
case of the plasticized polymer, by its poor heat
resistance and easy extractability of plasticizers
and stabilizers. These drawbacks were overcome by
various toughened materials (cf. Section 2.4).
Transparency was improved, and some food contact
acceptable stabilization systems were found on the
basis of synergistic mixtures of calcium and zinc
salts, esters of aminocrotonic acid, and certain
organotin stabilizers, permitted in several countries
(e.g. USA,   FRG). Now, transparent bottles of
various shapes and colours are produced. The use of
poly(vinyl chloride) increased sharply in 1969-1970;

in France, bottle production was almost doubled from
27.000 t to 47,000 t of polymer /51/, one third of
the production being for mineral water bottled by
the firm Vittel.

Another development is the application of poly(vinyl
chloride) for wine, due mainly to the reluctance of
most self-service stores to sell beverages in returnable containers. Poly(vinyl chloride) packages can
also be seen in the field of cosmetics (cremes, liquid
shampoos, bath preparations, etc.), coffee /51, 52/,
non-alcoholic beverages, etc.

The Swiss firm F. Baumann introduced a production
technique for bottles from unplasticized poly(vinyl
chloride) film of 0.35 to 0.45 mm thickness by means
of the *Renopac* system /2/.

Another application of poly(vinyl chloride) bottles
is in the field of carbonated beverages, especially
beer /53, 54/. The beer bottles made by the Swedish
firm *Rigello Pak AB* have an unconventional shape;
the bottle is designed as a pressure vessel and reinforced with a collar made of wound paper provided
with an aluminium label (see Fig. 12) /2/. Poly(vinyl
chloride) bottles for these purposes facilitate
collation packing and, hence, reduced the need for
transport crates.

Among the polystyrene hollow containers the most
important are the crystal-clear bottles with a threaded
neck which are usually produced by injection blow
moulding; toughened polystyrene is also used. Polystyrene bottles are used for packing pharmaceutical
products and liquid sauces, but the applications are
limited by the significant gas permeability. Polystyrene is also used for yogurt cups, fruit juices,
ice creames, etc., often closed with laminated aluminium foil /55/.

5.3.3 COMPRESSION MOULDING OF PLASTICS

Compression moulding is the oldest process for shaping polymers. Its beginning dates back to the last decades of the nineteenth century /26/. The plastics mainly in powder form are subjected to elevated temperatures and pressures, the moulds imparting the required shape. According to the pressure applied compression moulding is classified as high-pressure, low-pressure, or contact /1/. The high-pressure process uses pressures above 3 MPa, the low-pressure process operates at 0.5 to 3 MPa, and the contact process needs only pressures below 0.5 MPa. Compression moulding is largely restricted to thermosetting resins; thermoplastics are used only in specialized applications.

5.3.3.1 H i g h - P r e s s u r e   M o u l d i n g

High-pressure moulding is further classified into direct moulding, indirect moulding or transfer moulding, and impact moulding.

Direct compression moulding was very widespread in the plastics industry, being used initially for processing thermosetting resins and thermoplastic sheets.

It uses almost exclusively hydraulic presses built for moulding pressures from 30 to 500 t. Mechanical power presses, having many drawbacks, are hardly used any more, and only as automatic circular presses based on similar principles to tablet making machines.

The advantages of hydraulic presses include easier upkeep, lower weight and dimensions, and most of all the possibility to adjust ram speed and pressure according to the resistance offered by the moulding material. This means that full pressure can be obtained at any position of the plunger.

The positive mould and flash mould designs have
been superceeded by the semi-positive mould. One part
of the mould is a cavity and attached to the plunger
plate and called the punch, the plunger part, called
the force, is incorporated in the stable stand of the
press (Fig. 69).

Fig. 69. Schematic representation of a mould for direct
compacting /1/

a - loose plastic material, b - compact

The necessary amount of material is determined by
preliminary practical tests for each kind of moulding
composition. The amount must be sufficient to ensure
perfect filling of the mould with a small surplus to
produce flashes. The force of the mould is coated with
a thin layer of release agent prior to filling, so that
the product can be removed more easily from the mould.
Multiple moulds are used for smaller products.

Perfect filling of the mould as well as the properties
of the product depend on correct moulding procedure,
i.e.

- time to release entrained air from the mould
- moulding pressure (the pressure exerted on the
exposed resin surface, e.g. 20-60 MPa for phenolic
resins, 30-40 MPa for carbamide resins)
- time to cure a 1 mm layer at the moulding temperature. The cure is carried out at the maximum possible
temperature, compatible with product quality, to make
the moulding cycle as short as possible. Resol materials
of 1 mm thickness require 60 s, novolaks 30 s, amino
plastics 40-55 s. The maximum temperatures are about
$170^{\circ}C$ for phenolic resins and $145-155^{\circ}C$ for carbamide

resins

   - greatest thickness of the profile of the product
determines the time needed for cure throughout the
material

   - flow properties determined by resin type and the
thermoplastic content, i.e. resol. In the manufacture
of the moulding material, the resin is partially
converted from resole to resitol and resite. The ratio
of these three forms of the resin determines its fluidity. The lower the resitol + resite content the
higher is the fluidity of the resin. On the other hand,
more resite means less shrinkage, and faster cure.
Therefore a compromise between these factors is
unavoidable. The resol content may be determined by
extraction with alcohol or acetone.

   As soon as the material is cured, the mould is
opened, the product removed, and the mould blown with
compressed air. The adhering residues of material are
removed with wooden or copper tools.

   Indirect moulding or transfer moulding adopts the
same method of mould filling as with injection moulding.
The mould is closed, and the material flows rapidly
in through a relatively small channel. The two
variants   differ in construction of the pressure
chambers.

   In transfer moulding the pressure chamber, called
the transfer pot or "cup", forms usually part of the
heated mould (Fig. 70). After injection of each shot
it is completely emptied.

   In the injection moulding process the pressure
chamber is called a melting chamber and usually forms
a part of the injection machine, separated from the
mould which is cooled. In the melting chamber the
material is pre-heated, melted, kept at the injection
temperature, and continuously replenished. In productivity this method is better than transfer moulding,

especially with the incorporation of screw plasticizing.

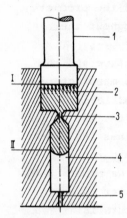

Fig. 70. Schematic representation of
a mould for transfer moulding /13/

1 - punch, 2 - the auxiliary pressure
chamber (transfer chamber, "cup"),
3 - gate channel, 4 - mould cavity,
5 - deareation joint

An important point of the transfer moulding process
is the reliable heating of the material to the correct
temperature. This is carried out in pre-heating units
and by transformation of mechanical energy into heat
during flow of the material through the injection channel.

The transfer pressure varies within broad limits,
from 50 to 150 MPa. Sometimes lower pressures are used,
e.g. in processing plasticized poly(vinyl chloride).

The maximum transfer pressure depends largely on
the cross-sectional flow in the mould, further influenced by product weight mould size, etc. Thus thin-
walled products, which are most important for packaging,
necessitate higher pressures, as the flow path is
relatively long in these cases. This contributes to the
reasons why compression moulding is not so important
in packaging, except perhaps for caps and closures.

Impact moulding represents a transition between
direct compression moulding and transfer moulding. It
differs from direct compression moulding by high
forming velocities and relatively cold moulds, and from
transfer moulding in that the dose of pre-heated
material is placed directly into the mould cavity, the

material being melted in a separate apparatus (chambers
with high-frequency heating, two-roll kneaders, extruders, etc.). The rapid forming velocities should
avoid undesirable cooling of the pre-heated material,
thus ensuring that the mould is filled properly. At
the same time, the residual stress in the moulding can
be greater, reducing the dimensional stability of the
product /13/. The technique is little used in packaging.

5.3.3.2 L o w - P r e s s u r e a n d C o n t a c t
         M o u l d i n g   P r o c e s s e s

Low-pressure moulding is mainly used for thermosetting
resins, especially phenolic resins and amino resins,
as well as reactive polyester resins, e.g. glycolmaleic
acid polyesters combined with styrene and with reinforcing materials (cf. Sect. 3.4.6). In the last
case contact moulding may also be used.

   Prior to application of resin, the reinforcing
material is arranged in such a way that it will be
uniformly distributed in the final article. This is
achieved either by pre-forming to the required shape
or by dispersing the fibrous material (cellulose,
cotton, glass fibers, etc.) in water and transferring
it by suction onto sieves which have the shape of
the final moulded piece. This procedure ensures uniform
distribution of the reinforcement, because, if the
suction produces a thinner layer at one place, water
penetrates more quickly through it and immediately
brings further fibers so that the non-uniformity is
compensated. This method of pre-forming is mainly
used for smaller deep profile products. In the production of large mouldings the reinforcement is placed
directly into the moulds and the contact resin is
applied by brush or spray gun.

In producing laminated materials from condensation resins, the reinforcement is impregnated with resin solution and after drying it is placed in moulds. Suitable resins are those with excellent flow which penetrate well into the reinforcement and with the ability to reach maximum strength under low pressure in a short time. Table 44 gives suitable conditions for low-pressure moulding of the resins most often used in this technology /1/.

The low-pressure and the contact moulding processes are carried out in the following ways:

In a press with stationary moulds which are usually open. Both surfaces of the moulding are smooth, and its wall thickness corresponds to the distance between the punch and force (Fig. 71). The presses for use in

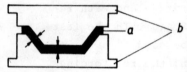

Fig. 71. Schematic representation of moulding of glass-reinforced laminates in a press with stationary moulds /1/

a - laminated resin, b - mould

these forming processes have large effective surface areas, i.e. large distances between the dowels. As the working pressures are low (0.1-0.7 MPa), the mould can be made of cast iron, reinforced concrete, light alloys, etc.

Moulding by means of elastic components, Fig. 72, needs only one stationary rigid component, the other part of the mould being formed by an elastic component (plunger) in a metal frame or by an elastic bottom (made of expanded rubber or poly(vinyl chloride)) pressed down by a solid punch. In the latter case, the elastic component usually has the shape of a tilting cover. Sometimes the elastic bottom can be replaced by an elastic tube. This method is used especially for long articles (Fig. 73).

Table 44. Conditions of low-pressure moulding of some resins /1/

Kind of resin	Type of resin	Curing temperature (°C)	Moulding time (min)	Moulding pressure (kPa)
polyesters	curing by heat	60 – 160	2 – 300	7 – 35
ureas	curing while cold	20 – 60	10 – 240	7 – 525
thioureas	curing by heat	60 – 105	5 – 15	7 – 1400
melamine	curing by heat	125 – 150	5 – 60	1050 – 1750
phenolic	curing while cold	20 – 60	30 – 720	7 – 525
	curing by heat	60 – 105	15 – 120	7 – 1750
	curing by heat	135 – 180	5 – 60	7 – 1750

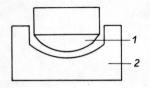

Fig. 72. Schematic representation
of a mould for moulding of glass-
reinforced   laminates by means
of elastic segment

1 - elastic segment, 2 - stationary
mould

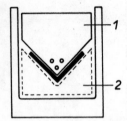

Fig. 73. Schematic representation
of moulding of glass-reinforced
laminates by means of an elastic
tube /1/

1 - the heated mould, 2 - the elastic
tube

Bag moulding, Fig. 74, uses an elastic bag filled
with air or steam for pressing the impregnated reinforcement onto the mould.

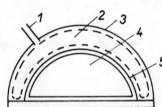

Fig. 74. Schematic representation
of bag moulding of glass-reinforced
laminates /1/

1 - steam admission, 2 - elastic
bag, 3 - covering of metal
sheet, 4 - mould, 5 - laminated
resin

Large-capacity vessels and giant containers and
reservoirs are produced by a newer technique which
involves winding continuous glass fibers impregnated
with polyester resin on a mandrel, either solid like
wood, or inflatable (Fig. 75).

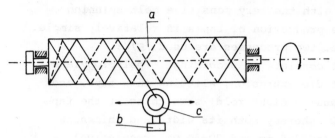

Fig. 75. One of the principles used for winding-up
of glass fibres in production of large-capacity
vessels and tubes of giant diameters /56/

a - shaping core (mould), b - base of the wind-up
equipment, c - roving guide

5.3.4 NETS, CIRCULAR KNIT GOODS, AND WOVEN SACKS

Nets and circular knit sleeves are usually made of
polyethylene or polypropylene fibres produced by melt
spinning possibly with further orientation. In spite
of relatively low heat resistance and moderate mechanical properties, polyethylene is used for high-
quality fibres, exhibiting high chemical resistance,
converted to nets and circular knit sleeves where its
mechanical properties and heat resistance are adequate.
Woven sacks, which must meet higher requirements,
are made of polypropylene.

Nets and circular knit sleeves are used especially
for short-term consumers packages of piece goods,
e.g. fruit, vegetables, household utensils, etc. The
circular knit sleeves are also suitable for protecting
glass bottles. Woven sacks are used for more exacting
industrial packaging (foodstuffs, fertilizers,
chemicals, textiles, etc.).

Formerly woven sacks were produced predominantly
from hard natural fibres especially from jute, but now
synthetic fibres and tapes made from thermoplastic
films (mainly polypropylene) are mainly used for these

products /57/.

Compared with the very sensitive melt spinning
process, the production of tapes is relatively simple.
In principle, two processes are used /57/:

The first consists in producing   polypropylene
film by slot die extrusion, slitting into tape, and
multiple (usually eight fold) stretching of the tape
at 120-200°C whereby both its width and thickness
are reduced (to 2-5 mm and 20-50 μm, respectively).
The tapes first pass through a hot air tunnel; at both
sides of the tunnel there are stands with rolls rotating
at different speeds according to the degree of stretch
required. The air temperature is adjusted to suit the
film and the degree and rate of stretching. The
material dwells in the tunnel for 2-3 seconds only.
The stretching causes orientation of the macromolecules
in the direction of elongation, which increases the
ultimate tensile strength in this direction to the
detriment of that in the transverse direction (cf.
Sect. 5.4.1). Prior to re-winding, the strips are
annealed at 90-130°C in another tunnel, which causes
a controlled shrinkage (by 1-2%). This annealing prevents
later uncontrolled shrinking of the strips during
further processing.

The alternative process starts with unidirectional
orientation of the polypropylene film before slitting.
Stretching is carried out on a series of heated
cylinders; the first pair heat the film to the required
temperature, and the second pair apply the stretching.
The film then passes through annealing and cooling
cylinders. It is advantageous to slit the film in the
weaving machines to obtain consistent warps for the
fabric.

Polypropylene woven sacks have a number of advantages
compared with jute sacks: they do not absorb water,
so they are ruptured neither by cooling below 0°C
(formation of ice) nor by drying at elevated temperatures;

they resist rotting and (practically) ageing, and are
acceptable for food contact. The chemical resistance
of polypropylene is important in packing chemicals.
The low density of polypropylene (0.90-0.91 g cm^{-3}),
even lower than that of polyethylene (0.92-0.96 g
cm^{-3}) /57/, means that polypropylene sacks are both
lighter and stronger than jute sacks of the same size.
Polypropylene woven sacks are useful also for packing
textiles and as mail bags. In comparison with sacks
made from thermoplastic films, the woven sacks are
much stronger and more resistant to bursting.

5.3.5 WELDING OF PLASTICS

The welding or heat-sealing of plastics is a method
of joining them by heat and pressure applied in the
area to be joined. The procedures include welding by
conduction, radiation, friction, hot gas, high-frequency
heating, impulse heating, ultrasonic irradiation and
more recently by lasers.

Conduction welding (Fig. 76) is used in packaging
to join films of thickness up to about 0.12 mm. Thicker
films can be welded by means of a soldering iron with
a longer heating arm (Fig. 77) with various shapes at
the end. The temperature can be varied by changing the
arm length. The areas heated with the soldering iron
are joined by pressing them together with a roller.

Radiation welding (Fig. 78) is used to join bulkier
objects made of the same material. The areas to be
joined are heated with a radiator (usually heated
electrically) and pressed together to give a very good
joint.

Sometimes, the material is melted by contact with
a high-gloss electrically heated metal mirror, or
plate, whereupon the mirror is taken away and the pre-
heated areas are pressed together (butt-welding). The

mirror surface is usually nickel-plated or coated with
a thin polytetrafluoroethylene (*Teflon*) layer to prevent
its sticking to the plastic. This method is called
polyfusion welding, or hot-plate welding.

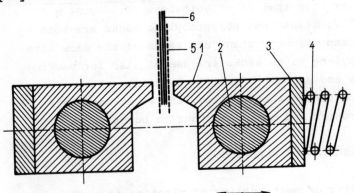

Fig. 76. Schematic representation of conduction
welding apparatus /1/

1 - jaw, 2 - heating unit, 3 - insulation, 4 - spring
with a stop, 5 - protective insert, 6 - the film
joined by welding

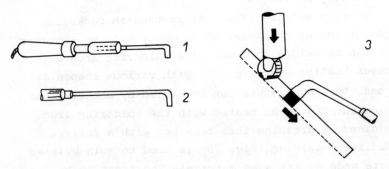

Fig. 77. Conduction soldering lamp with temperature
regulation /1/

1 - the arm is pushed in, its end is heated at higher
temperature, 2 - the arm is pulled out, its end is
cooled by the milieu, 3 - the working scheme

Friction welding is only applicable to semi-finished
products which can be clamped in a lathe. Heat is
liberated by friction at the surfaces counter-rotated

in contact, and the temperature depends on the pressure
and rate of rotation. The material is melted by the

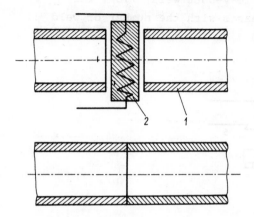

Fig. 78. Schematic representation of the radiation
welding process /1/
1 - the article to be welded, 2 - radiator

heat, and the contact surfaces are joined. The welded
surfaces must have conical shape, so that their centres
may be heated (Fig. 79). If such joints are made
properly, they are very strong.

Another method is welding with hot gas. It is
particularly useful for thick-walled materials. The
gaseous medium can be air, nitrogen, or carbon dioxide.
The surfaces to be joined and the material added (the
welding wire) to connect the surfaces are heated with
hot gas. The welding is carried out by means of a
welding gun (Fig. 80) which heats the gas and directs
it onto the material to be welded. Correct heating of
the area to be joined is achieved by suitable inclination
of the welding gun nozzle and its swinging. The welding
wire is introduced into the weld area in a direction
perpendicular to the plane of welding (Fig. 81). It
is pressed slightly into the welding bed so that the
"weld bead" is somewhat shorter than the welding wire

diameter. In this method, almost all the weld shapes
are applied which are known in metal welding (fillet
welds, V-butt weld, double-V-butt weld) /13/. The
strength of a weld decreases with the number of weld

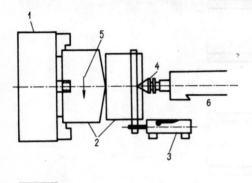

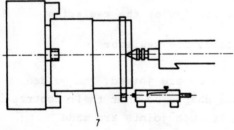

Fig. 79. Welding of plastics by friction with the use
of a lathe /1/

1 - jaw chuck, 2 - the components to be welded,
3 - the stop, 4 - live centre of the tailstock,
5 - direction of rotation, 6 - the stationary part,
7 - welding

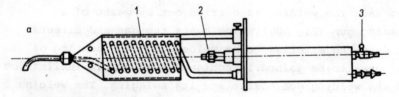

Fig. 80. Welding gun for welding of plastics with hot
gas /1/

a - heating with town gas: 1 - heating tube, 2 - the
burner nozzle, 3 - gas regulation valve

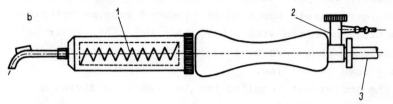

b - electrically heated gun: 1 - electric heating
coiling, 2 - gas inlet regulation valve, 3 - cable
attachment

beads needed. The inclination of the nozzle of the
welding gun (the α angle in Fig. 81) determines the
spot of most intense heating. If the α angle is smaller,
then the welding wire is heated more than the welding
bed, whereas for greater α angle the opposite is true.

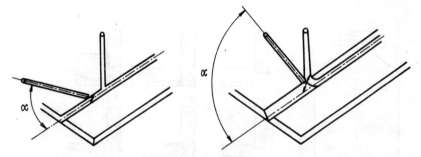

Fig. 81. Schematic representation of hot air welding /1/
(For thin materials the α value is lower than for
thicker materials. The arrows denote angle of oscil-
lation of the mouth of the welding gun)

This technique can also be used for production of
large-volume vessels and containers.

Large and deep plastic containers present considerable difficulties when produced by moulding, particularly in wall thickness uniformity. The body of the
vessel may be produced from film or sheet which is
wound and sealed by welding. Also the base and top
can be attached by welding or by adhesives. This
method is used for the production of vessels from

unplasticized poly(vinyl chloride) /13/.

Larger vessels can also be produced from polyethylene (possibly combined with other plastics) /15/. Their individual parts are usually made by injection moulding and joined by welding.

The individual sections can be chosen by dividing the vessel according to its axis or in the way represented in Fig. 82. In the A5 variant the base is usually made of a stiff material (e.g. HD polyethylene or polystyrene) so that there will be no leakage after pressing the base into the groove in the cylindrical wall of the body. The A6 + B3 variant is chosen for bottles made of HD polyethylene which is welded relatively easily.

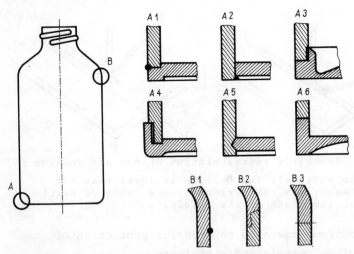

Fig. 82. Schematic representation of bottle production by combining its individual segments /15/

A - methods of connections at the bottom, B - methods of connections at the neck

Because of various drawbacks (difficult sealing of the joints, imperfect appearance) gas welding is used nowadays practically only for large-volume vessels.

The main advantage of the method is that it does not
require any intricate or unusual apparatus. The individual components of vessels can be connected in various
ways. According to the Van Leer method, for example,
pipe for Valasex HD polyethylene containers is hot-plate
welded to injection moulded base and top. A special method
of welding the components of polyethylene containers
has been developed in U.S.S.R. /58/. It is based on
joining the components by the heat and pressure of a
melt of the same polymer which is injected into the seam
cavity of a special casting mould. After cooling, the
injected melt forms the weld seam.

Some plastics having a sufficiently high dielectric
constant can be welded by high-frequency (dielectric)
heating. This method of heating is based on dielectric
losses of the polymeric material itself when exposed to
a high frequency electric field, which induces electronic
and orientation polarization in the material /1/.

Ideal adjustment to changes of the electric field at
increasing frequencies is exhibited by polymeric hydrocarbons, e.g. polyethylene, polystyrene, polyisobutylene,
and rubber, and by the polarly balanced polytetrafluoroethylene, which have low relative permittivities
(dielectric constants) ε (2.2 to 2.5) and practically
zero dielectric loss angles δ . The insulating factor,
ε tan δ , which determines the energy absorbed and
transformed into heat when the material is exposed to
an oscillating electric field, has practically zero
values for these polymers, and even the slight dielectric
losses observed are presumed to be due to the impurities
present. That is why it is impossible to heat or weld
these polymers with high-frequency alternating electric
current. A different situation is encountered with the
polymers whose molecules are polarized due to the
presence of such bonds as $C=N$, $C-N$, $C-O$, $N-O$, $C-Cl$, $C-O$,
and whose structure allows - at certain temperatures -

dielectric losses sufficient to be utilized at high
frequencies for heating or welding.

Plastics are welded by high-frequency heating in
welding presses, pliers, automatic presses, etc. The
jaws of the presses are in fact the electrodes of a
capacitor (Fig. 83), and the material placed between

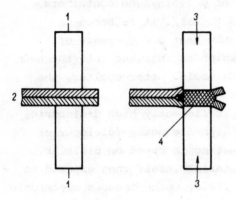

Fig. 83. Schematic representation of welding by means
of dielectric heating /1/

1 - electrodes, 2 - plastic film, 3 - pressure,
4 - welding

them is heated and melted due to dielectric losses.
The jaws are made of brass, which can be easily
repaired and is an excellent conductor. The profiled
jaws with cutting edges, which are designed to separate
the residual material from the article welded, are
made of steel. When these cutting electrodes are
applied, the weld must be supported by material having
a low relative permitivity and high dielectric strength
to prevent a puncture. Suitable supports are made of
mica, cellulose triacetate film, polyvinylcarbazole,
polypropylene, etc. The pressure between the jaws
(about 0.2-0.3 MPa) is achieved by mechanical,
hydraulic, or pneumatic means. Thermoplastic materials
are welded with frequencies from 5 to 200 MHz. As the
high frequency apparatus can interfere with radio-

communication, there are restrictions on which frequencies may be used in a given location.

Thin films, e.g. polyethylene, can be welded conveniently by a heat impulse which can be produced in various ways /15/. Impulse welding is the procedure in which a sufficiently large impulse current (up to 300 A) is applied to a resistance band to heat it to the temperature necessary for welding the film. Single-sided impulse apparatus is effective with films up to about 0.08 mm thickness; greater thickness, up to 0.12 mm, requiring double-sided impulse apparatus.

Another impulse method produces heat in a conductor by current induced by an electromagnetic field. In this case the conductor inserted between the components welded becomes part of the weld (e.g. joining poly-ethylene pipes by special weldable couplings) /17/.

Impulse welding by heat radiation is represented in Fig. 84. A permanently heated resistance wire is

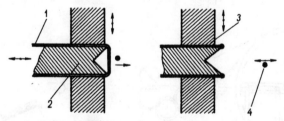

Fig. 84. Schematic representation of welding by radiation impulse /15/

1 - film, 2 - insert, 3 - clamping jaws, 4 - heated resistance wire

placed near two layer of stretched film so that the films are melted through, and two high-quality welds are formed simultaneously.

For welding bags, a highly efficient impulse welder is used, as shown for an automatic machine in Fig. 85. The heat impulse is supplied by a permanently heated

welding blade which moves with a drum during welding,
afterward returning to its original position. Sticking of material to the blade is prevented by a separation
belt. High outputs are reached with this method (up
to 24,000 welds per shift) /17/.

The application of ultrasound to the welding of
plastics is not very widespread in spite of having
indisputable advantages including cheapness, high
rate of bond formation (from 0.2 to 2 s), non toxicity,
and safe operation. Practically no jigs or couplings,
and no solvents are necessary, and even quite distant
areas of plastic components can be joined. As shown in
Fig. 86, the ultrasonic waves penetrate through the
material to the surface to be joined where they are
abruptly dispersed and refracted to produce the necessary
heat for welding. As soon as the ultrasonic waves begin

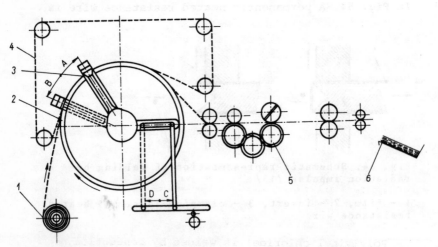

Fig. 85. Schematic representation of an automatic
machine for welding of bags /15/

1 - the material to be welded, 2 - rubber-coated
cylinder, 3 - tilting (swinging) welding knife,
4 - infinite separation belt, 5 - cutting device,
6 - stack equipment

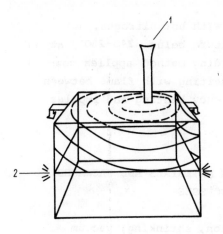

Fig. 86. Schematic representation of ultrasonic welding /17/

1 - the ultrasound source,
2 - the welded area

to penetrate through the joint (weld) into the other material, their source is switched off. The material joined remains quite cool except for the area welded.

This welding method has the limitation that it is only applicable to hard materials, e.g. acrylonitrile-butadiene-styrene terpolymers, both unmodified and toughened polystyrene, polyethylene, polypropylene, polyamide, etc.

The methods of welding of the most common polymers /1/ are given below for example:

Polyethylene is welded by the radiation or impulse methods at temperatures corresponding to red heat /15/, by conduction methods at 200-250°C, and with hot gas at 190-210°C /59/. Friction welding is also applicable /15/, but high-frequency welding is impossible due to the unsuitable dielectric properties of the polymer.

Poly(vinyl chloride) is welded by conduction, radiation, friction, and most often by high-frequency heating and with hot gas (240-270°C) /13/.

Poly(methyl methacrylate) is welded with hot air (320-350°C), or by friction using at least 1000 rpm /59/, whereas the conduction and high-frequency methods are very difficult. Another friction welding method is given in reference /60/.

Polyamides can be welded with hot nitrogen, but only if dry, the gas temperature being 240-280°C at the nozzle. Another easy welding method applies conduction-heated inserts and melting with flame between two metal plates. High-frequency heating and impulse welding are also used /61/.

5.4 PRODUCTION OF PACKAGING MATERIALS AND PACKAGES BY MEANS OF FORMING PROCESSES

Forming processes (orientation, shrinking, vacuum forming, sheeting) are applied in the region of viscoelastic deformation. They are either independent processes (vacuum forming, shrinking) treating the material in the form of a pre-fabricated intermediate, e.g. films, or parts of certain combined processes involving forming, i.e. treatment of polymers in their plastic state. This applies, e.g., to the production of biaxially oriented films by bubble extrusion. The material is oriented during blow-up and stretching while being formed in its viscoelastic state.

5.4.1 ORIENTED AND HEAT-SHRINKABLE MATERIALS

Plastic films are expected to have a number of properties, e.g. strength, tear resistance, stiffness, ductility, frost resistance, heat resistance, resistance to effect of weather and chemicals, printability, dyeability, transparency, electrical insulating properties, permanency of form, shrinkability, or certain barrier properties to gases, humidity, and aromatics. An individual plastic can satisfy only some of these requirements which can be mutually exclusive. The required properties can, however, be obtained or

improved by certain special modifications which extend
the application possibilities of plastics. Sometimes
various materials are combined (reinforced plastics,
toughened polymers, etc.), or the polymer is modified
physically (especially by orientation or stretching).

5.4.1.1 O r i e n t a t i o n   (S t r e t c h i n g)
o f P o l y m e r s

Both crystalline and amorphous polymer change their
shape and the arrangement of macromolecular chains
during large and permanent deformations, with ensuing
changes in their properties. The effect of such large
deformations on the polymer structure can be exemplified
by the generation of anisotropic properties in calandered
materials (calandering effect) /62/.

5.4.1.1.1 *Orientation of Crystalline Polymer*

If crystallization of a polymer takes place without
the influence of any outside forces, its crystalline
structure is unmodified. If such crystalline material
is exposed to stress, its crystalline phase will
rearrange (Fig. 87) /63/. Changes in X-ray diffraction
pattern provide evidence that the polymer chains became
oriented along the direction of the applied force.
The physical properties of the oriented polymer are
changed in the same direction.

The orientation is carried out by stretching the
polymer below the melting point of the crystals (T_m)
but above the glass transition temperature (T_g) of the
amorphous portion. The elongation of the material is
sudden, and the sample becomes thinner, initially at
one point (Fig. 88). This phenomenon is called necking.
All the thin elongated sections of the sample have the
same gauge, the same being true of the gradually

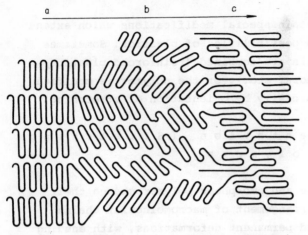

Fig. 87. Schematic representation or orientation of
polymeric chains of a crystalline polymer /63/

a - crystalline phase of the polymer before orientation
(fibrillae), b - changes in arrangement of the chains
during orientation (stretch in the direction of the
arrow), c - the newly-formed (oriented) crystalline
structure (fibrillae)

disappearing unstretched sections.

As stretching continues, the thin section extends
to the whole length of the film. For most polymers
(e.g. LD polyethylene, polyesters, polyamides) the
ratio of the length of the elongated material to the
original length is 4:1 to 5:1. In some cases, however,
it can be much higher, e.g. up to 10:1 with HD polyethylene.

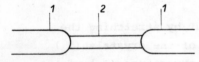

Fig. 88. Schematic rep-
resentation of necking during
fiber stretching

1 - the original fibre section
(before elongation), 2 - cross
section of the stretched
(elongated) fibre (neck)

If the polymer is sufficiently crystalline before
orientation, the process will not change its crystallinity

significantly. As the rearrangement mainly involves
the whole crystalline system, the orientation of a
crystalline polymer will stabilize its crystalline
state. The crystallinity of mainly amorphous or only
partly crystalline polymers seems to be increased
during stretching.

Spherolitic aggregates forming the crystalline
phase are elongated elliptically during drawing, and
if rupture occurs, then it is observed at their inter-
faces /63/.

5.4.1.1.2 *Orientation of Amorphous Polymer*

The stretching of polymers does not necessitate the
presence of a crystalline phase. About the glass transition temperature amorphous polymers can be oriented,
too. Two types of orientation are possible with an
amorphous polymer: orientation of whole molecules and
that of their segments /62/. Thus the orientation of
sections of macromolecules is facilitated by the
mobility of the segments, i.e. it proceeds more easily
if the macromolecular segments are shorter. This
orientation can take place even during the reversible,
highly elastic deformation of the polymer, and can
be fixed by cooling the polymer below its glass transition temperature. The reversible deformation of the
polymer then remains in a "frozen state". The polymer
exhibits enhanced strength and tendency to shrinking
along the direction of chain orientation, and pronounced
and rapid shrinking takes place on heating the polymer
above its glass transition temperature.

The forces necessary for elongation of a polymer
are changed during the course of orientation of the
chain segments of the macromolecules. As elongation
increases the length of the segments, which is a measure
of chain mobility, the polymer becomes more and more

stiff. This stiffness retards both further orientation
and disorientation. Thus high orientation can transform
a soft, rubber-like polymer into hard, glassy material.
It is, however, very difficult to obtain such glassy
polymer by orientation, because the viscosity increase
accompanying chain stretching is so great that further
orientation results in rupture of the material. A
higher degree of orientation could only be reached at
a lower viscosity which can be realized by increasing
temperature or by applying swelling agents which are
removed after the orientation process is finished. The
lowering of viscosity, however, must not be too great,
because it would permit increased thermal motion with
consequent disorientation of the polymer structure. The
drawing of amorphous polymers connected with the
orientation of whole molecules (at flow) can result in
a more stable oriented state.

The oriented amorphous and crystalline polymers
are different. First of all they differ in their phase
state /62/. Also the courses of orientation and disorientation in crystalline polymer are different from
those in amorphous polymer. Whereas the latter are
continuous, the former are discontinuous, being
connected with necking, and the oriented state is stable
up to the melting temperature which destroys the
crystalline structure.

The properties of these two oriented systems also
change with time. The oriented crystalline polymers
show a tendency to reach still higher organisation
until the disorientation tendency existing in the
amorphous phase prevents this. The oriented amorphous
polymers in the stretched condition can crystallize
if thermodynamic prerequisites are satisfied, but in
most cases gradual disorientation occurs, which can
be accelerated by viscosity decrease. In this way an

equilibrium, non-oriented, isotropic state is obtained.
This procedure carried out practically in the production
of "oriented" films is known under the name of thermo-
fixing (thermostabilizing or "annealing"). Kargin and
Slonimskiy /62/ say: "The erroneous idea that oriented
polymer must be in a crystalline state has unpleasant
consequences in practice, e.g., spontaneous deformation
during application of the anisotropic films which should
have maintained precise dimensions".

For evaluation of phase state of an oriented polymer
it is necessary to apply structural analysis, but thermo-
dynamical methods are also necessary if the history of
the application of force during orientation is not known.
The methods based on measurement of changes of properties
in the region of the glass transition temperature (e.g.
determination of the shrinkage force) can also be useful
occasionally.

5.4.1.2 Methods of Drawing

Various possibilities follow from the above discussion
for the practical realization of drawing and the application of oriented polymers. Drawing is a normal part
of modern processes for manufacturing films. The film
is stretched by the action of external forces either
immediately after the extrusion die or after production
of the pre-film /64/. In this way the area yield is
increased with simultaneous reduction of thickness. It
is recommended to carry out the drawing with materials
of viscosity within the limits 10^4 to 10^9 Pa s.

Film orientation by drawing can be accomplished
either in one direction (longitudinal or transverse)
or in two directions (biaxial).

There are two main advantages of transverse orientation
as compared with longitudinal orientation /61/:

1. The width of the oriented film can be 1/3 of that
of the final film, so the extrusion slot die is less

expensive, its manufacture is easier, and its dimensional
requirements can be met with higher precision.

2. Transverse non-uniformities in thickness of the
unoriented film are diminished by this mode of orientation.

Various techniques have been developed for drawing
films. Good practical results of the drawing were
obtained with the expansion frames (tenters) used in
the textile industry /61/. Depending on the particular
construction chosen, these frames allow both uniaxial
and biaxial (simultaneous or stepwise) orientation of
films.

Occasionally the oriented films are produced with
extrusion and drawing combined in a single production
line, the film being oriented immediately after extrusion
at the cooling drum while still in the viscoelastic
state. This method requires relatively small elongation
forces, and the orientation proceeds easily and uniformly. If the interval between extrusion and drawing
is longer, the orientation with crystalline polymers
is more difficult and less uniform due to the formation
of this crystallinity, and the elongation forces
needed are much greater. There are several established
procedures /23/:

For the production of flat film the first process
is usually longitudinal orientation by elongation at
rollers followed, if need be, by transverse orientation
in the extension frame.

The film extruded from a circular die can be oriented at once, biaxially, if the blow-up ratio is large
and, simultaneously, the tubular film is pulled away
at increased speed at a temperature close to the melting
temperature of the crystallites /65-67/.

The macromolecular chains of oriented thermoplastic
materials tend to restore the original non-oriented state
at increased temperatures, which results in shrinkage

of the material. This phenomenon, originally considered
quite undesirable, is often termed elastic (plastic)
memory or recovery. However, in the fifties it found
an advantageous application in USA in the field
of packaging /65/ which spread to Europe in the
sixties /68/. Thus a quite new, but now widespread
and popular packaging technique was created: packaging in shrinkable films.

5.4.1.3 P r o p e r t i e s   o f   S h r i n k a b l e
        F i l m s   (Shrink-Wrap)   C o m p a r e d
        w i t h   T h o s e   o f   O r i e n t e d
        F i l m s

Oriented films are monoaxially or biaxially oriented
thermoplastic materials which were subjected after
orientation to a fixation procedure ("annealing") in
which the material in its stretched condition was
exposed to a temperature lower by several tens of
degrees than that used for drawing (orientation). The
fixation causes considerable relaxation of the tension
introduced by the drawing, and the material thus
acquires considerable dimensional stability even at
higher temperatures.

The main aim in the production of oriented films
is to enhance some of the properties of the packaging
material. Properties are improved in the direction
of orientation (drawing) to the detriment of the
properties in the transverse direction. Thus the
monoaxial orientation of the packaging materials is
particularly important in the cases where especially
good mechanical properties are required in a given
direction (e.g. filaments, strings, packaging tapes).
However, for films, biaxial orientation is increasingly
important and brings more versatile improvements of
the properties of packaging materials, especially if

the procedure is properly balanced in the two directions.
Properties particularly improved are:

- tensile strength /13, 61/
- impact resistance (toughness) /60, 69/
- stiffness /69/
- resistance to tear initiation and propagation
  /26, 69/
- flexibility and frost resistance /60, 69/
- solvent resistance /26/
- vapour barrier performance /69/
- transparency and brilliance /69/

On the contrary, the orientation of thermoplastic
material in the production of "shrinkable" films brings
a quite new quality - shrinkability at increased temperatures. As a rule, the film shrinkage is expressed as a
percentage of the original linear dimension of the
oriented film. At increased temperature, the film shrinkage is accompanied by relaxation of the tension
introduced during its elongation. For example, Fig. 89
gives the temperature dependence of tension and shrinkage of a shrinkable polyethylene film /66/.

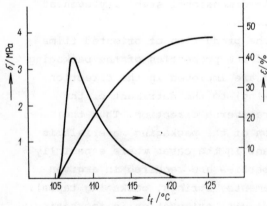

Fig. 89. Temperature dependences of strain δ and
shrinkage ε of a shrinkable polyethylene film /66/
t_f - the film temperature

The practical application of the shrinkable films in packaging technology aims at the following properties:

a) the maximum possible shrinkage of the film in order that it may fit snugly to the goods packed, and at the same time

b) the maximum possible shrinking force (tension) in order that the packed goods may be firmly held.

With respect to the temperature dependence of the shrinking (which is increased with increasing temperature) and the tension (which, on the contrary, decreases with increasing temperature), it is always necessary to find a compromise between the two.

The shrinkable packaging materials are manufactured in the form of single or double flat films, half-sleeves, sleeves, casings, sacks, and bags. Use of shrinkable films necessitates well-balanced shrinkability and physical properties. The physical properties can often be affected favourably by cross-linking the polymer, which is mostly achieved by exposure to ionizing radiation /65/. Shrinkability depends on the polymer type and the degree of orientation (extent of elongation). Some characteristic properties of shrinkable films based on various polymeric materials are given in Table 45.

The first plastics from which shrinkable films were produced were poly(vinyl chloride) and, later, nylon 6 /64/. Nowadays, the most widespread material is low-density polyethylene (LDPE) (cf. Sect. 4.2.1). Its favourable properties are low processing temperature, high impact resistance (toughness), excellent tear strength (resistance to tear propagation), good weldability, and cheapness. Poly(vinyl chloride) and polypropylene, however, exhibit better transparency and provide a greater shrinking force (tension). Other polymers applicable to the production of shrinkable films are copolymers of ethylene with vinyl acetate,

Table 45. Characteristic properties of shrinkable films /71/

Polymer	Tensile strength (MPa)	Permeability for water vapour) P_{H_2O} (g h^{-1})	Permeability for oxygen$^{a)}$ P_{O_2} (mm^{3}h^{-1})	Shrinking (%)	Shrinking temperature (°C)	Welding temperature (°C)
vinyl chloride-vinylidene chloride copolymer	42-140	0.8-22	0.8-25	30-60	65-110	80-120
branched polyethylene	12	0.5	417	15-40	100-120	150-200
cross-linked branched polyethylene	56-90	0.5	312-400	70-80	70-115	150-200
cross-linked linear polyethylene	98-130	0.1	146	70-80	90-140	150-200
polypropylene	100-190	0.1-0.2	63-71	70-80	100-165	175-200
polyester	118-250	0.1-0.9	0.6-13	25-40	70-120	-

$^{a)}$The film thickness of 25 μm, surface area of 1 m^2, and pressure difference of 101 kPa.

poly(vinylidene chloride), copolymers of vinyl chloride
with vinylidene chloride, polystyrene, and polyesters
/65, 68, 70/.

Three basic parameters affect the suitability of a
polymer for the production of shrinkable films:

   1. the mean molecular weight expressed usually by
the melt flow index,
   2. the molecular weight distribution,
   3. the crystallinity expressed usually by density.

Certain optimum values of these parameters along
with the necessary orientation of the material represent
the necessary conditions for the production of first-rate
shrinkable films.

For example, we give dependence of properties on the
characteristics of LDPE /68/: Increasing molecular
weight results in increasing toughness and decreasing
melt flow index. The toughest films are obtained from
polymers with the melt flow index from 0.02 to 0.1 g min^{-1}.
The density (crystallinity) of polyethylene affects
the stiffness, toughness, tear strength, weldability,
and transparency of films; the most favourable properties
correspond to a density in the range 0.918 to 0.923
g cm^{-3}. The distribution of molecular weight and branching of the main chain of the polymer affect the processing and shrinkability of film. These parameters,
depending on the polymerization process, mean that
only some types of LDPE can be used in the production
of shrinkable films.

The degree of orientation also depends on several
parameters:

   1. temperature of drawing (elongation),
   2. extent of elongation,
   3. ratio of longitudinal to transverse elongation,
   4. cooling rate.

Increasing drawing temperature results in decreasing

viscosity of the material, therefore the tension
decreases faster. The elongations of the film in the
longitudinal and transverse directions are chosen
according to the required packaging type and purpose of
the shrinkable film. They vary within relatively broad
limits. If the orientation (elongation) is the same in
both directions, then the shrinkability of the film
is balanced. The cooling rate of the oriented film
directly determines the tension fixed in the polymer
and, hence, the shrinkability of the material.

5.4.1.4 Processing of Shrinkable and Oriented Films

The packaging methods using shrinkable films are of
recent introduction. They penetrated into European
packaging technology from USA at the beginning of
the sixties /68/, being applied, at first, to the packaging of consumer foodstuffs without or with supporting
rigid trays (meat, poultry, fruit, bread, etc.) and
for multiple packs during clearance sales or for use
in larger families (e.g. 4 to 6 bottles or tins or
cartons). Thus for collation packaging of several units,
the assembly is covered with shrink-film instead of
paper or cellophane. The shrinkage of the film also
provides good compactness for handling. This is why
its application also penetrated into the field of
primary packaging where it successfully replaces
paperboard wrappings. Furthermore, the application of
shrinkable films also spread to securing goods on
loading pallets.

The technology of machine packaging is relatively
simple, being slightly different for small and medium
packagings (individual products or groups, and transport
packaging) and for the large ones (palletization).

Besides the type of machinery with a very broad field

of application, some arrangements are specially devised for packaging of certain products or group of products (meat, poultry, bottles, textiles, newspapers, journals, furniture, etc.). On the whole, however, these means of mechanization represent a line combination of two basic units:

1. the actual packaging machine and
2. the tunnel in which film is shrunk by heating.

An example is the arrangement represented in Fig. 90 /68/.

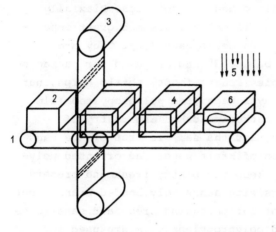

Fig. 90. Schematic representation of production of packages with open ends (with the so-called open handle) /68/

1 - supply conveyer, 2 - set of the chosen number of items to be packed, 3 - unwinding and welding equipment, 4 - the wrapped set, 5 - shrinking tunnel, 6 - final package

The advantages of shrink-wrap packaging responsible for its rapid growth are: increased protection from moisture, dust, and contamination, transparency of the packaging and its good appearance (the content is easily seen and checked), better utilization of loading capacity and easier handling (such packaging does not increase the volume of the packed goods and

the weight of the film is negligible). The method of
packaging in shrink-wrap produces a minimum of waste
packaging material which is easily disposed of /68/.
Experience has also shown that transparent wrapping
has a psychological effect during handling, leading to
more care. Also this method of packing goods often
facilitates palletization.

Oriented films are produced from HDPE, polypropylene
/69/, polyamides /61/, acrylate polymers /26, 61/, and
homopolymers and copolymers of vinyl chloride /13/.
They are used both alone and as laminates. Biaxially
oriented polypropylene films have assumed great importance, largely displacing cellophane /69/. They are
mainly produced as "balanced" and are used for packaging
dried fruit, vegetables, paste foods, potato chips, nut
kernels, etc. Multilayer bags from polypropylene film
are widely used for packaging cakes and pastry of
long-term shelf-life, and the same is true of the bags
made of a combination of kraft paper and oriented polypropylene film. They keep the pastry fresh and preserve
its aroma. Paper bags with adhesively bonded transparent
windows are important for packaging bread and sandwiches.
Blanks from oriented polypropylene film are used for
the manual or semi-automatic packing of flowers and a
number of other products.

However, in some cases the orientation of thermoplastic polymer materials has undesirable effects on
their properties. Oriented polypropylene films no longer
have good welding characteristics. Therefore, for
universal application, oriented polypropylene film is
coated with a suitable heat-sealable layer.

Surface modifications of oriented polypropylene
film can be achieved in various ways conferring greater
or smaller differences in properties, which determine
its suitability for a particular use. The following
methods are known:

1. production by circular or slot die coextrusion
with an olefinic copolymer
2. production with a slot extrusion die and subsequent
coating with polyolefin (predominantly LDPE)
3. incorporation of a resin into the polypropylene,
which enables welding of the oriented film
4. production with a circular or slot extrusion die
and subsequent coating with poly(vinylidene chloride)
dispersion
5. production with a circular or slot extrusion die
and subsequent coating with poly(vinylidene chloride)
lacquer.

The films produced by the methods 1 and 2 are suitable
for forming, filling, and sealing machines. The film
produced according to method 3 is usually highly
transparent, but its processing is made difficult due
to the presence of high electrostatic charge.

The films made according to method 4 have the broadest
field of application. They are suitable for packed
goods requiring protection from oxygen, water vapour,
and also preservation of aroma. They are used, e.g.,
for packaging of cigarettes. Similar properties are
encountered also with the films produced according to
the method 5 which, however, provide less protection
from oxygen and preservation of aroma.

A number of unmodified uniaxially or biaxially
oriented polyethylene films are produced under various
commercial names, e.g. *Plastotrans X* (Folienfabrik,
Forschheim), *Suprotherm S* (Kalle), *Walotherm* (Wolf Walsrode), *Alkoron SH* (Alkor-Oerlikon Plastics GmbH.), etc.
Their applications are very varied at present.

Oriented polystyrene films are distinguished by
their crystal clarity, high gloss, good weldability,
and printability; they are highly permeable to gases
and resistant to fats and to temperatures from -60°C
to + 100°C. They are produced in USA mainly by the

firm Monsanto Chemicals Ltd. under the name *Polyflex*
and in Europe by the firms Norddeutsche Seekabelwerke
AG (FRG) and BX Plastics Ltd. (GB) under the commercial
names *Norflex* and *Bexphane S,* respectively. They are
used for packing fresh meat (particularly the types with
anticondensation modification), fruit, vegetables, as
supporting trays for meat and fruit, as window films
for envelopes and boxes, and they are laminated to paper,
metal foil, etc.

Oriented polyamide films, which are chiefly used
for their enhanced impermeability to gases, serve mainly
as supporting materials for laminating. Therefore, they
have not yet found such broad application as the non-
oriented types which can be vacuum formed.

Among the shrinkable film materials so far the most
widespread are based on LDPE. The heat-shrink films
Cryovac L and *Cryovac XL* (W.R. Grace and Co.) are
cross-linked by irradiation which improves their physical
properties. They have high tensile strength (60-90 MPa)
with a maximum shrinkage of 70-80%, whereas the
non-irradiated ones have tensile strengths about 12 MPa
and maximum shrinkages of 15-40%. Also the shrinking
tension of the cross-linked films is 10 times higher
than that of the non-irradiated ones, and is maintained
with time. The shrinking temperatures vary from 110 to
315°C. *Cryovac XL* is further modified with an anticondensation coating on one side.

Another important producer of shrinkable polyethylene
films (called *Lexel*) is the Dow Chemical Corp.

The shrinkable "ionomer" film *Surlyn A* (Du Pont de
Nemours) is an ethylene copolymer found useful in packing metal products (even those with sharp edges); its
high gloss improves the overall appearance of the goods.

Shrink-wraps based on ethylene-vinyl acetate copolymers, compared with LDPE, are characterized by higher
permeability to gases, high toughness even at very

low temperatures, greater elasticity and flexibility, better printability, better weldability (also in high-frequency welding), gloss and transparency.

Many applications were found copolymers, e.g. *Alathon E/VA 3120* and *Alathon E/VA PE 6646* (Du Pont de Nemours, USA), *Urathene UE 630-81* and *Urathene UE 637-80* (National Distillers Chemical Corp.), *Alkathene YJF 502* (Imperial Chemical Industries Ltd.), *Montothene G* (Monsanto Chemicals Ltd.). The film *Meatwrap* produced by I.C.I. is used for packing fresh meat.

Polystyrene shrinkable films, whose shrinking tension is maintained with time, have good permeability to oxygen, carbon dioxide, and water vapour, and are used for packing fresh fruit and vegetables. They possess crystal clear transparency and good printability.

Polypropylene shrinkable films have not yet found any broad application to packing of fresh foodstuffs with short-term shelf-life, because they have relatively low permeability to water vapour and gases. Besides various technical applications they are used for packing freshly baked bread. Commercial names are *Cryovac Y* (W.R. Grace and Co.) and *Propafilm S* (Imperial Chemical Industries, Ltd.).

Films based on poly(vinyl chloride) are produced with or without plasticizers. The latter are used for packing foodstuffs which necessitate a wrap of low gas permeability, e.g., bread and pastry, and they are produced by the Italian firm Fiap S.P.A. under the name *Termovir TA*.

Plasticized poly(vinyl chloride) shrinkable films are made by W.R. Grace and Co., and Goodyear Tyre and Rubber Co., but are not very widespread because they have poor dimensional stability and are not permitted for packing foodstuffs due to the plasticizers used. The latter problem is solved by using special harmless plasticizers for the drawable films *Vitafilm* (Goodyear

Tyre and Rubber Co.) and *Resinite* (The Borden Chemical Co. Ltd.). Resinite films are produced in a number of modifications for packing various foodstuffs:

Resinite RMF for fresh meat, poultry and fish,

Resinite VF-70 for bacon, boiled meat, dried and smoked fish, and delicatessen,

Resinite VF-71 for fresh fruit and vegetables, and

Resinite AF-50 for cheese, pastry, and sandwiches.

Of the vinyl copolymers, shrinkable packaging materials based on vinyl chloride-vinylidene chloride have increasing importance; they are produced by Dow Chemical Co. (*Saran Wrapfilm*), W.R. Grace and Co. (*Cryovac S* film and bags) and Kureha Chemical Industry Co. Ltd. (*Krehalon*).

They are characterized by very low permeability to oxygen, carbon dioxide, and water vapour. Hence, they are particularly useful for the vacuum packing of poultry and ham, for the ripening of hard cheese, and for packing dough and pastry on supporting trays. They are also used in the form of shrinkable sleeves for packing dried fruit. They are shrinkable at $100^{\circ}C$, which allows the use of a hot water bath, often more economical than shrinking in hot-air tunnels.

Shrinkable films based on rubber hydrochloride have been known for several years under the name *Pliofilm* (Goodyear Tyre and Rubber Co.). They exhibit high permeability to oxygen and carbon dioxide, lower permeability to water vapour, but have poor light resistance. They are used for packing some.kinds of fruit, vegetables, and fresh meat.

High impermeability to oxygen, carbon dioxide, and water vapour is exhibited by shrinkable polyester films. They are shrunk at about $100^{\circ}C$, as with vinyl chloride-vinylidene chloride copolymers films. A commercial example is *Hostaphan* (Kalle AG) in the form of sleeves *Hostaphan Strumpfschlauch* used for vacuum packing ham, poultry, ripening of hard cheese, and as casing under the name *Nalophan*.

5.4.2 PRODUCTION OF PACKAGES BY FORMING

In contrast to compression moulding of plastics in presses, injection moulding machines, or screw extrusion machines, where the materials are treated in the plastic state, the starting material being in the form of powder, granules, agglomerates, etc., the technology of forming uses pre-fabricated films or sheets. Change of shape is achieved by die-press, compressed air or vacuum, either in the viscoelastic (rubber-like) state after heating, or by "cold" forming.

5.4.2.1 H e a t F o r m i n g P r o c e s s e s

The thermoplastic film or sheet fixed in a holding frame is heated to a suitable temperature and pressed to adopt the shape of a mould, in which it is quickly cooled before removing.

Mechanical die forming is relatively insignificant. Forming with compressed air gives more precise profiles than those from vacuum forming, but economic aspects led to greater development of automatic vacuum-forming machines.

The vacuum forming process adopts the negative or positive methods. In the positive method the punch penetrates the pre-heated film which adopts its shape assisted by the applied pressure difference (Fig.91) /17/, whereas in the negative method the film adopts the shape of the force (Fig. 92) /1/. As follows from the figures, the slip technique with application of air is often used for bulky and more complicated articles to prevent irregularity in wall thickness. In principle, the pre-heated film is pre-formed air pressure to give a large bubble, so that it is

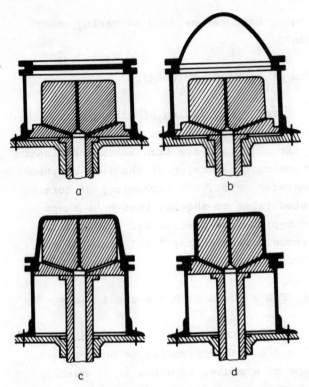

Fig. 91. Positive method of vacuum forming process of a thermoplastic film by drawing with the use of air /17/

a - heating of the film by means of preheating plates,
b - prestressing of the film by air pressure,
c - drawing of the film over the positive mould (the beginning of the proper forming process), d - final forming of the film by introduction of underpressure between the positive mould and the film

uniformly stretched. This also gives a better yield from a given film area, because it does not necessitate a broad rim.

The increase of surface area during the forming process varies with different thermoplastic materials /45/: polycarbonates by a factor of 2, HD polyethylene 2.5, polypropylene 3, poly(vinyl chloride) 4, acrylonitrile-butadiene-styrene terpolymer 4, polystyrene 5.

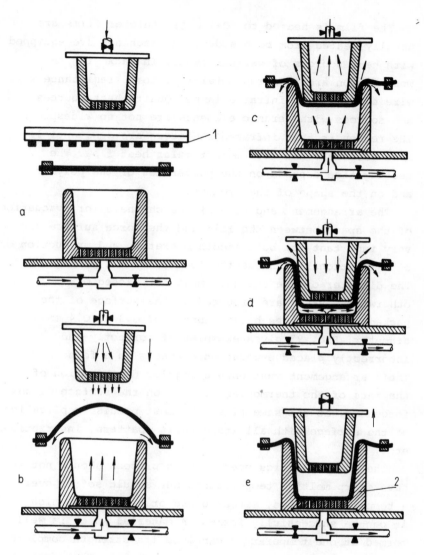

Fig. 92. Negative method of vacuum forming of a
thermoplastic film by the slip method with the use of
air /1/

a – heating of the film by an infrared radiator 1,
b – pre-stressing of the film by air pressure, c, d –
the beginning of the film forming by the slip,
e – finishing of the forming of the film by introduction of underpressure between the negative mould 2
and the film

The film is heated to 90-110°C; thicker films are
usually heated from both sides. The machines are equipped
with heat sources of various types. In packaging, the
most common are infrared metal radiators, resistance
wire sources, and infrared lamps. Quartz heat sources
and sources with ceramic elements are not so widespread.
The reason is that infrared radiators are substantially
more efficient and give more regular heat flow. The
heating time depends on the properties of the polymer
and on the shape of the product.

The arrangement and size of the channels for evacuation
of the space between the film and the force surface are
very important for both smooth operation and production
of a functionally and aesthetically correct package.
The diameter of the channels must not exceed 0.8 mm,
otherwise "marks" are produced on the surface of the
package. Most forces have channels of 0.3 to 0.4 mm
diameter. Fig. 93 gives examples of correctly and
incorrectly placed evacuation channels in a force.
Their arrangement must ensure regular distribution of
the mass of the thermoplastic film on the surface of the
force and, at the same time, the most precise replication
of the surface with all its details (pattern, inscriptions,
etc.) /14/.

The thermoplastics used in this process should not
have sharp melting temperatures but should soften over
a temperature range, either broad or narrow, in which
it can be successfully formed. A material which is well
formed at one temperature can lose its strength completely
at a temperature by several degrees higher. Therefore,
the choice of temperature range is of no less importance
than the selection of plastic. The closer is the forming
temperature to the thermoplastic transition temperature,
the more thermostable are the products; i.e. only a small
reversible deformation is possible after cooling, or the
products have only small "elastic memory". On the other

hand, too high a forming temperature decreases the
ductility of the material.

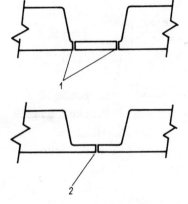

Fig. 93. Examples of location
of evacuation channels in a
force of a mould for vacuum
forming process /14/

1 - correct location,
2 - incorrect location

Some materials become so yielding at the forming
temperature that a minimum pressure or underpressure
suffices for perfect copying of every detail of the
force. Others are considerably resistant and necessitate
heavier equipment and tools, because the application
of pressure or underpressure is limited, and often it
is impossible to ensure copying of more complex details.
This behaviour, which is called hot strength, can
sometimes be related to the ability of the material to
be stretched at enhanced temperature (hot ductility),
but the relation need not be parallel. The pressures
and underpressures used reach the values of 1 MPa and
120 Pa, respectively.

The pressure forming process is used to treat sheets
of 6-10 mm thickness to obtain products up to 500 mm
in depth. The drawing rate can reach 0.5 m s^{-1}. Forming
by atmospheric pressure necessitates evacuation of the
air between the plate and the form within 0.5 s.

The low pressures used in the hot forming processes
make it possible to use various materials for the moulds,

e.g. wood, plaster, cement, casting resins, light
metals, etc. The choice also depends on the length of
the production run:

- plaster for a few tens of prototypes
- wood for up to several hundreds of pieces
- zinc-aluminium alloy for up to 10,000 pieces
- glass-reinforced polyesters or epoxide resins

for over 10,000 pieces /72/.

The moulds must have 2° minimum bevels and rounded
edges of minimum radius 1.5 mm. For sheets thicker than
3 mm this radius of edges must be at least equal to
the sheet thickness. Multicavity moulds are also used,
e.g. with 12 or more units, and can produce up to
8000 pieces per hour.

Both continuous and discontinuous processes are
used for forming. The discontinuous apparatus is fed
with cut sheets whose dimensions correspond to those
of the holding frame. The capacity of the equipment is
increased by rotary arrangement of the individual
operations fixing in the frame - heating - forming -
removing the product. The discontinuous process is
chosen especially for large-area products (e.g. covers)
and smaller series.

The continuous process (Fig. 94) can utilize the
sterility of the reeled film surface after its extrusion
heating, and can include the filling operation, which
is especially advantageous in the production of cups.
For these purposes it is also possible to use combinations with other processing technologies, e.g. the
film or sheet is annealed to a suitable forming temperature immediately after extrusion.

The thermoplastics used for hot forming include
polystyrene and its toughened modifications, polyethylene,
polypropylene, acrylonitrile-butadiene copolymers,
poly(methyl methacrylate), cellulose acetate and acetobutyrate, and polycarbonate.

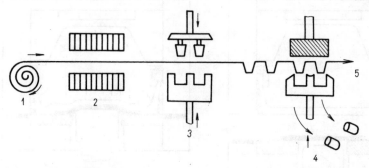

Fig. 94. Schematic representation of a production
line for continuous production of hollow packages from
a thermoplastic film /72/

1 - unwinding equipment, 2 - heating unit, 3 - shaping
unit, 4 - cutting unit, 5 - recycle of technological
wastes

5.4.2.2 C o l d   F o r m i n g   P r o c e s s

Cold forming is one of the more recent technologies
used in processing of thermoplastics. The starting
materials are films, and the main field of application
is in packaging (i.e. for production of cups and tubs).

The cold forming process, i.e. forming at about $20^{o}C$,
consists in three-dimensional deformation of a plastic
film by the action of pressure of a liquid in an elastic
bag with simultaneous drawing of the material by the
forming plunger. The process is thus analogous to
forming in a mould whose one part (usually the matrix-
force) is the filled elastic bag, and the other part
(usually the punch) is replaced by the forming plunger.

The forming cycle is shown schematically in Fig. 95
/73/. The thermoplastic film is placed on the thrust
ring so that it overlaps the forming plunger (Fig. 95a).
After the mould is closed, the film is exposed to pressure
from the side of the force (matrix) made of a rubber
membrane (the chamber above the membrane is filled
with pressure oil), and from the other side it is
pressed by the approaching plunger (Fig. 95b). Under

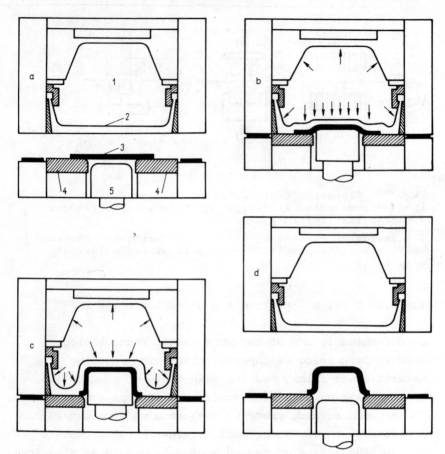

Fig. 95. Schematic representation of the procedure of cold forming of a plastic material /73/

a - loading of the plastic film 3 into the open mould between the holding-down ring 4 and the rubber membrane 2; b, c - simultaneous filling of the chamber 1 with pressure liquid and lifting of the plunger 5 which shapes the film 3 by drawing, d - opening of the mould and taking out of the product

these conditions the film is drawn and forced to replicate the plunger shape (Fig. 95c). In the last phase of the forming cycle, the pressure is released by venting the oil from the chamber, the forming plunger is returned to its initial position, the mould is opened, and the product removed (Fig. 95d).

Cold forming has the advantage over hot forming, that the polymeric material maintains its characteristic strength and therefore its integrity during the process, and avoids the residual tensile strain that can occur through heating and cooling and the consequent dimensional instability.

The cold forming process utilizes the cold flow which is associated with forced shifts of whole macro-molecules, and any slight elastic reversible deformation is removed either immediately after the forming pressure is released, or additionally by heating the material. The absence of residual strain in the product is particularly valuable for the base of cup type containers. Also the mechanical properties of the container walls are homogeneous.

Further development of this technology requires development both of the equipment and of the understanding of the stress-strain relationship at room temperature to aid the selection of suitable materials. It is obvious that a polymer for use in the cold forming process must exhibit high tensile strength and high modulus and at the same time be capable of deformation to a high degree at room temperature. It is important that no residual strain remains in the material, which requires high values of the modulus of elasticity in both tension and compression.

The materials fulfilling these requirements and used industrially include polycarbonates, cellulose derivatives, and acrylonitrile-butadiene-styrene terpolymers.

5.4.2.3   F o r m i n g   P r o c e s s   o f   F i l m s
          f o r   t h e   " S k i n - P a c k "   a n d
          " B l i s t e r - P a c k "   P a c k a g e s

Important methods of packing piece articles are known under the names "Skin-Pack" and "Blister-Pack".

The "Skin-Pack" method consists in a skintight closing of the articles to be packed, which are placed on a support (mostly of cardboard or paperboard), using a soft or hard thermoplastic film. During the packing process, the articles are covered with the film which is then heated until it softens; simultaneous exhaustion of air causes the film to collapse tightly around the foods and, at the same time, the film is bonded to the support (Fig. 96) /14/. After cooling, the film tightly holds the goods packed, so the method is suitable for packing both consumer and industrial products, e.g. tools, glass object, toys, etc. Sometimes the film is directly extruded on a group of articles placed on the support with simultaneous exhaustion of air.

In principle, the "Skin-Pack" method is simple, nevertheless, its application, which should produce cheap and useful packages, can meet a number of problems /14/:

1. the film can adhere to the goods packed,

2. unpacking of the goods can be difficult,

3. some articles can be damaged or degraded by heat or mechanically.

4. the goods can shift position during the packing operation.

Until recently, only two types of film were available

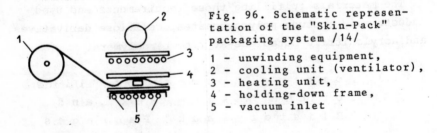

Fig. 96. Schematic representation of the "Skin-Pack" packaging system /14/

1 - unwinding equipment,
2 - cooling unit (ventilator),
3 - heating unit,
4 - holding-down frame,
5 - vacuum inlet

for the "Skin-Pack" system, viz polyethylene and poly (vinyl chloride). They are still used most frequently, but films based on ionomers also proved useful during the last decade.

Also suitable for the "Skin-Pack" system are modified polyethylene films with an oxidized surface usually obtained by exposing one side of the film to electric discharge. This treatment improves adhesion of the film to the paper support, and facilitates direct printing and bonding to printed supports, but involves additional cost.

In this application, polyethylene film has the advantage of being cheap, strong, and resistant to cold and heat. However, film shrinkage can distort the paper support after cooling.

Polyethylene film is used mainly for packing industrial products, and here the main factors are its cheapness and good protective function; the fact that the film is translucent rather than clear is no disadvantage.

Poly(vinyl chloride) films can have various properties ranging from high flexibility (plasticized) to crystal transparency and hardness (unplasticized), so more variants are possible than with other plastics. For the "Skin-Pack" system, producers of poly(vinyl chloride) recommend the soft or medium-soft films with high impact strength, and good resistance against water vapour, solar radiation, and ageing, and good weldability. They are heated through more quickly than polyethylene films. The shorter welding cycle compensates for the higher price of the material. The plasticized poly(vinyl chloride) films, however, cannot be used for packing pharmaceuticals or foodstuffs. For these purposes only the unplasticized films are suitable.

Compared with polyethylene, the poly(vinyl chloride) films are advantageous in being transparent, easily formed, having excellent adhesion to the support and less tendency to cause distortion, but exhibiting lower strength and lower resistance to low temperatures.

The ionomer copolymer films developed by Du Pont and

marketed since 1965 under the name *Surlyn A* combine
many of the good properties of polyethylene and
poly(vinyl chloride). The film is as transparent as the
poly(vinyl chloride) films and even stronger than the
polyethylene films. It is chemically inert, and resists
oils, fats, and solvents.

The forming process uses infrared heating as these
films absorb the radiation well. The material is heated
very quickly and so cycle times are very short. Another
advantage lies in the good physical properties of the
ionomer which allow the use of much thinner films
than in the case of polyethylene or poly(vinyl chloride).
The films shrink very little and therefore exhibit
better dimensional stability, and little distortion
of the support. However, the adhesion of the film to
the support needs to be improved by coating the support
with an adhesive layer (*Surlyn D Primer*).

The "Blister-Pack" method uses a transparent hard
film pre-formed to the shape of the main contour of
the article to be packed. This transparent film, the
blister, is attached to support (mostly of cardboard)
by heat sealing, adhesive bonding or stapling, or it
is inserted into a connecting slot (this method is
also known as "Bubble-Pack").

The "Blister-Pack" method is used especially in the
foodstuff industry for packing sweets and pastry, and
in the pharmaceutical industry for packing tablets, etc.

The film packages fulfil primarily a protective
function, whereas blister packaging, which is equally
cheap, also lends itself to printing information and
decoration on the support while the transparency of the
blister allows easy inspection of the article packed.
Another advantage for the consumer lies in easy removal
of the goods if the support is provided with a perforated
section or consists of a thin film or metal foil only.
A functional scheme of blister packaging is shown in

Fig. 97 and 98.

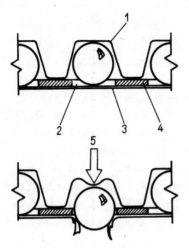

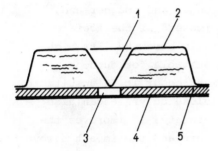

Fig. 97. Functional scheme of a blister-type package for solid articles /14/

1 - blister, 2 - perforation in the support,
3 - the closing film,
4 - support, 5 - the pressure needed for opening of the blister-type package by bursting the closing film

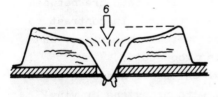

Fig. 98. Functional scheme of the blister-type package for loose materials /14/

1 - the opening element of the blister,
2 - the proper blister, 3 - perforation in the support,
4 - closing film, 5 - support, 6 - pressure

Most problems connected with blister packs concern the production of the blisters themselves (mostly by vacuum forming) which in turn depends on the selection

of suitable plastic film. The following materials /14/
are those most often used:

1. cellulose derivatives (acetate, propionate,
butyrate),

2. vinylic polymers,

3. polystyrene.

The cellulose acetate, propionate, and butyrate are
formed very easily, are transparent, have good impact
strength and excellent strength at low temperatures,
resist oils and fats, and exhibit good weldability,
but are fairly permeable to water vapour.

Cellulose acetate, which was one of the first plastics
used for the production of film, is substantially cheaper
and also exhibits better weldability than cellulose
propionate and butyrate.

Cellulose butyrate is stronger and tougher and,
therefore, it is often used instead of the acetate
for heavier articles, especially hardware. Its high
weather resistance makes it suitable for outdoor
applications. Due to its slight odour it is unsuitable
for packing foodstuffs. Cellulose butyrate can also
be used to modify cellulose acetate: it improves the
resistance to light and moisture and the impact strength.

The properties of cellulose propionate are between
those of the acetate and the butyrate. In contrast to
cellulose butyrate, the propionate is odourless, and
so it can be used for packing foodstuffs.

Films based on poly(vinyl chloride) and some copolymers of vinyl chloride e.g. with propylene or vinyl
acetate are odourless and tasteless, strong, transparent,
and resistant to oils, fats, and many chemicals. They
are easily formed. In packages they fulfil the function
of a barrier to water vapour, oxygen, carbon dioxide,
and aromatics. They are self-quenching in contrast to
mildly but still flammable cellulose derivatives.
The films can be easily bonded by means of dielectric

or heat welding or with adhesives.

Unoriented polystyrene films are some of the cheapest, but have a tendency to distort the paper support and are fragile and not very resistant to low temperatures.

Biaxially oriented polystyrene films (cf. Sect. 5.4.1) properly stabilized exhibit dimensional stability, are transparent, odourless, tasteless, resistant to oils and fats, and food compatible. Their water absorption is low, they do not absorb flavour, odours, and dyestuffs, and they are therefore suitable for packing foodstuffs. They are widely used in blister packaging.

REFERENCES

/1/ Štěpek J., Zpracování plastických hmot, SNTL, Prague, 1966.

/2/ Čurda D., Obalová technika v potravinářství, Texts of Prague Institute of Chemical Technology, Prague, 1971.

/3/ Vajgent B., Prům. potrav., 1966, 17, 30.

/4/ Schieber R., A publicity booklet of the firm Chemische Fabrik, Bohfinden, Württemberg, FRG.

/5/ Lambert P., Adhesives Age, 1973, 16 (7), 22.

/6/ Anonym, Packaging, 1974, 45 (534), 40.

/7/ Vajgent B., Obaly, 1968, 14, 131.

/8/ De Fries J., Neue Verpack., 1971, 24, 460.

/9/ Hicks E., Shellac, Chemical Publishing, New York, 1961.

/10/ Švastal S., Lím D., Kolínský M., Úvod do chemie a technologie plastických hmot, Práce, Prague, 1954.

/11/ Martens C.R., Alkyd Resins, Reinhold, New York, 1961.

/12/ Stevens M.P., Polymer Chemistry. An Introduction, Addison-Wesley, Massachusetts, 1975.

/13/ Kubík J., Gřundel F. et al., PVC, výroba, zpracování a použití, SNTL, Prague, 1965.

/14/ Farnham S.E., A Guide to Thermoformed Plastics Packaging, Cahners, Boston, 1972.

/15/ Tomis F. et al., Polyethylen, SNTL, Prague, 1961.

/16/ Planeta A., Planeta M., Manipul. sklad. bal., 1972, 1 (2), 15.

/17/ Kovačič L., Bína J., Plasty - vlastnosti, spracovanie, využitie. Alfa, Bratislava, 1974.

/18/ Pokrovskiy L.I., Murashov J.S., Plast. Massy, 1971, (5), 33.

/19/ Šimoník J., Vilím O., Plast. kauč., 1973, 10, 232.

/20/ Anonym, Package Eng., 1968, 13 (5), 120.

/21/ Anonym, Kunststoffe, 1966, 56, 361.

/22/ Visnovsky L., Plast. Mod. Elast., 1970, 22, 103.

/23/ Visnovsky L., Brit. Plast., 1970, 43, 90.

/24/ Anonym, Verpack. Wirtschaft, 1968, 16 (3), 16.

/25/ Anonym, Package Eng., 1968, 13 (3), 15.

/26/ The Society of the Plastics Industry, Plastics Engineering Handbook, Van Nostrand-Reinhold, New York, 1960.

/27/ Zelinger J., Zařízení závodů gumárenského a plastikářského průmyslu. Texts of Prague Institute of Chemical Technology, Prague, 1964.

/28/ Schildknecht C.E., Polymer Processes, Interscience, New York, 1956.

/29/ Anonym, Kunststoffe, 1962, 52, 478.

/30/ Breitbach K., Droyer W., Predöhl W., Schotte K., Steinau P., Trausch G., Werner E., Kunststoffe, 1971, 61, 356.

/31/ Banning H.J., Diener H., Freudenberger F., Rickling E., Schotte K., Witt W., Kunststoffe, 1971, 61, 337.

/32/ Prokop J., Obaly, 1969, 15, 97.

/33/ P. Dixon and Sons, Ltd., Especially Transparent Polyethylene Film, A publicity booklet of the firm.

/34/ Anonym, Packaging, 1968, 39 (460), 59.

/35/ Anonym, Plast. Rubber Weekly, 1968, 246, 1.

/36/ Grey P.J., Mod. Plast., 1965, 42 (5), 173.

/37/ Nippon Gohsei (Osaka, Japan), Water-Soluble Poly(Vinyl Alcohol) Film, A publicity booklet of the firm.

/38/ Anonym, Emball. Digest, 1967, 10 (95),281.

/39/ Anonym, Emballages, 1967, 37 (248), 22.

/40/ Anonym, Emballages, 1968, 38 (259), 22.

/41/ Anonym, Canning Packing, 1968, 38 (445), 11.

/42/ Anonym, Plastics, 1966, 31 (340) 117.

/43/ Anonym, Intern. Bottler Packer, 1967, 41 (12), 120.

/44/ Montecatini-Edison SpA. Moplefan - Unoriented Polypropylene Film, Information Bulletin DIES No. 1, 1063-Z.

/45/ Trávníček D., Způsob zpracování thermoplastů pro aplikace v obalové technice. The Prague Institute of Chemical Technology, Prague, 1974.

/46/ Anonym, Neue Verpack., 1968, 21, 932.

/47/ Anonym, Kunststoffe, 1968, 68, 222.

/48/ Zippel B., Kunststoffe, 1966, 56, 361.

/49/ Eckert G., Verpackung, 1967, 8 (1), 26.

/50/ Greiner G.J., Neue Verpack., 1967, 20, 1800.

/51/ Pardos-Jacques F., Plast. Mod. Elast., 1971, 23 (1), 94.

/52/ Anonym, Neue Verpack., 1968, 21, 708.

/53/ Deutsch K., Neue Verpack., 1968, 21, 1259.

/54/ Anonym,Verpack. Rdsch., 1968, 19, 647.

/55/ Anonym, Internat. Bottl. Packer, 1967, 41 (9), 70.

/56/ Saechtling H., Bauen mit Kunststoffen, Hanser,
     Münich, 1973.

/57/ Papst E., Ropa uhlie, 1973, 15, 276.

/58/ Luganov G.E. et al., Plast. Massy, 1974, (12), 29.

/59/ Schrader W., Zpracování a svařování plastických
     hmot, SNTL, Prague, 1962.

/60/ Marek O., Tomka M., Akrylové polymery, SNTL,
     Prague, 1964.

/61/ Veselý O., Sochor M. et al., Polyamidy, jejich
     chemie, výroba a použití, SNTL, Prague, 1963.

/62/ Karghin V.A., Slonimskiy G.L., Úvod do fyzikální
     chemie polymerů, SNTL, Prague, 1963.

/63/ Billmeyer F.W. Jr., Textbook of Polymer Science,
     Wiley, New York, 1971.

/64/ Breitbach K., Dreyer W., Predöhl W., Schotte K.,
     Steinau P., Trausch G., Werner E., Kunststoffe, 1971,
     51, 356.

/65/ Hager J., Joung D.R., Chem. Techn., 1974, 10, 677.

/66/ Müller J., Werner W., Plastverarbeiter, 1971, 22,
     461.

/67/ Müller J., Werner W., Plastverarbeiter, 1971, 22,
     570.

/68/ Červenka M., Revue prům. obch., 1973, 10 (3), 37.

/69/ Siebrecht M., Kunststoffe, 1975, 65, 229.

/70/ Anonym, Offic. Plast. Caoutch., 1972, 19 (36), 153.

/71/ Kapel L., Müan. Gumi, 1971, 8, 257.

/72/ Reyne M., Plast. Inform., 1968, 19 (416), 7.

/73/ Smoluk G.R., Klaus M.B., Mod. Plast., 1968, 45
     (15), 240.

6

Evaluation of Hygienic Properties of Plastic Packaging Materials

J. HORÁČEK

6.1 HYGIENIC ASPECTS OF PLASTICS

Food contact acceptability is a very important requirement in packing foodstuffs besides protective properties
and functional suitability. A package must meet a
series of specific requirements in order that it may
be used without any unfavourable effects on the goods
packed and, hence, on the health of the consumer. When
evaluating the hygienic aspect of a package, nowadays
the main attention is focused on microbiological, and
organoleptic aspects, but primarily on the risk of
contamination of the packed goods with substances
present in the packaging material.

The material must be completely inert from the microbiological point of view. This means that it not only
must protect the foodstuff against the penetration of
microorganisms from outside but also itself must not
be a source of microbial contamination and must resist
attack by all the normal microorganisms present in the
foodstuff. Besides that it must not contribute to the
creation of a medium supporting growth and propagation
of undesirable microflora in the closed package. Also

inadmissible is any contamination of the foodstuff
with the toxins which would be produced by microorganisms in contact with the package or which would
already come from the use of unsuitable raw material
in the production of the plastics. Some carcinogenic
mycotoxins are so heat resistant that they are not
destroyed even at the temperatures used in processing
plastics. Otherwise, thermal treatment during
processing can, in most cases, supply a satisfactorily
sterile product (except in the case of strong contamination or microbial attack of the raw material). In
the secondary contamination, however, some components,
e.g. plasticizers, can become sources of nutrients
(carbon, nitrogen) supporting the growth and effect
of various microorganisms, until, in extreme cases,
so-called "microbial corrosion" of plastics can take
place.

Another requirement, which both plastics and final
packages for foodstuff applications must meet with
regard to hygiene, is organoleptic neutrality. The
material must cause no change in the consistency,
colour, odour and taste characteristics of the foodstuffs
with which it comes into contact: such changes could
be due to transition of some substances from the packaging material into the foodstuff or to absorption of
some components of the foodstuff by the plastics (e.g.
absorption of dyestuffs) or to escape of e.g. specific
odours to the environment. These sensory effects are
connected with some of the functional properties of
plastics. Thus e.g. long-term storage of foodstuffs
in packages with high permeability to oxygen can
result in fats turning rancid, oxylabile vitamins being
degraded, etc., so that the quality of the foodstuff
is impaired or even completely destroyed. In most
countries foodstuffs with changed sensory properties

are considered to have inferior quality, even causing
disgust and, hence, to be inedible although the changes
could be tolerated from the toxicological point of view.

The greatest attention, however, is paid to packaging materials with respect to the presence of
secondary contaminants. In this respect plastics are
investigated for the release of various substances
due to their volatility or to extraction. Not only are
raw materials investigated but also the compounds which
can be formed during their processing or by interactions
between the individual components or with food components.
The information obtained about the identities, amounts
of the substances released under various conditions,
and the results of toxicological studies, together
serve as a basis for the evaluation of individual
additives and, after generalization, for proposing
legislative measures.

6.2 PLASTICS AS A SOURCE OF SECONDARY CONTAMINANTS

The introduction of more sensitive analytical methods
made it possible to prove that contact of foodstuffs
with various materials results practically always in
migration of certain amounts (sometimes quite slight)
of various substances into the foodstuffs. These
substances have neither nutritive value nor flavouring
character and are added to foodstuffs either intentionally to improve their sensory properties (dyestuffs,
aromatics), to affect technological parameters
(emulsifiers), or to increase their storage stability
(preservatives, antioxidants), or are added accidentally during primary production, food preparation,
packing, storage, distribution, or by some external
effect /1-4/. According to whether they are added
intentionally or occur accidentally, the substances

are classified as additive or contaminating, respectively
/1-5/. They cannot be entirely excluded from foodstuffs, but their content must be kept within physiologically acceptable limits. First of all it is necessary
to eliminate the substances of high toxicity and those
not easily metabolized, those deposited permanently
in the organism and, of course, carcinogens, terato-
gens, and mutagens; also the less toxic or non-toxic
substances which do not belong to the foodstuff must
be excluded.

When evaluating plastics for use in the foodstuff
industry or for foodstuff packages it must be determined whether some contaminating substances may be
released. In systems of such complexity as are foodstuffs,
it would be very difficult, and in some cases practically
impossible, to follow negligible amounts of the contaminating substances; therefore, it is usual to carry out
model migration experiments. The investigated material
is treated with selected liquids at a given temperature
for a given time at a constant ratio of the extraction
liquid volume to the amount of the sample (which is
usually defined by its surface area). The conventional
conditions not only simplify the chemical determination
but make the results obtained in different materials
comparable.

The experts of the British Plastics Federation,
the first to deal systematically with the relations
between the surface area, thickness, and weight of
the sample and the amount of the migrating substances,
assumed that in the case of very thin materials the
amount of the substances extracted depends primarily
on the weight, and not on the thickness and surface
area of the wetted surface of the material /6, 7/.
Only above a thickness of 0.5 mm does the surface area
of the sample become determining. However, the results
of numerous investigations disagree with these assump-

tions /8, 9/. In ČSSR, practically since the beginning
of systematic scientific research of extractability
from plastics, the results have been related to the
surface area of the specimen tested under the standard
conditions of the model migration experiment /10/.

Although in practice the migration is never so high
that even a sparingly soluble substance would saturate
the extraction medium, the ratio of the volume of the
extracting liquid to the surface area of the sample
is also defined. This ratio as well as the other
experimental conditions   are given by a convention.
In ČSSR the ratio usually adopted is 100 ml dm^{-2} /10/,
in other countries it is 20-200 (or even several
hundreds) ml dm^{-2}. This conventional ratio is not
observed in special cases such as the evaluation of
filter materials, various vessels, surface treatment,
etc.

The test liquids are defined liquids and solutions
which make it possible to determine in the minimum
number of model experiments the effects of a range
of foodstuffs and beverages, and also cleansing agents
in the case of returnable packages designed for re-use.
The test liquids must be easily prepared, and their
composition must interfere as little as possible with
the analytical determination of the substances tested.
Therefore, more complicated solutions were abandoned
(as e.g. the "artificial wine" prepared from distilled
water, ethanol, sugar, and organic acids /11/). Such
mixtures are used only exceptionally nowadays - e.g.
in evaluation of a certain substance for a specific
application.

The selection of acids and choice of concentrations
of the solutions are not universal. Often, the use
of more aggressive compounds or higher concentrations
of the solutions make the conditions stricter and
thereby a certain safety factor is created. Although

some test liquids have been changed and unification
of methods has not been reached yet, the original idea
of simulating the effects of acid foodstuffs, alcoholic
beverages, fats, or alkaline purifying agents has
remained /12-14/. Therefore, a number of test liquids
contain distilled water, acetic or other organic acid,
and ethanol. It is more difficult to simulate the
effects of fats. Vegetable oil is used most often for
this purpose; later it was replaced by heptane which
is chemically pure and offers advantages during
estimation of the extracted substances. Numerous studies
in the field of the extraction tests using natural
fats and heptane as the extraction liquids gave comparable results, but other studies indicate significantly
greater extractability of the plastic components into
heptane than into oil /8/. Particularly with polyolefins,
some stereoisomeric forms are more soluble in heptane
than in oil, which facilitates extraction of the additives
present in the plastic during the extraction experiments,
so that the migration into heptane is greater than into
fats. Therefore, it is often more advantageous to use
a defined mixture of synthetic glycerides of fatty
acids and alcohols, e.g. *Fettsimulant HB 307* (Natec,
Hamburg, FRG). This mixture has the advantage that it
facilitates some analyses (e.g. spectrophotometry)
without previous isolation of the compounds migrated.

A great number of liquids for the migration tests
are used in France. These include: distilled water;
10, 50, and 95% ethanol solutions; 10 and 50% acetic
acid solutions; 1 and 5% sodium chloride solutions;
10% saccharose solution, and 2% citric acid solution
with pH adjusted to 4.5. Fats are simulated with lard
and peanut oil.

In Czechoslovakia, the following series /10/ of model
test liquids is mostly used at present: distilled water;
8% aqueous solution of acetic acid; solution with 2%
tartaric acid and 2% sodium chloride; 50% (v/v)

aqueous ethanol; 5%   aqueous sodium carbonate; table oil
(according to ČSN 58 0220 standard).

However, very often the reduced series of liquids
which can be easily evaporated without residue is used:
distilled water; 8%   acetic acid; 50%   ethanol; heptane
(instead of oil).

In connection with the test liquids it should be
noted that some components of plastics can migrate also
into "dry" foodstuffs; it is not justified to assume
that packaging materials for such foodstuffs need not
be tested /12/. It was found that dibutyl phthalate
can migrate into sugar and flour from paper bags with
an inner layer of plasticized poly(vinyl acetate) and
apples can be contaminated with 6-caprolactam from
polyamide bags /15/.

The amount of substance migrating from plastic packages depends on several factors. Temperature plays a
very important role. Most extraction tests are carried
out at elevated temperatures. However, a higher temperature means more severe conditions although one can
reduce the duration of the test. The testing of some
materials at higher temperatures, however, has the
disadvantage that certain physical properties of the
material can be changed, which increases the migration
of some components. The published testing procedures
use different temperatures from about $20^{o}C$. to $121^{o}C$
(extractions in autoclaves). In West European countries,
these extraction tests are carried out at 40 or $60^{o}C$
(see below), and only in the determination of overall
migration is a short-term test at higher temperatures
used /13, 16, 17/.

The US legislation prescribes the temperatures of
the extraction tests according to the principle that
the tests should be carried out at a temperature close
to that maximum temperature at which the material is
practically in contact with the foodstuff. This procedure

is advantageous for evaluation of specific cases,
but, generally, all applications of the material
tested are not known in advance.

In Czechoslovakia, migration tests are often carried
out at 70°C for 2 hours or at room temperature (about
20°C) for 10 days, and, if possible, the course of
migration is followed by withdrawing samples at intervals, for analysis and to determine the trend.

Time represents another important factor of the
model extraction experiments. If the amount of the
substance released is plotted against time, the usual
result is a typical exponential dependence resembling
an isotherm, i.e. at the beginning, the increase of
the extracted substance per time unit is substantially
greater than that in later stages. Hopf /18/ found
practically no further extraction increase after 5
days at 60°C. Hence, a number of reports which try
to describe the migration during a long-term contact
of plastics with foodstuffs show that the extraction
at 60°C for 10 days (or, according to some authors,
at 45°C for 10 days) is similar to six months at
room temperature, i.e. at 20-25°C /8/.

The results obtained in the standard extraction
tests and expressed in milligrams of the substance
extracted from 1 dm^2 of the wetted surface area of
the sample are sufficiently clear for comparison of
the results on various materials and procedures. In
spite of that, in some countries the term contamination
is introduced. It is expressed as the weight of the
component extracted from a plastic and present in
1 kg (or 1 l) of the foodstuff which was in direct
contact with the material tested. The results are
given in ppm. In these tests it is assumed that 1 kg
of foodstuffs is in contact with 6 dm^2, 8 dm^2, 10 dm^2,
or even 40 dm^2 surface area of the plastic (e.g. in
GDR it is 20 $dm^2 kg^{-1}$).

Besides these painstaking and exacting model migration tests, the following methods are also used for check purposes: determination of the global migration (e.g. the evaporation residue of the extract) or general tests (extraction of oxidable substances /19, 20/, changes in the surface tension of the extracting liquid /21/, effects on the polarographic maximum /22/, changes in the cryoscopic constant of the extract /21/, changes in UV and IR spectra of the extract, etc.) which were shown to be related to the amount and character of the components released from plastics during extraction. The extraction experiments designed for checking variants of the basic materials and finding the optimum composition are restricted to analyses of selected compounds. In such cases the compounds followed are either the most toxic ones or those whose content in the material is intentionally varied or whose extractability depends largely on the quality of the material.

Laminated materials are evaluated as a whole, although in their applications the decisive layers are those in contact with the foodstuffs packed. If the combinations of plastics are correctly chosen, the extractability of the upper layer does not differ much from that of the same plastic alone, the migration of components from inner layers being only rarely significant.

One of the health hazards connected with the application of plastics cannot usually be revealed by extraction experiments, viz. that due to reinforcing materials and some fillers. If such plastic packages are damaged or are of poor quality, the foodstuffs packed can be contaminated with solid particles. So e.g. fragments of glass fibres from polyester glass laminates can enter the digestive tract with the contaminated food and can cause irritation or even serious damage to mucous membranes. Therefore, the glass laminates used in the food industry must have

their reinforcement covered with a polymer layer on
all surfaces and edges.

Evaluation of contaminants takes into account their
pathological effect (acute toxicity given by the LD_{50}
value, chronic effects, irritation of the epidermis,
conjunctiva tests, accumulation in the organism, route
of metabolism, carcinogenity, mutagenity, and teratogenity
tests, etc.) and the assumed daily exposure of the
organism. The estimates of admissibility are based on
results of the extraction tests and the highest allowed
daily dose determined on the basis of chronic experiments
(three generations of test animals) or at least subacute
experiments (90 to 120 days) with animals. According
to the WHO proposals these experiments should be carried
out with three different species and at least one should
be different from rodents. Most often used are mice,
rats, and dogs. Several groups of the experimental
animals are exposed daily (e.g. by means of food) to
a constant amount of the compound investigated, and
from several such experiments with different amounts
of the compound, the highest concentration is determined
which has still no effect on the animals ("No Effect
Level" /NEL/ or also "No Toxic Effect Level"). This
value along with average food consumption and average
weight of the animal are used for calculation of average
weight of the investigated substance accepted per 1 kg
body weight without any negative effects on the experimental animal ("Animal Acceptable Daily Intake" /AADI/).
The resulting value is divided by a safety coefficient
(100 to 500) to give the ADI ("Acceptable Daily Intake")
for a man. The ADI value can be different for a healthy
adult and for ill people and children with increased
sensitivity. In connection with packaging materials,
another value was suggested, viz. PADI ("Packaging
Acceptable Daily Intake") /23/. In contrast to ADI,
the latter is not determined from chronic but from

subacute 90 days experiments with the application of
a higher safety coefficient (200 to 1000).

A long-term uncontrolled intake of various kinds
of contaminants (which often cannot be easily determined)
always means a load on the organism and can even cause
serious damage. Therefore, a number of countries accepted
legislative measures regulating the use of plastics in
the food industry. On the basis of the migration experiments, toxicological data, generalization of individual
findings, evaluation and solution of particular problems,
it is then possible to admit certain materials or those
of a certain composition for food contact applications.
These results are most usually presented in the form
of lists of the admitted starting materials and auxiliary
additives with specified purity requirements which must
be met by the ingredients. Some of these positive lists
also give the concentration limits of some substances
or restrictions to applications of materials of a certain
composition for some purposes (usually contact with
fats or alcoholic beverages). It is assumed that materials
produced by correct processing from the polymers and
additives given in a valid positive list will present
no risk (in normal applications) of release of
contaminants above the tolerated limits. This means
that according to available knowledge the health hazard
will be minimal, and the material can be denoted as
acceptable in this respect. Most lists and proposals
also present variants of the requirement for organoleptic neutrality of plastics.

The best-known positive lists of plastics - a special
list for each material - were compiled in GDR (as a
law /24/) and FRG (as Recommendation of Bundesgesund-
heitsamt in the sense of food law /17/), as well as in
the Netherlands /25, 26/. In France /14/ there exists
a positive list of admitted compounds which can be used
in the production and processing of plastics for the
purposes of the food industry. The admitted compounds

are not grouped according to the individual materials
or classes, but a single list was compiled for all
raw materials and auxiliary additives used within
the whole scope of plastics technology. In some cases,
purity of the admitted compound is specified, and for
some additives their maximum acceptable concentration
in the material (or different concentrations for
different purposes) is stated.

Similarly in Italy, an appendix of the food law
/13/ lists the plastics which can be used in contact
with foodstuffs and the additives which can be used
in the production and processing of these plastics.
Foodstuffs are divided into six classes according
to their water and fat content and acidity, and
according to the assumed use, the limit values of the
extraction tests are determined.

In the United States of America, the requirements
on plastics and acceptable raw materials are published
in the law digest /12/ sometimes with specification of
the maximum concentrations and scope of applications.
Besides that, further criteria are prescribed for
some materials as well as general extraction tests with
specified maximal limiting values.

In Great Britain, The Committee of Toxicology Experts
of the British Plastics Federation periodically published
their accounts /6, 7/. These documents list alphabeti-
cally the compounds with their toxicity factors (which
are, in fact, the rounded-off ADI values). From these
factors (T) and extractability (E) it is possible to
calculate the "toxicity quotient" ($Q = 1000E/T$) which -
calculated from all the migration experiments and for
all substances present in the material - must not
exceed 10.

Recently, however, the British Plastics Federation
began to publish (in cooperation with the BIBRA)
positive lists of raw materials and auxiliary additives

for individual plastics and lists of admitted compounds
along with their trade names, specification of the
material type, the maximum concentration acceptable in
the material, and types of foodstuffs which can come
into contact with materials containing the given in-
gredient /27/.

The Czechoslovak laws No 20/1966 /28/ and 45/1966 /29/
state that permission of use materials which can affect
human health (particularly in foodstuffs) can only be
granted by the Chief Hygiene Officer. In the sense
of these laws, instructions were published /30/ which
specify general demands on plastics, principles of
their application, purity of dyestuffs, and principles
regarding new plastics and additives. They summarize
the admitted starting monomers and additives for the
individual materials, and also specify limits of some
general tests.

6.3   EVALUATION OF HEALTH ASPECTS OF RAW MATERIALS
      FOR PLASTICS

In theory, the attitude of hygienists towards the use
of plastics in the food industry should be the same
in all countries. In practice, however, considerable
differences exist between the individual positive
lists of raw materials for these purposes. The principles
of evaluation of polymers and their additives are given
in this section along with notes about some less
investigated or not acceptable additives as well as
relatively safe additives. Clearly enough, further
development will bring new materials and new additives
which will have to be evaluated with respect to health.
New findings will also affect attitudes towards current
compositions.

Monomers are reactive substances, also with respect
to living organisms and, hence, more or less toxic

(except for caprolactam). Therefore, hygiene regulations usually restrict the content of monomers in starting raw materials, plastics, and articles made thereof. In some cases these limitations are dictated primarily by unfavourable organoleptic effects, the biological activity being of lower importance (styrene).

Olefins used in the plastics industry have no serious physiological effects except for narcotic effects observed at higher concentrations.

As vinyl chloride has been found responsible for liver angiosarcoma /31, 32/, the content of this monomer in the materials used has been substantially restricted in recent years. Its limit concentrations are 10 ppm to 1 ppm, and also limited is its maximum admissible content in foodstuffs (10 ppb), and some uses for the polymer are altogether prohibited (e.g. poly(vinyl chloride) bottles for spirits) /17, 30, 31, 32/.

Against expectation, any carcinogenic activity of vinyl acetate (which is less toxic than vinyl chloride) has not been proved. It has a slight narcotic effect on the organism, and larger doses irritate the eyes, mucous membranes, and respiratory organs /33/.

The acute toxicity of styrene (vinylbenzene) is not high. The LD_{50} value (per os) usually given is about 5000 mg per 1 kg of body weight /34/. The perception threshold is given between 20 and 190 mg m^{-3} /33/. Its inhalation slightly irritates the conjunctiva and mucous membranes. At high concentrations it irritates eyes and respiratory organs and has also a slight narcotic effect. Delayed effects cannot be excluded. Newer investigations show that it probably does not affect haematopoiesis. Lazarev /33/ describes changes in ovaries and menstrual cycles of women after exposure to styrene, but this finding was not confirmed /35/. Some individuals working in an atmosphere containing styrene vapours for a long period suffered from mild

changes of liver function, slight lowering of white
blood cell count, and decreased blood pressure. All
these changes spontaneously receded, if the exposure
was interrupted.

Lower esters of acrylic and methacrylic acids
also have no high acute toxicity. After intake they
irritate the digestive organs and cause changes in
liver and kidneys /33/. A very dangerous poison is
acrylonitrile, its LD_{50} being 93 mg per 1 kg of body
weight. The action of acrylonitrile in organisms is
usually explained by its releasing hydrogen cyanide;
some authors /36/ ascribe the toxic effect to the
nitrile group. Clinical symptoms of chemical action
of acrylonitrile on man are: slight anaemia, nausea,
headaches, vomiting, and leucocytosis /37/. As acrylonitrile is assumed to have carcinogenic effects, its
acceptable content in plastics is restricted in some
countries, e.g. to a value of 0.0001% in Czechoslovakia /30/.

Isocyanates (even very small doses) irritate mucous
membranes, eyes, upper respiratory organs, and skin.
Their oral administration was reported to damage the
oesophagus and stomach, but the toxicity is not high.
Long-term inhalation exposure causes development of
asthmatic symptoms, and sensitive individuals show
allergic reactions /38, 39/. The highest admissible
concentrations in the atmosphere are relatively low;
it is this fact which probably evokes care in evaluation
of residua of these compounds in package materials.

6-Caprolactam is not especially toxic on oral administration. Its low toxicity is probably due to its
ready solubility, rapid excretion from the body, and
fast metabolism.

The metabolism of caprolactam proceeds most probably
via aminocaproic acid. Ceresa and Grazioli /40/ carried
out subacute experiments administering caprolactam

to guinea pigs and found anaemia, leucocytosis, eosi-
nophilia, decreased calcium level in blood, increased
level of urea nitrogen, changes in liver, kidneys, and
hypertrophy of spleen. Non-lethal doses caused cramps
of striated muscles, peripheral vasodilation, increased
secretion of glands (salivary glands) and had a minor
protracted effect on thermoregulation. The tolerated
doses of 6-caprolactam only cause reversible changes
and no permanent deviations of biological functions
/41, 42/. The limits of the caprolactam content in
polymers thus take into account its good solubility
and disagreeable bitter taste.

Formaldehyde belongs to substances of medium to
strong toxicity with cumulative effects. It damages
mucous membranes and respiratory organs. Symptoms
of chronic intoxication include loss of appetite,
perspiration troubles, headaches, decrease of body
weight, and weakness. Among children who consumed
milk containing a small amount of formaldehyde, 25%
excreted protein in urine after several days, but this
ceased when the contaminated milk was replaced by normal
milk /43/. Metabolism of formaldehyde proceeds probably
via formic acid which is further oxidized by dehydro-
genases (to carbon dioxide). Carcinogenic effects of
formaldehyde were not found; mutagenic effects are
ascribed to formic acid /44/. Extraction of formaldehyde
from cured resins is limited.

Phenols and cresols have sharp characteristic odours.
They are neurotoxic substances. They damage mucous
membranes and skin. Concentrated phenols cause (even
in small amounts) serious or even lethal damage to the
digestive organs /45/. Chronic intoxication is characterized by irritation of respiratory organs, malaise,
muscular weakness, insomnia, increased perspiration,
itching, and digestive troubles. In some countries, the
use of phenol-formaldehyde moulding products (in contact

with foodstuffs) is allowed, but in Czechoslovakia it
is strictly forbidden because of possible extraction
of phenol and formaldehyde especially in acid media.

Urea is a metabolic product of nitrogen compounds,
hence its toxicity is very low. An increase of the level
of urea nitrogen in blood is accompanied by loss of
appetite, fatigue, malaise, headache, vomiting, etc.
/46/. Urea has distinct diuretic effects which are made
use of therapeutically. Low resistance of moulding
products from urea-formaldehyde resins even to weakly
acidic foodstuffs (release of formaldehyde) precludes
any practical applications of these materials in the
food industry.

The toxicity of melamine (triamide of cyanuric acid)
is also low, although it is higher than that of urea.
Its oral intake has diuretic effects on the organism.
Moulding products made of melamine-formaldehyde resins
belong to the most resistant among the thermoreactive
resins containing formaldehyde. In spite of this fact,
in many countries there is a proposal to test the
formaldehyde release from these articles during exposure
to acetic acid solutions (a Czechoslovak proposal gives
the limit of 0.10 mg formaldehyde per 1 dm^2 after 24 h
action of 8% acetic acid /30/).

Macromolecular compounds which form the basis of
packaging materials are usually insoluble, remain intact
after oral administration, and, hence, are practically
non-toxic. Nevertheless, they are undesirable as
contaminants in foods. The content of soluble polymers
(poly(vinyl alcohol)) and soluble components of plastics
(e.g. low-molecular fractions in polyamides, atactic
polypropylene, etc.) is usually limited.

Some polymerization additives are also followed,
especially the non-decomposed peroxides or peroxo-
dicarbonates. Their incorporation in the positive lists
depends on the toxicity of their transformation products.
Azo-bis(isobutyronitrile) can be decomposed to relatively

non-toxic compounds, but can also give the toxic nitrile
of tetramethylsuccinic acid. Therefore, every new
technological procedure must be evaluated. The finding
concerning changes in liver due to chloroformates prompt
carefulness, in spite of the fact that chloroformates
are decomposed with water and bases to hydrochloric
acid.

In the cases of polyesters, epoxide plastics,
silicones, and plastics based on polyphenylene oxide,
it is recommended to examine the content of amines used
as catalysts. Aliphatic amines affect the nervous
system, one of the mechanisms of their action consisting in cholinesterase inhibition /33/. The toxicity
increases with increasing molecular weight. Diamines
are more toxic than monoamines. Chronic intoxications
are accompanied by degenerative changes in the kidneys,
liver damage, impaired vision, and, sometimes, also
changes in the lungs. Aromatic amines are very toxic;
they affect the nervous system and react with haemo-
globin to give methaemoglobin /47/.

Considerable amounts of surfactants are used in
plastics production as emulsifiers, protective colloids
in polymerizations, and also as antistatic agents and
antifog agents (i.e. preparations which decrease
electrostatic charges on plastics and those preventing
the formation of condensed water droplets in the package, respectively). These substances do not usually
have a high toxicity (except for cation-active tensides
based on quaternary ammonium salts), but they can
increase considerably the reabsorption of toxic substances in the digestive tract. Some anion-active
tensides irritate the digestive tract, and, sporadically,
changes in liver were observed /48/. Shaffer et al.
/49/ carried out a toxicological study from which it
follows that non-ionic ethylene glycol emulsifiers
pass through the digestive tract without being reabsorbed.
With respect to the LD_{50} values, the non-ionic emulsifiers

belong to the least toxic class /14/. They are, however, disadvantageous in being less biodegradable. The cation-active preparations cause hyperplasia of the gastric mucous membranes and affect the nervous system; higher doses act haemolytically and often evoke haemoglobin-uria /50/. If the principle is observed that even the content of relatively non-toxic contaminant surface-active compounds in foodstuffs should not exceed that permitted for drinking water (0.2 ppm) /51/, then it will be difficult, especially with antistatic and antifogging agents, to meet this requirement at concentrations in plastics which would be technically effective.

Stabilizers present a complicated problem with respect to the hygienic evaluation of plastics. They constitute a large and varied class of compounds, and their content in the materials varies from 10^{-2}% to 3% . The variety of this category of additives is reflected in the variety of their physiological effects and toxicity.

Of the broad selection of thermal stabilizers, carbonates and phosphates are acceptable. Due to their lower efficiency they are used predominantly as pre-stabilizers. Soaps of fatty acids are also acceptable because of their low toxicity and slight solubility in water. The acceptability of these two groups of stabilizers depends significantly on the cation; generally accepted are sodium, calcium, and magnesium salts, but strontium compounds are unaccept-able and, in many cases, also barium, cadmium, and lead derivatives. There are different views about zinc compounds; the toxicity of zinc dioctoate is comparable with that of common inorganic zinc salts.

Of the organometallic stabilizers, the most widespread are tin(IV) derivatives, which are among the best-known additives to plastics /52/. Isotope techniques were

used /53/ in the respective extraction experiments and
toxicological tests. The toxicity of organotin(IV)
compounds depends considerably on the substituents
and molecular structure. Highly toxic are the trisubstituted derivatives which are nerve poisons.
Barnes and Stoner /54/ observed penetration of tri-
alkyltin(IV) compounds into erythrocytes. The dialkyl
derivatives are used for stabilizing poly(vinyl chloride), and it was found (although with some exceptions)
that their solubility, and, hence, reabsorption and
toxicity decrease with increasing molecular weight
/54, 55/. Experiments with animals showed that their
intoxication with dialkyltin(IV) derivatives is characterized by damage of the gall-bladder and hepatic duct.
A case was described /54/ of intoxication of a group
of people caused by administration of a dialkyltin(IV)
compound instead of the usual medicine: cerebral oedema
was found to be the most marked damage in lethal cases.
A number of organotin(IV) compounds cause severe damage
to skin. The compounds containing sulphur bridges in
their molecules are less toxic on oral administration,
but they attack mucous membranes and skin much more
than analogous compounds without sulphur /21/. The
possibility of formation of the dangerous trialkyl
derivatives on heating the dialkyltin(IV) compounds
(described by Franzen /56/) was not proved in practical
application of organotin(IV) stabilizers. In many
countries it is permitted to use the dioctyltin(IV)
stabilizers of prescribed purity in materials used in
the food industry. Intensively studied are the recently
suggested dimethyltin(IV) derivatives which can exhibit
low extractability when combined with plastics of
suitable composition. The development of polymeric
organotin(IV) stabilizers is directed to the suppression
of extractability.

Significant among thermal stabilizers are the
epoxidized oils. Their toxicity is affected by their

purity, the residual ethylene oxide being quite toxic.
The toxicity of pure epoxidized oils (oxyethylated oils)
decreases with increasing molecular weight (i.e. with
decreasing solubility) /57/. A long-term intake of food
with a high content of oxyethylated soya bean oil by
experimental animals resulted in no pathological changes
/14, 58/. Large doses increased the level of fats in
blood and caused disturbances in the metabolism of fats
and atrophy of testicles, but no changes in the organs
investigated were found by histological analysis at
the end of the experiment.

Glycol esters of 3-aminocrotonic acid are efficient
stabilizers whose use is allowed by medical authorities.
Their toxicity is low. A 90 days´ experiment with
animals whose food contained 5% thiodiethylene glycol
aminocrotonate resulted in weight increase of their
liver and kidneys. However, biochemical and haemato-
logical tests as well as urine tests showed no changes
except for a decrease in leucocytes of males in the
seventh week of the experiment /58/. The given ADI for
man /59/ is 9.5 mg ethylene glycol ester per 1 kg body
weight.

2-Phenylindole has a low acute toxicity /11/, and
even a long-term intake of small amounts by rats
resulted in no abnormal response. The number of tumours
did not increase either /60/. Similarly favourable
results were also obtained with the glycol esters of
aminocrotonic acid. Therefore, both 2-phenylindole
and some aminocrotonates are included in the positive
lists of some European countries including ČSSR.

Arylsubstituted urea derivatives were also suspected
to possess carcinogenic activity. However, the results
of tests with diphenylthiourea made it possible to
include these substances in the positive lists (FRG up
to 1%, France up to 0.5%, ČSSR up to 0.7%).

Esters of gallic acid are considered hygienically
safe but technologically rather unsuitable. Some of

them (if they fulfil the purity requirements) are even
allowed for use in preservation of fats in analogy to
ditert. butyl-p-cresol (BHT) which belonged among the
most widespread antioxidants for polyethylene and
polystyrene. Hygienic acceptability of most substituted
phenols can be assessed from their structure. There
are numerous of them which are very little toxic and
have good stabilization effects and other effects;
hence, they are included in the positive lists.

Aromatic amines, on the contrary, have unfavourable
parameters due to their high toxicity and possible
carcinogenic activity. As to sulphur-containing compounds
favourable data were obtained for thiodipropionic acid
and particularly its dialkyl esters (diethyl, dipropyl,
dilauryl, distearyl, and dipalmityl esters). Diamino-
diaryl sulphides are unsuitable for plastics use in the
food industry. Very controversial are the attitudes
towards thiophenolic substances. 4,4´-Thiobis(6-tert.
butyl-m-cresol) is allowed in most countries as a
hygienically unobjectionable additive to plastics.
However, in ČSSR it is not admitted in these applications.
A subacute experiment showed effects on growth and
an increase of liver weight.

Arylsubstituted phosphites also are efficient
antioxidants. Generally, they are considered toxic /14/,
nevertheless, some of them (especially tris(phenylnonyl)
phosphite) are found in the positive lists of many
countries, often accompanied by strict purity requirements.
This is due to the trisubstituted derivative being much
less toxic than the mono- and disubstituted ones.
Triphenyl phosphite can be included among highly toxic
substances. The antioxidants and light stabilizers based
on nickel(II) phenolic derivatives have not been allowed
yet, due to the presence of toxic nickel.

Out of many types of light stabilizers (UV absorbers,
which include salicylates, resorcinols, coumarine

derivatives, substituted acrylonitriles, benzophenones, and benzotriazoles) the only admitted additives to the materials for hygienically inobjectionable packages are higher derivatives of mono- and dihydroxybenzophenones and some benzotriazoles. Their toxicity is not high, but, with regard to their structure, they should be tested for possible photosensitizing effects and benzo- triazoles with respect to possible genetic effects. The admitted compounds include most often octylbenzo- phenone and 2-(2´-hydroxy-3´-tert. butyl-5´-methyl- phenyl)-5-chlorobenzotriazole. 2,4-Ditert. butylphenyl- 3,5-ditert. butyl-4-hydroxybenzoate is allowed, unless the respective packaging material is in contact with fat-containing foodstuffs.

Fluorescent brightening agents represent a question- able category. Hygiene specialists consider these sub- stances to be dispensable in the packaging materials used in the food industry. All these substances must be tested for photosensitizing and genetic effects and carcinogenic activity (besides the usual tests).

Among the additives used to modify the properties of plastics, the plasticizers cause considerable troubles from the hygienic point of view. Generally, the low- molecular plasticizers are not very resistant to ex- traction from plastics, and the extractability depends on the nature of the plasticizers, especially in case of highly plasticized materials (lipophile substances are more easily extracted by fatty foods, whereas the hydrophile plasticizers are more readily extracted by food with a higher water content). Migration of plasti- cizers generally increases the extractability of other additives. Suitable plasticizers with low toxicity include: butyl stearate, acetyltributyl citrate, alkyl sebacates and adipates, and - to a limited extent - higher dialkyl phthalates. Newer toxicological studies /61/ can be expected to bring further restrictions to

the use of the phthalate plasticizers. In some countries
the positive lists of allowed plasticizers include:
phenyl alkanesulphonates, substituted glycols, glycolates,
and azelaic esters. The toxicity of glycol plasticizers,
given by the LD_{50} value, decreases along the homologous
series. Toxicity of the ester plasticizers depends on
both the alcoholic component (it usually decreases
with increasing chain length) and acid component of the
molecule. Many esters of phosphoric acid are highly toxic
and not permitted in the packaging technology of food
industry. Tritolyl phosphate and tricresyl phosphate
(especially the ortho-isomer) are dangerous nervous
poisons responsible for many thousand cases of lethal
intoxication in North Africa in the fifties of this
century. The symptoms of intoxications by these substances are very varied. The main feature, which is
common to all organophosphates, is the cholinesterase
inhibition. There exist, however, also little toxic
phosphate plasticizers, out of which tripropylglycol
phosphate is allowed in the USA.

With respect to easy extraction of the above-mentioned
types of plasticizers, some specialists /14/ consider
polymeric plasticizers to be the only suitable types
especially for packing fatty foodstuffs and alcoholic
beverages. These types include polyesters (e.g. poly-
sebacates and polyadipates), condensates of glycols with
dicarboxylic acids, and some polymers denoted as
modifiers. These compounds are usually characterized
by low solubility, considerable resistance to extraction,
and low toxicity.

Lubricants are added to practically all types of
plastic films. These substances include additives of
varied nature, and their content in plastics varies
within the limits from $10^{-2}\%$ to 1%. Many positive lists
include stearin, waxes, paraffins, paraffin oils,
silicone oils, metal soaps, higher aliphatic alcohols,

pentaerythritol and its derivatives, glycerol and
its esters with higher fatty acids, amides of higher
fatty acids, etc. Most of the lubricants given are not
very dangerous to health, a certain care being
necessary in case of amidic types. They are presumed
to be hydrolyzed in organism, probably in the liver /14/,
with simultaneous liberation of the corresponding acid,
which is, however, insignificant except for the undesirable erucic acid. The situation is not quite clear
(due especially to the impurities present) in the case of
higher chlorinated paraffins; the pure derivatives are
little toxic, and no reabsorption by experimental
animals was observed even after repeated intake of
small doses. Thermal treatment in production of films
and packages thereof can result in formation of a thin
film of these substances at the surface of plastics;
so in spite of their low solubility, these additives
can contaminate the foodstuffs they are in contact with.
Some of them can affect unfavourably the sensory
properties of the foodstuffs packed. Lubricants have
the disadvantage (similarly as plasticizers) that they
enable migration of ingredients from plastics whereby
the extractability of these ingredients is increased.
Therefore, health regulations restrict the content of
lubricants.

The most suitable blowing agent (from the point of
view of hygiene) is carbon dioxide. Azo-bis(formamide)
(azodicarboxamide) which is decomposed at about $190^{\circ}C$
to give nitrogen, carbon monoxide, carbon dioxide,
and water is very favourable. The kickers (i.e. compounds
decreasing the decomposition temperature) which can be
used in acceptable plastics are zinc oxide and exceptionally barium sulphate of the purity given in
pharmacopoeia. In the United States of America,
azo-bis(formamide) is allowed as food additive even in
flour. Residual compounds of difluorodichloromethane

type (so-called Freons) and petrol residuals in cellular
materials contaminate foods and are undesirable due to
both toxicological aspects and sensory effects. A quite
unsuitable blowing agent is azo-bis(isobutyronitrile)
which can be decomposed to toxic dinitrile of tetramethylsuccinic acid (see above). The other blowing
agents suggested (Porofors), which are decomposed to
nitrogen and carbon dioxide, as e.g. diphenyl-
sulphone-3,3´-disulphonyl hydrazide, are used predominantly for technical products and are not included into
the positive lists, because their decomposition products
are not sufficiently investigated with respect to
toxicological effects.

Various modifiers serve for obtaining tough plastics
and improving processability of mixtures: often they
are mixed polymers containing, sometimes, acrylate and
methacrylate esters or vinyl acetate. The requirements
which must be met by the polymeric modifiers are proposed
on the basis of the same principles as those applied
to current plastics. In addition, their presence must
not affect extractability of other components of the
modified plastic.

A very much discussed group of plastics additives
are dyestuffs and pigments. The principles underlying
their admission vary in different countries. The general
requirement is that the compounds must not be acutely
toxic or carcinogenic. In some countries it is not
allowed to use benzidine dyestuffs and pigments containing cadmium for dyeing plastics used in the food industry.
The legislation concerning use of dyestuffs to plastics
is practically based on two principles. One of them
postulates "zero migration" (in FRG, see ref. /17/),i.e.
not even traces of dyestuffs and/or their components
must be extracted from the materials dyed. The other
principle includes regulations concerning purity of the
dyes and pigments used. These requirements are similar

in many European countries, the differences existing
predominantly in acceptable values of cadmium, chromium,
and antimony compound content. In the Netherlands /62/
the regulations prescribe both purity of the dye and
extractability of dangerous components of the plastic
dyed. In ČSSR it is not allowed to use pigments containing antimony, arsenic, cadmium, chromium, selenium,
lead, and uranium as basic components (they only can
be present as impurities, provided, of course, that
the pigment meets the purity requirements). The pigments
must not contain more than 0.01% lead, 0.005% arsenic,
and 0.05% aromatic amines; 0.2% zinc compounds,
0.01% barium, 0.01% chromium, 0.01% cadmium, 0.01%
selenium, and 0.005% mercury compounds soluble in 0.1N
hydrochloric acid are the highest acceptable amounts.
It cannot be excluded that, in some countries, the
regulations concerning purity of dyestuffs and pigments
will be supplemented by polychlorinated biphenyls.

Special purity requirements are prescribed for carbon
black limiting the content of polycyclic aromatic
hydrocarbons and, in some cases, also the whole content
of substances extractable by organic solvents (toluene,
benzene). Fritz /63/ found for linear and branched PE,
respectively, that 0.2 and 0.7 μg benzopyrene was
extracted into 200 ml milk from 10 dm^2 polyethylene
containing 2% carbon black (with 110 mg 3,4-benzopyrene
per 1 kg carbon black) during 24 h. On the basis of
these findings it was proposed in the GDR that the
maximum acceptable content be 0.5 mg benzopyrene per
1 kg carbon black for use in plastics applied in the
food industry.

Printing of transparent packages on the inner surface
improves their appearance, but, because of possible
contamination of foodstuffs with the printing inks,
this method is refused by most hygienists.

REFERENCES

/1/ WHO/FAO Ref. No. CAC/FAL-1-1973.

/2/ WHO/FAO Ref. No. CAC/FAL-2-1973.

/3/ Rozsíval L., Szokolay A., Cudzorodé látky v poží-
    vatinách, Osveta, Martin, 1969.

/4/ Halačka K., Čs. hyg., 1959, 4, 365.

/5/ Janíček G., Čs. hyg., 1959, 4, 375.

/6/ Anonym, Report of the "Toxicity" Sub-Committee of
    the Main Technical Committee of the British Plastics
    Federation, British Plastics Federation, London,
    1958.

/7/ Anonym, Second Report of the "Toxicity" Sub-Committee
    of the Main Technical Committee (publ.45), British
    Plastics Federation, London, 1962.

/8/ Van der Heide R.F., The Safety for Health of Plastics
    Food Packaging Materials, Kemink en Zoon N.V.,
    Utrecht, 1964.

/9/ Phillips I., Marks G.C., Brit. Plast., 1961, 34,
    319, 385.

/10/ Houška M., Horáček J., Čs. hyg., 1961, 6, 38.

/11/ Klimmer O.R., Nebel L.V., Arzneimittel-Forsch.,
     1960, 10, 44.

/12/ Code of Federal Regulations, Title 21. Food and
     Drugs, Part 170-178, U.S. Government Printing
     Office, 1978.

/13/ Decreto Ministeriale - Disciplina igienica degli
     imballaggi recipienti, utensili destinati a venire
     contatto con le sostanze alimentari con sostanze
     d'uso personale, Gazzetta Ufficiale, 1963.

/14/ Lefaus R., Practical Toxicology of Plastics,
     Iliffe, London, 1968.

/15/ Horáček J., Malkus Z., Ernährungsforsch., 1966,
     11, 409.

/16/ Woggon H., Uhde W.J., Nahrung, 1967, 11, 359.

/17/ Franck R., Mühlschlegel H., Kunststoffe im Lebens-
     mittelverkehr (Empfehlungen der Kunststoff-Kommission
     des BGA), Heymanns Verlag, Köln-Berlin-Bonn-München,
     1962-1987.

/18/ Hopf P.P., Plast. Inst. Trans. J., 1961, 29, 2.

/19/ Staub M., Mitt. Gebiet. Lebensmittelunters. Hyg., 1958, 49, 1.

/20/ Kiermeier F., Wildbrett G., Schaftefroh G., Z. Lebensmitt. Unters., 1959, 109, 1.

/21/ Pokorný F., Prac. lékař., 1961, 13, 439.

/22/ Horáček J., Malkus Z., Nahrung, 1963, 7, 173.

/23/ Leimgruber R., Chap. 18 in the book: Taschenbuch der Kunststoff-Additive, Carl Hansen Verlag, München, Wien, 1983.

/24/ Anordnung über Plaste für Bedarfsgegenstände, Gesetzblatt der DDR, No. 1, 1964, No. 2, 1967, No. 3, 1968, No. 4, 1970, No. 5, 1971, No. 6, 1972, No. 7, 1977.

/25/ Tentative Packaging Decree (Food Law), Netherlands Stat. Public., No. 61, 1964.

/26/ Tentative Decree on Food Packaging, Netherlands Offic. Gazette No. 189, 1965.

/27/ Anonym, Plastics for Food Contact Application (A Code of Practice of Safety in Use)(publ. 45/4), British Plastics Federation, London 1981 (Revision).

/28/ Zákon č. 20/1966 Sb. o péči o zdraví lidu (Health Law).

/29/ Regulation of Ministry of Health on the creation and protection of healthy living condition.

/30/ Záväzné opatrenia 36/1977. Hygienické požiadavky na plastické látky prichádzajúce do styku s poživatinami, Vestník ministerstva zdravotníctva SSR, r. 25, čiastka 21, 15.12.1977. Směrnice č. 49/1978 Sb. Hygienické předpisy o hygienických požadavcích na plasty a předměty z plastů přicházející do styku s poživatinami, Prague, 1978. (Czechoslovak hygienic regulations for plastics for food contact application).

/31/ WHO, IARC Report of Working Group on Vinyl Chloride, Int. Techn. Report No. 74/005, WHO, Lyon, 1974.

/32/ WHO, IARC Report of Working Group on Epidemiological Studies on Vinyl Chloride, Int. Techn. Report 75/001, WHO, Lyon, 1975.

/33/ Lazarev N.V., Chemické jedy v průmyslu, Vol. 1,
Státní zdravotnické nakladatelství, Prague, 1959.

/34/ Wolf M., Rowe V.K., McCollister D.D., Hollings-
worth R.L., Oyen F., Amer. Med. Assoc. Arch.
Ind. Health, 1956, 14, 387.

/35/ Bardoděj Z., Čs. hyg., 1964, 9, 224.

/36/ Levina E.N., Gig. Sanit., 1951, 2, 34.

/37/ Willson R.H., J. Amer. Med. Assoc., 1944, 124, 701.

/38/ Fuchs S., Volade P., Arc. Mal. Prof., 1951, 12, 191.

/39/ Munn A., Ann. Occup. Hyg., 1965, 8, 163.

/40/ Ceresa G., Grazioli C., Med. Lavoro, 1952, 43, 124.

/41/ Pokorný F., Sbor. lékař., 1952, 54, 28.

/42/ Goldblatt M.W., Farquharson M.E., Bennet G.,
Askew B.M., Brit. J. Ind. Med., 1954, 11, 1.

/43/ Chrustalev, Gig. Sanit., 1946, 11, (10), 30.

/44/ Anonym, Food. Cosm. Toxicol., 1965, 3, 515.

/45/ Fabre R., Truhaut R., Précis de Toxicologie,
Sedes, Paris, 1960.

/46/ Sollmann T. A Manual of Pharmacology and its
Applications to Therapeutics and Toxicology,
Saunders, Philadelphia, London, 1944.

/47/ Fabre R., Actual. Pharmacol., 1957, 10, 117.

/48/ Happer S.S., Hulpien H.R., Cole V.V., J. Amer.
Pharm. Assoc., 1949, 28, 408.

/49/ Shaffer C.B., Critchfield F.H., J. Amer. Pharm.
Assoc., 1947, 36, 152.

/50/ Lehman A.J., Assoc. Food Drug Off. USA, 1954,
18, No. 2.

/51/ ČSN 83 0611. Pitná voda (Czechoslovak Standards:
Drinking Water), 1974.

/52/ Luijten J.G.A., Bibliography of Organotin Toxicity,
TNO, Utrecht, 1971.

/53/ Seidler H., Woggon H., Härtig M., Uhde W.J.,
Nahrung, 1969, 13, 257.

/54/ Barnes J.M., Stoner H.B., Brit. J. Ind. Med.,
     1958, 15, 15.

/55/ Franzen V., Neubert G., Chem. Z., 1965, 89, 801.

/56/ Franzen V., Ernährungsforsch., 1966, 11, 368.

/57/ Hine C.H., Kodoma J.K., Anderson H.H., Simsonson
     D.W., Wellington J.S., Amer. Med. Assoc. Arch.
     Ind. Health, 1958, 17, 129.

/58/ Larson P.S., Finnegan J.K., Haag H.B., Blackwell
     S.R., Hennigar G.R., Toxicol. Appl. Pharm., 1960,
     2, 649.

/59/ De Groot A.P., Til H.P., Feron V.J., Huismans
     J.W., Food Cosm. Toxicol., 1969, 7, 473.

/60/ Mallette F.S., von Haam E., Arch. Ind. Hyg.
     Occup. Med., 1952, 5, 311.

/61/ U.S. Department of Health and Human Service
     "Environmental Health Perspectives", Vol. 45,
     Phthalate Esters, November 1982.

/62/ Anonym, Translation of the Netherlands Packaging
     and Food-Utensil Regulation (Food Law), Netherlands
     Plastics Federation (NVFK), Haarlem, 1975.

/63/ Fritz W., Ernährungsforsch., 1971, 16, 547.

APPENDIX

Survey of basic polymers used in packaging technology

Name of the polymer	Structural formula	Abbreviation
polyethylene	$\left[-CH_2-CH_2-\right]_n$	PE
polypropylene	$\left[CH_2-\underset{CH_3}{CH}-\right]_n$	PP
polyisobutylene	$\left[\underset{CH_2}{\overset{CH_3}{\underset{\vert}{C}}}-CH_3\right]_n$	PIB
polytetrafluoroethylene	$\left[-CF_2-CF_2-\right]_n$	PTFE
polytrifluorochloroethylene	$\left[\underset{F\ \ F}{\overset{F\ \ Cl}{C-C}}\right]_n$	PTFClE
polystyrene	$\left[-CH_2-CH\!\!\left\langle\!\!\bigcirc\!\!\right.\right]_n$	PS

APPENDIX - continued

Name of polymer	Structural formula	Abbreviation
poly(vinyl chloride)		PVC
poly(vinylidene chloride)		PVDC
poly(vinyl acetate)		PVAC
poly(vinyl alcohol)		PVAL
poly(vinyl acetals) poly(vinyl formal), R = H poly(vinyl butyral), R = C_3H_7		PVF PVB

APPENDIX - continued

Name of polymer	Structural formula	Abbreviation
poly(vinyl ethers)	$\left[\begin{array}{c} H\ H \\ -C-C-O-R \\ H\ H \end{array}\right]_n$	
polyacrylates poly(methyl acrylate), R = CH_3	$\left[\begin{array}{c} -CH_2-CH- \\ \|\\ COOR \end{array}\right]_n$	PMA
polymethacrylates poly(methyl methacrylate) R = CH_3	$\left[\begin{array}{c} CH_3 \\ \| \\ -C- \\ \| \\ CH_2 \end{array}\quad COOR\right]_n$ R = CH_3	PMMA
poly(methyl α-chloroacrylate)	$\left[\begin{array}{c} -CH_2-CCl- \\ \| \\ COOCH_3 \end{array}\right]_n$	
polyacrylonitrile	$\left[\begin{array}{c} H\ H \\ -C-C- \\ H\ CN \end{array}\right]_n$	PAN

APPENDIX - continued

Name of polymer	Structural formula	Abbreviation
polyamides		PA
polycaprolactam (x = 5)	$[-NH-(CH_2)_x-CO-]_n$	PA 6
poly(hexamethylene-adipamide) (y = 6, z = 4)	$[-NH-(CH_2)_y-NH-CO-(CH_2)_z-CO-]_n$ $x = 5$ $y = 6 \quad z = 4$	PA 66
polyurethanes		PU
polyesters	$[-O-R-O-CO-R'-CO-]_n$	PES
poly(ethylene terephthalate)		PET
phenolic plastics		

APPENDIX – continued

Name of polymer	Structural formula	Abbreviation
aminoplastics	$\left[-NH-R-CH_2- \right]_n$	
epoxides		
trans-1,4-polyisoprene (gutta-percha)		
α-modification		
β-modification		
polysulphide rubbers	$\left[-CH_2-CH_2-S_4-CH_2-CH_2-S_4- \right]_n$	TM, TOM

APPENDIX - continued

Name of polymer	Structural formula	Abbreviation
polycarbonates		PC
polyformaldehyde (polyoxymethylene)	$\left[-O-CH_2-\right]_n$	POM
poly(ethylene imine)	$\left[-NH-CH_2-CH_2-\right]_n$	
cis-1,4-polyisoprene (natural rubber) (synthetic isoprene rubber)		NR IR
cis-1,4-polychloroprene (chloroprene rubber)		CR
cis-1,4-polybutadiene (butadiene rubber)	$\left[-CH_2-CH=CH-CH_2-\right]_n$	BR

Index